Igneous Rocks of South-West England

THE GEOLOGICAL CONSERVATION REVIEW SERIES

The comparatively small land area of Great Britain contains an unrivalled sequence of rocks, mineral and fossil deposits, and a variety of landforms that span much of the earth's long history. Well-documented ancient volcanic episodes, famous fossil sites and sedimentary rock sections used internationally as comparative standards, have given these islands an importance out of all proportion to their size. The long sequences of strata and their organic and inorganic contents have been studied by generations of leading geologists, thus giving Britain a unique status in the development of the science. Many of the divisions of geological time used throughout the world are named after British sites or areas, for instance, the Cambrian, Ordovician and Devonian systems, the Ludlow Series and the Kimmeridgian and Portlandian stages.

The Geological Conservation Review (GCR) was initiated by the Nature Conservancy Council in 1977 to assess, document and ultimately publish accounts of the most important parts of this rich heritage. Since 1991, the task of publication has been assumed by the Joint Nature Conservation Committee on behalf of the three country agencies, English Nature, Scottish Natural Heritage and the Countryside Council for Wales. The GCR series of volumes will review the current state of knowledge of the key Earth-science sites in Great Britain, and provide a firm basis on which site conservation can be founded in years to come. Each GCR volume will describe and assess networks of sites in the context of a portion of the geological column, or a geological, palaeontological or mineralogical topic. The full series of approximately 50 volumes will be published by the year 2000.

Within each individual volume, every GCR locality is described in detail in a self-contained account, consisting of highlights (a précis of the special interest of the site), an introduction (with a concise history of previous work), a description, an interpretation (assessing the fundamentals of the site's scientific interest and importance), and a conclusion (written in simpler terms for the non-specialist). Each site report is a justification of a particular scientific interest at a locality, of its importance in a British or international setting and ultimately of its worthiness for conservation.

The aim of the Geological Conservation Review series is to provide a public record of the features of interest in sites being considered for notification as Sites of Special Scientific Interest (SSSIs). It is written to the highest scientific standards, but in such a way that the assessment and conservation value of the site is clear. It is a public statement of the value given to our geological and geomorphological heritage by the Earth-science community which has participated in its production, and it will be used by the Joint Nature Conservation Committee, English Nature, the Countryside Council for Wales and Scottish Natural Heritage in carrying out their conservation functions.

All the sites in this volume have been proposed for notification as SSSIs; the final decision to notify, or renotify, lies with the governing Councils of the appropriate country conservation agency.

Information about the GCR publication programme may be obtained from:

Earth Science Branch,
Joint Nature Conservation Committee,
Monkstone House,
City Road,
Peterborough PE1 1JY.

Titles in the series

1. **Geological Conservation Review**
   An introduction

2. **Quaternary of Wales**
   S. Campbell and D. Q. Bowen

3. **Caledonian Structures in Britain**
   South of the Midland Valley
   Edited by J. E. Treagus

4. **British Tertiary Volcanic Province**
   C. H. Emeleus and M. C. Gyopari

5. **Igneous Rocks of South-West England**
   P. A. Floyd, C. S. Exley and M. T. Styles

6. **Quaternary of Scotland**
   Edited by J. E. Gordon and D. G. Sutherland

7. **Quaternary of the Thames**
   D. R. Bridgland

Igneous Rocks of South-West England

P. A. Floyd
Department of Geology,
University of Keele.

C. S. Exley
Department of Geology,
University of Keele.

M. T. Styles
British Geological Survey,
Keyworth,
Nottingham.

GCR Editors: W. A. Wimbledon and P. H. Banham

JOINT NATURE CONSERVATION COMMITTEE

CHAPMAN & HALL
London · Glasgow · New York · Tokyo · Melbourne · Madras

Published by Chapman & Hall, 2–6 Boundary Row, London SE1 8HN

Chapman & Hall, 2–6 Boundary Row, London SE1 8HN, UK

Blackie Academic & Professional, Wester Cleddens Road, Bishopbriggs, Glasgow G64 2NZ, UK

Chapman & Hall, 29 West 35th Street, New York NY10001, USA

Chapman & Hall Japan, Thomson Publishing Japan, Kirakawacho Nemoto Building, 6F, 1–7–11 Hirakawa-cho, Chiyoda-ku, Tokyo 102, Japan

Chapman & Hall Australia, Thomas Nelson Australia, 102 Dodds Street, South Melbourne, Victoria 3205, Australia

Chapman & Hall India, R. Seshadri, 32 Second Main Road, CIT East, Madras 600 035, India

First edition 1993

© 1993 Joint Nature Conservation Committee

Typeset in 10/12pt Garamond by Columns Design & Production Services Ltd, Reading
Printed in Great Britain at the University Press, Cambridge

ISBN 0 412 48850 7

Apart from any fair dealing for the purposes of research or private study, or criticism or review, as permitted under the UK Copyright Designs and Patents Act, 1988, this publication may not be reproduced, stored, or transmitted, in any form or by any means, without the prior permission in writing of the publishers, or in the case of reprographic reproduction only in accordance with the terms of the licences issued by the Copyright Licensing Agency in the UK, or in accordance with the terms of licences issued by the appropriate Reproduction Rights Organization outside the UK. Enquiries concerning reproduction outside the terms stated here should be sent to the publishers at the London address printed on this page.

   The publisher makes no representation, express or implied, with regard to the accuracy of the information contained in this book and cannot accept any legal responsibility or liability for any errors or omissions that may be made.

A catalogue record for this book is available from the British Library

Library of Congress Cataloging-in-Publication data available

Contents

Acknowledgements		x
Access to the countryside		xi
Preface		xii

1 The igneous rocks of south-west England — 1

    Introduction and site synthesis — 3

2 Geological framework — 9

    The regional setting – Variscan Orogen — 11
    The local setting – Rhenohercynian zone of south-west England — 13
    Devonian volcanic activity — 16
    Carboniferous volcanic activity — 21
    Carboniferous–Permian plutonic activity — 22
    Post-orogenic volcanic activity — 27

3 Lizard and Start Complexes (Group A sites) — 31

    Introduction — 33
    List of sites — 33
    Lithological and chemical variation — 33
    A1 Lizard Point (SW 695116–SW 706115) — 44
    A2 Kennack Sands (SW 734165) — 46
    A3 Polbarrow–The Balk (SW 717135–SW 715128) — 49
    A4 Kynance Cove (SW 684133) — 53
    A5 Coverack Cove–Dolor Point (SW 784187–SW 785181) — 54
    A6 Porthoustock Point (SW 810217) — 58
    A7 Porthallow Cove–Porthkerris Cove (SW 798232–SW 806226) — 61
    A8 Lankidden (SW 756164) — 66
    A9 Mullion Island (SW 660175) — 70
    A10 Elender Cove–Black Cove, Prawle Point (SX 769353–SX 769356) — 73

4 Pre-orogenic volcanics (Group B sites) — 77

    Introduction — 79
    List of sites — 79
    Lithological and chemical variation — 80
    B1 Porthleven (SW 628254–SW 634250) — 84

Contents

B2	Cudden Point–Prussia Cove (SW 548275–SW 555278)	86
B3	Penlee Point (SW 474269)	89
B4	Carrick Du–Clodgy Point (SW 507414–SW 512410)	91
B5	Gurnard's Head (SW 432387)	95
B6	Botallack Head–Porth Ledden (SW 362339–SW 355322)	97
B7	Tater-du (SW 440230)	102
B8	Pentire Point–Rumps Point (SW 923805–SW 935812)	106
B9	Chipley Quarries (SX 807712)	109
B10	Dinas Head–Trevose Head (SW 847761–SW 850766)	112
B11	Trevone Bay (SW 890762)	114
B12	Clicker Tor Quarry (SX 285614)	117
B13	Polyphant (SX 262822)	119
B14	Tintagel Head–Bossiney Haven (SX 047892–SX 066895)	122
B15	Brent Tor (SX 471804)	127
B16	Greystone Quarry (SX 364807)	130
B17	Pitts Cleave Quarry (SX 501761)	132
B18	Trusham Quarry (SX 846807)	135
B19	Ryecroft Quarry (SX 843847)	137

5 Cornubian granite batholith (Group C sites) — 141

Introduction — 143
List of sites — 143
Lithological and chemical variation — 146
Petrogenesis — 151

C1	Haytor Rocks area (SX 758773)	162
C2	Birch Tor (SX 686814)	165
C3	De Lank Quarries (SX 101755)	167
C4	Luxulyan Quarry (Golden Point, Tregarden) (SW 054591)	169
C5	Leusdon Common (SX 704729)	172
C6	Burrator Quarries (SX 549677)	173
C7	Rinsey Cove (Porthcew) (SW 593269)	176
C8	Cape Cornwall area (SW 352318)	178
C9	Porthmeor Cove (SW 425376)	182
C10	Wheal Martyn (SX 003556)	185
C11	Carn Grey Rock and Quarry (SX 033551)	187
C12	Tregargus Quarries (SW 949541)	188
C13	St Mewan Beacon (SW 985534)	191
C14	Roche Rock (SW 991596)	192
C15	Megiliggar Rocks (SW 609266)	194
C16	Meldon Aplite Quarries (SX 567921)	198
C17	Praa Sands (Folly Rocks) (SW 573280)	200
C18	Cameron (Beacon) Quarry (SW 704506)	202
C19	Cligga Head area (SW 738536)	205

6 Post-orogenic volcanics (Group D sites) — 209

Introduction — 211
List of sites — 211
Lithological and chemical variation — 211

D1	Kingsand Beach (SX 435506)	214
D2	Webberton Cross Quarry (SX 875871)	217
D3	Posbury Clump Quarry (SX 815978)	220
D4	Hannaborough Quarry (SS 529029)	222
D5	Killerton Park and quarries (SS 971005)	224

Contents

References 227
Glossary 245
Index 249

Acknowledgements

Work on this volume was initiated by the Nature Conservancy Council and has been seen to completion by the Joint Nature Conservation Committee on behalf of the three country agencies, English Nature, Scottish Natural Heritage and the Countryside Council for Wales. Since the Geological Conservation Review was initiated in 1977 by Dr G.P. Black, then Head of the Geology and Physiography Section of the Nature Conservancy Council, many specialists in addition to the authors have been involved in the assessment and selection of sites; this vital work is gratefully acknowledged.

Over many years the authors have benefited from the guidance of the following: past and present members of the Publications Management and Advisory Committees for their support and advice; Dr R.P. Barnes, Dr A.J.J. Goode and Dr R.T. Taylor of the British Geological Survey; Dr B. Rice-Birchall of Keele University and British Ceramic Research Ltd.; Professor C.M. Bristow of the Camborne School of Mines; Mr P. Hawken of the China Clay Museum, St Austell; Mr D.C. Methven, Manager of the De Lank Quarries; Mr A.D. Francis and Mr J. Howe of English China Clays International Ltd.; Dr M. Stone of the University of Exeter; Mr C.V. Smale of the Goonvean and Rostowrack China Clay Co. Ltd.; Mr G.J. Lees and Dr R.A. Roach of Keele University; Mr J. Symonds, Manager of Luxulyan Quarry.

Dr A.J. Barber, Dr A. Hall and Dr M.F. Thirlwall of Royal Holloway and Bedford New College, University of London, kindly made many helpful comments when they refereed the text.

Thanks are also due to the GCR publication production team: Dr D. O'Halloran (Project Manager); Valerie Wyld (Sub-editor); Nicholas D.W. Davey (Scientific Officer and Cartographic editor); and Chapman and Hall for their help and advice at the final stages of publication. Cartographic drafting was done by Silhouette (Peterborough).

Access to the countryside

This volume is not intended for use as a field guide. The description or mention of any site should not be taken as an indication that access to a site is open or that a right of way exists. Most sites described are in private ownership, and their inclusion herein is solely for the purpose of justifying their conservation. Their description or appearance on a map in this work should in no way be construed as an invitation to visit. Prior consent for visits should always be obtained from the landowner and/or occupier.

Information on conservation matters, including site ownership, relating to Sites of Special Scientific Interest (SSSIs) or National Nature Reserves (NNRs) in particular counties or districts may be obtained from the relevant country conservation agency headquarters listed below:

English Nature,
Northminster House,
Peterborough PE1 1UA.

Scottish Natural Heritage,
12 Hope Terrace,
Edinburgh EH9 2AS.

Countryside Council for Wales,
Plas Penrhos,
Ffordd Penrhos,
Bangor,
Gwynedd LL57 2LQ.

Preface

This volume illustrates some of the significant aspects of magmatic activity from Devonian (408 million years ago) to early Permian (270 million years ago) times in SW England. This period covers the progressive development of the Variscan mountain-building episode, from initial basin formation to final deformation and the subsequent development of a fold mountain belt – the Variscan Orogen. Both extrusive (volcanic) and intrusive (plutonic) rocks are found in the orogen, and chart the various stages of its magmatic development.

The sites described in this volume are key localities selected for conservation because they are representative of the magmatic history of the orogen from initiation to stabilization. Some of the earliest volcanic activity in the Devonian is represented by submarine basaltic and rhyolitic lavas developed in subsiding basins, caused by the attenuation of the existing continental crust. In some cases, extensive rifting and attendant magmatism produced narrow zones of true oceanic crust, whereas elsewhere basaltic volcanism is related to fractures in the continental crust at the margins of the basins. After the filling of the sedimentary basins, and their deformation caused by crustal shortening (late Carboniferous Period), further activity is manifested by the emplacement of the Cornubian granites and later minor basaltic volcanism in the early Permian.

Accounts of the constituent parts of this history have enriched geological literature from the nineteenth century onwards, and have contributed to the advancement and understanding of magmatic and tectonic processes. South-west England contains examples of the composition and emplacement of ancient ocean crust (ophiolites), the diversity and formation of submarine lavas, the emplacement of multiple granite intrusions and their effects on the surrounding rocks, and of the nature of economically important post-magmatic alteration processes and mineralization.

P.A. Floyd, C.S. Exley and M.T. Styles

Chapter 1

The igneous rocks of south-west England

Introduction and site synthesis

INTRODUCTION AND SITE SYNTHESIS

The magmatic rocks of south-west England fall within the northern European Variscan fold belt; they are dominated by pre-orogenic basic–acid volcanics and post-orogenic granites, together with minor volcanics, that span the Devonian and Carboniferous systems. These major magmatic groups have played their part in the evolution of petrogenetic theory, but on a more limited scale than, say, the igneous rocks of the British Tertiary province, and generally relative to the development of the Variscan fold belt. For example, the small-volume, effusive volcanic rocks of Devonian–early Carboniferous age in south-west England were identified in the early European literature as representative of the so-called 'spilite–keratophyre geosynclinal association'; that is, the association of basic and acidic volcanics in a deep basinal setting. They have their temporal counterparts throughout the Variscan Orogenic Belt of Northern Europe and provided the scientific battleground for argument over the primary versus secondary origin of spilitic rocks (for example, Vallance, 1960; Amstutz, 1974), rocks which we now recognize as metamorphosed basalts.

Volcanic activity in Britain during the Devonian–Carboniferous can be broadly divided into two geographically separate areas that show contrasting eruptive and tectonic settings. The volcanic rocks of south-west England are dominated by medium- to deep-water submarine extrusives, shallow intrusives and volcaniclastics generated within rifted ensialic troughs and narrow ocean basins which appear to characterize the Variscides as a whole. Subsequently, they were extensively tectonized and metamorphosed during the different stages of the Variscan Orogeny and are thus characteristic of pre-orogenic volcanism. On the other hand, the foreland continental environment to the north in central-northern England and southern Scotland was outside the active orogenic belt and, as a consequence, deformation of volcanic rocks was relatively limited. The eruptive setting was also different. The calc-alkaline Old Red Sandstone volcanics of southern Scotland are dominated by subaerial lavas and volcaniclastics interbedded with thick sequences of intermontane sedimentary debris. Similarly, the extensive basaltic volcanics of Carboniferous age in northern England and the Midland Valley of Scotland are characterized by subaerial lavas and shallow, but often thick, intrusive complexes. Another significant difference is that volcanism continued throughout the Carboniferous in the northern area, whereas in south-west England it terminated in the Viséan in response to thrust-generated crustal shortening (Floyd, 1982a, fig. 15.2).

The post-orogenic granite batholith volumetrically dominates the magmatic rocks found in south-west England. The granite batholith and its associated metalliferous ore bodies have provided the type area for fractionated, high-level, high heat-flow granitic terranes and models for hydrothermally induced zonal mineralization. From the economic viewpoint, the special character of the granite batholith has been used as a model for tin mineralization and late-stage alteration associated with acidic magmatism throughout the world. Direct comparisons can be made with the chemically distinct Caledonian granites, some of which also feature high heat flow, but lack the extensive mineralization and late-stage effects exhibited in south-west England. Recently the radioactive-element-enriched Cornubian granite batholith has also attracted national attention as a hot, dry rock energy source and as a potential environmental hazard due to the emission of seeping radiogenic radon gas.

Of no less importance is the Lizard Complex in south Cornwall, and the local problem of its age and tectonic significance. Recent work has firmly placed it in the Variscan tectonic regime, as a fragment of obducted ophiolite with an attendant sedimentary *mélange*. If the early Devonian age for this dismembered ophiolite is correct, then it has European significance as being one of the few remnants of ocean crust exposed in the external zone of the Variscides.

The above brief introduction places the south-west England Devonian–Carboniferous igneous rocks in their regional and tectonic context relative to contemporaneous, often more extensive, magmatic episodes in Britain. However, compared with the magmatic character of the northern 'stable' foreland, the rocks of south-west England are characteristic of the various stages of evolution of an orogen and, in particular, illustrate the spectrum of magmatic events relative to tectonic features within the external zone of the Variscan fold belt (Figure 2.1).

The general localities of the sites that have been selected are shown in Figure 1.1 (numbered as listed in Table 1.1) and an outline of the significant features exhibited by each group of sites is given below. Details of the spatial and

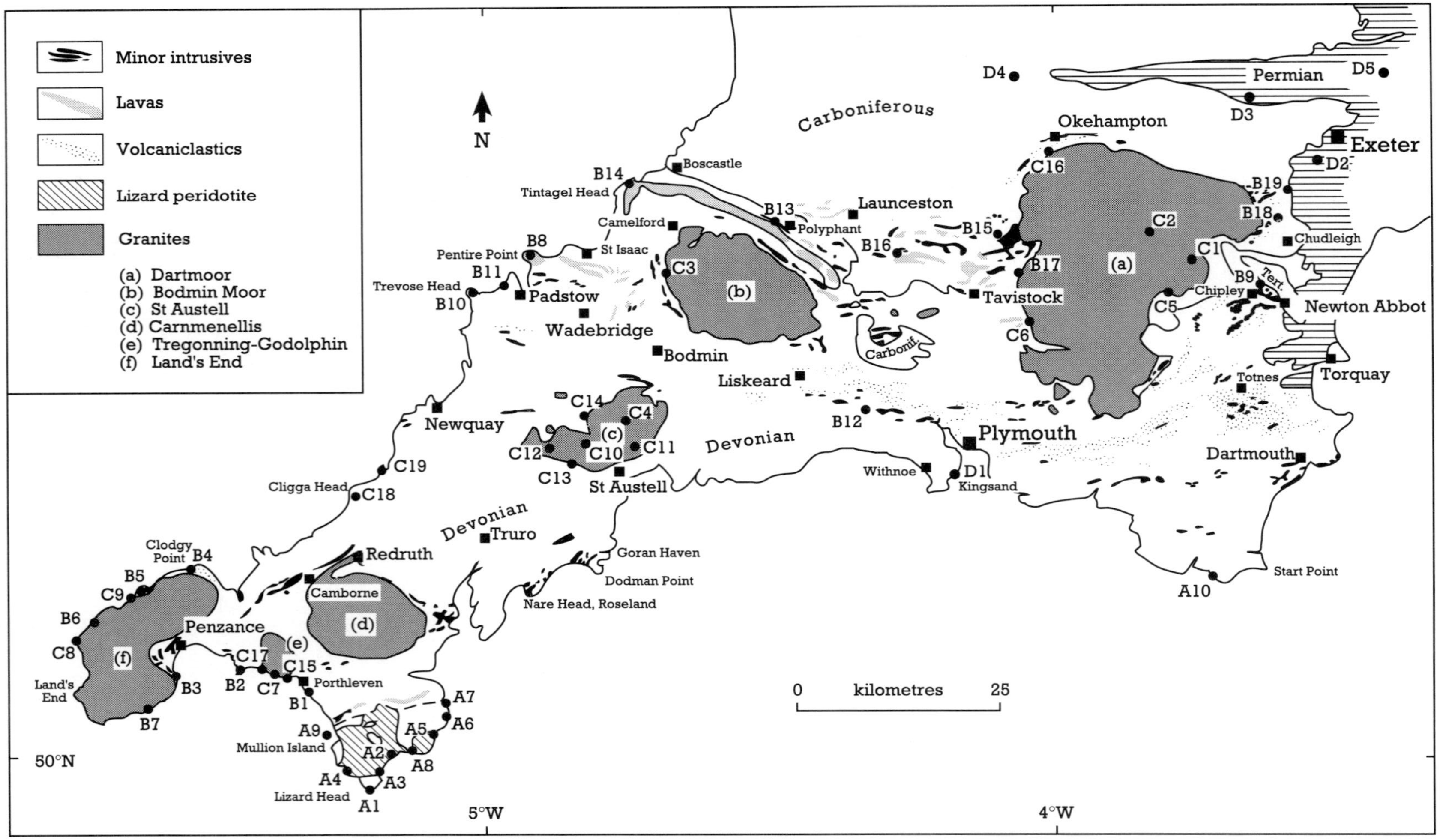

Minor intrusives
Lavas
Volcaniclastics
Lizard peridotite
Granites
(a) Dartmoor
(b) Bodmin Moor
(c) St Austell
(d) Carnmenellis
(e) Tregonning-Godolphin
(f) Land's End
N
Carboniferous
Permian
Devonian
Devonian
Tert.
Carbonif.
Exeter
Okehampton
Launceston
Tavistock
Plymouth
Newton Abbot
Torquay
Chudleigh
Chipley
Totnes
Dartmouth
Start Point
Kingsand
Withnoe
Liskeard
Bodmin
Wadebridge
Padstow
St Isaac
Camelford
Boscastle
Tintagel Head
Polyphant
Pentire Point
Trevose Head
Newquay
St Austell
Cligga Head
Truro
Redruth
Camborne
Penzance
Porthleven
Goran Haven
Dodman Point
Nare Head, Roseland
Clodgy Point
Land's End
Mullion Island
Lizard Head
0 kilometres 25
50°N
5°W
4°W

Introduction and site synthesis

temporal location, emplacement environment and origin of the volcanic and plutonic rocks within the tectonic zones of the Variscan Orogen and the local geological framework of south-west England are given in Chapter 2.

Two interrelated criteria were used for the selection of sites in south-west England:

1. to provide a full stratigraphical coverage of different magmatic activity throughout the Hercynian fold belt;
2. to illustrate the special or unique petrological and chemical characteristics of different magmatic units and their petrogenesis.

This allowed the continuum of magmatic activity within a fold belt to be documented, as well as highlighting special features best displayed in this region relative to elsewhere in the United Kingdom. The justification for choosing these specific sites, rather than others showing similar features, often rested on a combination of adequate geological exposure, lithological freshness and accessibility.

The sites can be conveniently grouped into four main units (A–D) that roughly relate to stratigraphical age and major magmatic events within the Variscan. Some significant geological features exhibited by these units are summarized below:

Lizard ophiolite, *mélange* and Start Complex

The plutonic complex of the Lizard ophiolite includes the serpentinized peridotite, gabbro and basaltic dykes, together with heterogeneous acid–basic gneisses. As representatives of oceanic crust, these units play an important role in the interpretation of early Variscan basins in south Cornwall; they also provide evidence for subsequent northward obduction. Although a volcanic carapace to the ophiolite is not present in sequence, tectonically associated, metamorphosed lavas (now hornblende schists) chemically similar to mid-ocean ridge basalts are consistent with a Lizard ocean-crust model. Metamorphism and tectonism possibly occurred both in a suboceanic setting and during obduction. The metavolcanic greenschists of the Start Complex also exhibit mid-ocean ridge chemical features and may represent another tectonized segment of the early Variscan ocean floor in this region.

Pre-orogenic volcanics

This unit comprises various stratigraphically localized volcanic rocks which were erupted contemporaneously with basinal sedimentation. They range in age from Devonian to early Carboniferous, but culminated in late Devonian–Viséan times. Although they represent a bimodal basic–acid suite (the old 'spilite-keratophyre association'), the volcanics are dominated by basaltic pillow lavas and high-level intrusives of both tholeiitic and alkaline character. Basic and acidic tuffaceous volcaniclastics are also common throughout the Upper Palaeozoic. The volcanics invariably have been altered subsequent to consolidation and deposition, exhibiting secondary assemblages indicative of the prehnite–pumpellyite and lower greenschist facies of metamorphism.

Cornubian granite batholith

The culmination or late stages of the Variscan Orogeny were marked by the emplacement of the Cornubian batholith at the end of the Carboniferous. This body is often interpreted as the product of the melting of sialic crust induced by continent–continent collision. Although predominantly a two-mica calc-alkaline granite, the batholith is composed of multiple intrusions, ranging in age from about 300 to 270 Ma and encompasses a number of highly fractionated acidic members including Li- and F-rich variants. Late-stage alteration effects are well displayed and include extensive tourmalinization, greisenization and kaolinization. The granite was also the source of the hydrothermal Sn–W mineralization as well as the heat engine for associated Cu–Pb–Zn–Fe–As deposits within the margins of plutons and their aureoles.

Figure 1.1 Simplified geological map of south-west England showing the distribution of magmatic rocks and the approximate location of sites described in the text (modified from Floyd, 1982b). Sites are numbered and grouped as in Table 1.1.

Group A sites: Lizard ophiolite and mélange	Group B sites: Pre-orogenic volcanics	Group C sites: Cornubian granite batholith	Group D sites: Post-orogenic volcanics
A1 Lizard Point (SW 695116 - SW 706115)	B1 Porthleven (SW 628254 - SW 634250)	C1 Haytor Rocks area (SX 758773)	D1 Kingsand Beach (SX 435506)
A2 Kennack Sands (SW 734165)	B2 Cudden Point-Prussia Cove (SW 548275 - SW 555278)	C2 Birch Tor (SX 686814)	D2 Webberton Cross Quarry (SX 875871)
A3 Polbarrow-The Balk (SW 717135 - SW 715128)	B3 Penlee Point (SW 474269)	C3 De Lank Quarries (SX 101755)	D3 Posbury Clump Quarry (SX 815978)
A4 Kynance Cove (SW 684133)	B4 Carrick Du-Clodgy Point (SW 507414 - SW 512410)	C4 Luxulyan (Goldenpoint, Tregarden) Quarry (SW 054591)	D4 Hannaborough Quarry (SS 529029)
A5 Coverack Cove-Dolor Point (SW 784187 - SW 785181)	B5 Gurnards Head (SW 432387)	C5 Leusdon Common (SX 704729)	D5 Killerton Park (SS 971005)
A6 Porthoustock Point (SW 810217)	B6 Botallack Head-Porth Ledden (SW 362339 - SW 355322)	C6 Burrator Quarries (SX 549677)	
A7 Porthallow Cove-Porthkerris Cove (SW 798232 - SW 806226)	B7 Tater-du (SW 440230)	C7 Rinsey Cove (Porthcew) (SW 593269)	
A8 Lankidden (SW 756164)	B8 Pentire Point-Rumps Point (SW 923805 - SW 935812)	C8 Cape Cornwall area (SW 352318)	
A9 Mullion Island (SW 660175)	B9 Chipley Quarries (SX 807712)	C9 Porthmeor Cove (SW 425376)	
A10 Elender Cove-Black Cove, Prawle Point (SX 769353 - SX 769356)	B10 Dinas Head-Trevose Head (SW 847761 - SW 850766)	C10 Wheal Martyn (SW 003556)	
	B11 Trevone Bay (SW 890762)	C11 Carn Grey Rock and Quarry (SX 033551)	
	B12 Clicker Tor Quarry (SX 285614)	C12 Tregargus Quarries (SW 949541)	
	B13 Polyphant (SX 262822)	C13 St Mewan Beacon (SW 985534)	
	B14 Tintagel Head-Bossiney Haven (SX 047892 - SX 066895)	C14 Roche Rock (SW 991596)	
	B15 Brent Tor (SX 471804)	C15 Megiliggar Rocks (SW 609266)	
	B16 Greystone Quarry (SX 364807)	C16 Meldon Aplite Quarries (SX 567921)	
	B17 Pitts Cleave Quarry (SX 501761)	C17 Praa Sands (Folly Rocks) (SW 573280)	
	B18 Trusham Quarry (SX 846807)	C18 Cameron (Beacon) Quarry (SW 704506)	
	B19 Ryecroft Quarry (SX 843847)	C19 Cligga Head area (SW 738536)	

Introduction and site synthesis

Post-orogenic volcanics

Shortly after consolidation of the batholith and regional uplift, a post-orogenic volcanic episode began in the late Carboniferous–early Permian. This comprised both suprabatholithic acid volcanism fed by late granite-porphyry dykes, and mafic extrusives and intrusives related to fault-bounded troughs. The latter group include various lamprophyres which often characterize the last stages of magmatic activity, apparently associated with granites of continental origin.

Table 1.1 List of GCR igneous rock sites in southwest England (see Figure 1.1 for locations)

Chapter 2

Geological framework

THE REGIONAL SETTING – VARISCAN OROGEN

A large part of the geology of Europe from Poland to Iberia is represented by the broad, sinuous, roughly E–W-trending Variscan fold belt which has affinities with the Appalachian Belt on the other side of the Atlantic Ocean and the Uralide Belt in Russia. This section provides a very generalized (and over-simplified) picture of the European Variscides within which to locate the particular magmatic rocks from south-west England which form the substance of this volume. Detailed reviews of the tectonic framework and development of the Variscan fold belt and its different segments may be found in Krebs (1977), Zwart and Dornsiepen (1978), Franke and Engel (1982), Behr *et al.* (1984), Lorenz and Nicholls (1984), Matte (1986), Ziegler (1986) and Franke (1989). The following outline largely follows Franke (1989).

The Variscan Orogen of northern Europe consists of a series of Ordovician to Carboniferous rift-generated basins, separated by metamorphosed crystalline ridges, which were progressively closed by the northward migration and subsequent collision of African Gondwanaland with northern Baltica. Early work by Kossmat (1927) subdivided the Variscan orogenic belt into a number of tectonic zones separated by major thrusts, ranging from the northern (external) Rhenohercynian zone to the (internal) Saxothuringian and Moldanubian zones of central Europe (Figure 2.1). The Rhenohercynian and Saxothuringian zones include basinal sedimentary sequences and volcanics indicative of initial rifting and later synorogenic sandstones heralding subsequent closure. Although much of the basic–acid

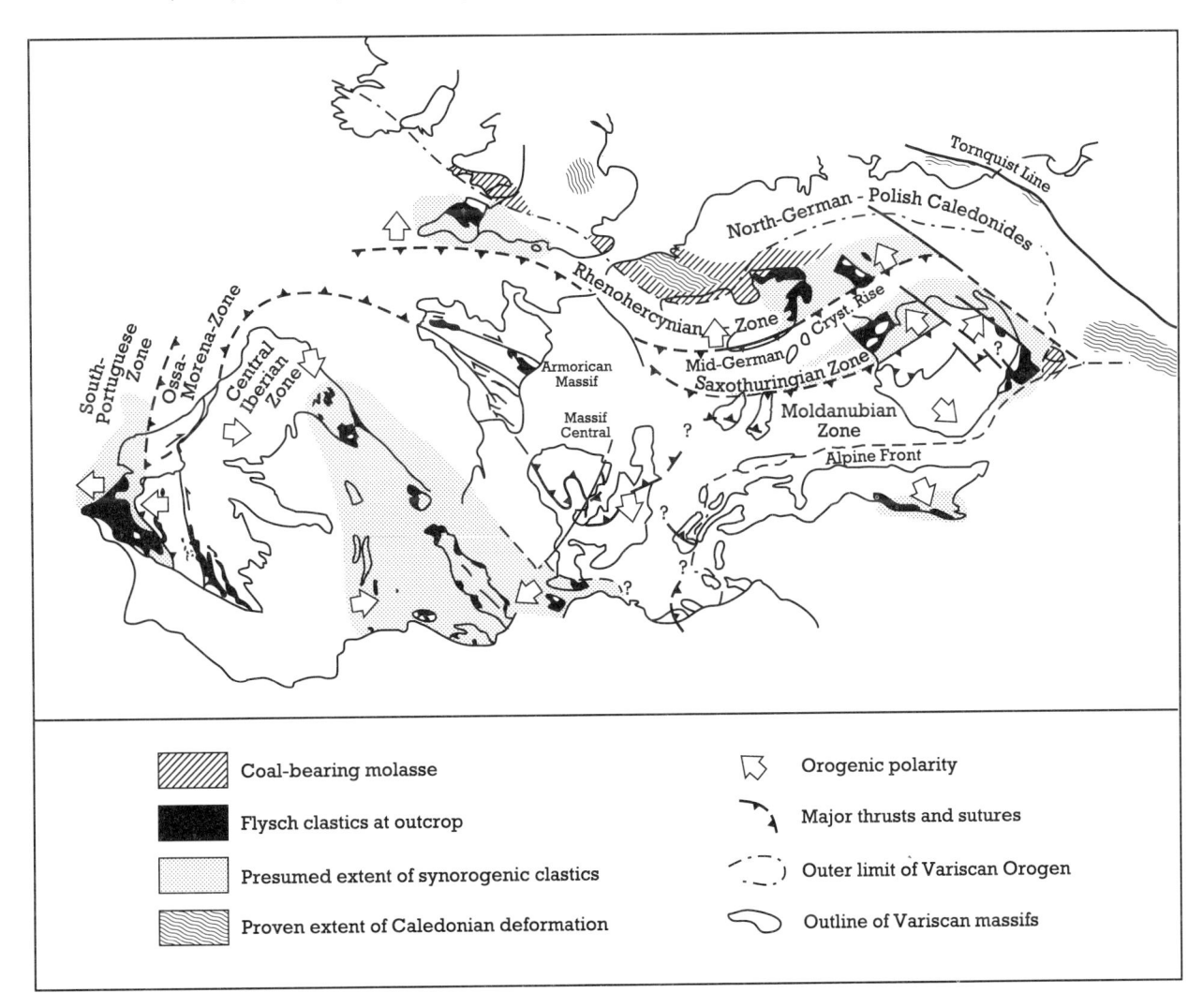

Figure 2.1 Distribution of tectonic zones in the Variscan Orogen of Europe (modified from Franke, 1989).

volcanism of these basins is typical of rifted continental crust, basaltic lavas with mid-ocean ridge (MORB) chemical characteristics are present and suggest that narrow oceans floored by oceanic crust were generated. Between these two zones in central Europe is the Mid-German Crystalline Rise largely composed of pre-Devonian magmatic and high-grade metamorphic rocks that have their counterpart in Armorica as the crystalline Normannia High. The Moldanubian zone is dominated by largely Precambrian magmatic and metamorphic rocks overprinted by Variscan tectonometamorphic events.

Essential to the Variscan story is the staged closure of the sedimentary basins by subduction of narrow oceanic segments and northwards-directed thrusting from late Ordovician through to the Carboniferous, such that many sequences are thrust bound and parautochthonous or allochthonous in character. In contrast to the general northwards convergence and progressive closure throughout much of the Palaeozoic, the basinal region that formed in the Rhenohercynian zone was not only late in development, but rapidly closed – opening in early Devonian and closing by the end of the late Devonian.

The magmatic rocks described in this volume are all within the external Rhenohercynian Zone of the Variscides, some of which have features in common with their counterparts in the rest of the orogen. The sedimentary and volcanic record of the Rhenohercynian Zone reflects an early Devonian rifting event that was rapidly followed by late Devonian to late Carboniferous crustal shortening that produced stacks of northwards-converging nappes and accompanying low-grade regional metamorphism. Relatively high-level, post-orogenic, granite magmatism is a common feature of this zone. There is a difference of opinion on the plate-tectonic situation of the initial rifting episode in the Rhenohercynian zone, but it is generally agreed that crustal attenuation was sufficient for the production of some oceanic crust. Four models have been suggested:

1. back-arc basin related to northerly directed subduction of an oceanic area to the south of Armorica (Floyd, 1982b; Leeder, 1982; Ziegler, 1986);
2. intracratonic strike-slip pull-apart basin (Badham, 1982; Barnes and Andrews, 1986);
3. small ocean basin related to southerly directed and although the older intrusions originated by subduction under an active arc to the north of Armorica (Holder and Leveridge, 1986);
4. small ocean basin generated as continental crust overrode a relict Caledonian spreading axis (Matthews, 1978; Franke, 1989).

Basic–acidic volcanism is a characteristic of all Variscan tectonic zones, although volcanic sequences are generally dominated by basalts ('spilites') with minor trachytes and rhyolites ('keratophyres' and 'quartz keratophyres'). Early Devonian rhyolites and late Devonian–Dinantian basalt pillow lavas and intrusives in south-west England have their temporal counterparts throughout the Rhenohercynian and Saxothuringian Zones, although the Moldanubian zone also exhibits late Dinantian calc-alkaline andesite-dominated volcanism. Throughout the Rhenohercynian zone most of the basalts are incompatible element-enriched, intraplate tholeiites and alkali basalts – the latter of which are particularly characteristic of south-west England (Floyd, 1982a; Wedepohl *et al.*, 1983). However, Middle and Upper Devonian basalts within stratigraphically restricted *mélange* sequences have chemical affinities to MORB (Floyd, 1984; Grosser and Dorr, 1986) and, together with the Lizard ophiolite and Start Complex in Cornubia, they provide evidence for the existence of Rhenohercynian oceanic crust. Another chemical characteristic that appears to be a common feature of the basalts within the Rhenohercynian (and possibly Saxothuringian) Zones is the change in degree of incompatible-element enrichment from the Devonian to the Carboniferous (Floyd, 1982a). This is probably a reflection of the change from generally depleted MORB-type basalts in mid-late Devonian rift zones to the more enriched intraplate basalts of late Devonian and early Carboniferous age on the margins of basins.

As far as the Variscan granites are concerned, all the tectonic zones contain granitic massifs, many of which include types resembling those of the Cornubian batholith. In addition to older pre-Variscan intrusives and metamorphics, two or three distinct intrusive phases are found, as illustrated below.

The Krušné hory Mountains granitoids (northwest part of the Bohemian Massif) in the Saxothuringian Zone, have ages ranging from 340 to 260 Ma (K/Ar method). They are grouped into an Older Intrusive Complex, with biotite and I-type granite characteristics, and a Younger Intrusive Complex, with Li-mica and the chemical attributes of S-type granites. The youngest rocks also have high concentrations of Rb (c. 500 ppm), Sn (c. 38 ppm) and F (c. 2900 ppm) (Štemprok, 1986). The granites of Normandy and northern Brittany group into suites with ages ranging from 340–300 Ma and 300–280 Ma (both K/Ar and Rb/Sr methods). Where these have been characterized as 'S' or 'I' types, the latter tend to be the older, and the younger group tends to be relatively enriched in Sn and other metals (Chauris et al., 1969; Adams, 1976; Peucat et al., 1984; Georget et al., 1986).

In the Schwarzwald, granites (dated by the Rb/Sr method) form three series ranging from about 363 Ma through 330–310 Ma (the majority) to 290–280 Ma. They lie in the Moldanubian zone and although the older intrusions originated by crustal melting, the younger show evidence of fractional crystallization and magmatic differentiation, together with high-level intrusive characteristics. With decreasing age, overall Sr contents fall (245 to ?24 ppm), whereas both Rb and F rise (228 to 483 ppm Rb and 618 to 1575 ppm F) (Emmerman, 1977). Also in the Moldanubian zone are the granitoids of southern Brittany and the Massif Central and here, too, there are 'older' and 'younger' generations dated at about 310–300 Ma and 290–280 Ma (K/Ar and Rb/Sr methods) (de Albuquerque, 1971; Reavy, 1989). In both cases there is evidence of differentiation sequences, but whereas de Albuquerque (1971) supposes that the parent rocks were pelites and greywackes, Pinto (1983) suggests that the younger granites have I-type characteristics.

Throughout the Variscides of northern Europe, therefore, granites of approximately the same age as those in Cornubia are found and they often display similar chemical characters, such as being enriched in Li, Rb, Sn and F. There are two important differences, however. On the continental mainland, they lie overwhelmingly within the Saxothuringian and Moldanubian tectonic zones, whereas the Cornubian rocks are in the Rhenohercynian zone where rather few granitoids (for example, Harz Massif) occur relative to the mainland. Secondly, in Cornubia, only the younger (300–280 Ma) suite is found, the older generations apparently being absent. These features tend to support the suggestion, discussed elsewhere in this volume, that the Cornubian batholith may be detached from its original roots.

The local setting – Rhenohercynian zone of south-west England

We now turn to south-west England, which is representative of the western end of the external Rhenohercynian zone. This section provides a simplified geological framework for the volcanic and plutonic events outlined in Chapter 1, and from which the sites have been chosen. Recent work and summaries of the geology of south-west England may be found in Durrance and Laming (1982), Hancock (1983), Hutton and Sanderson (1984) and Dineley (1986). An outline geology of south-west England and the distribution of magmatic rocks is shown in Figure 1.1.

Throughout the Devonian and Carboniferous periods, south-west England was mainly characterized by the successive northward development and subsequent closure of small, often deep-water, basins (Matthews, 1977). They were bordered to the north by shallow coastal shelves and initially received debris from the Old Red Sandstone continent in South Wales and the Bristol Channel (Freshney et al., 1972, 1979; Tunbridge, 1986). Later, a southern source was also important, with flysch and sedimentary *mélange* material derived from northwards-advancing nappes filling both late Devonian and Carboniferous basins (Isaac et al., 1982; Holder and Leveridge, 1986). During mid-Devonian times, basins in the east were segmented by 'rises' that developed shallow-water carbonate banks possibly capping volcanic seamounts (Edmonds et al., 1969).

Structurally, the Variscan fold belt is dominated by thrust ('thin-skin') tectonics with piles of southerly derived nappes characterizing Cornwall and central south-west England (for example, Matthews, 1981; Isaac et al., 1982; Shackleton et al., 1982; Leveridge et al., 1984; Selwood and Thomas, 1986a and 1986b). Figure 2.2 (from Dineley, 1986) summarizes the main structural features of Variscan South-west England. From a broad correlation of K/Ar age dates on slates and phyllites (Dodson and Rex, 1971) with structural zones (Dearman et al., 1969; Dearman, 1969, 1971; Sanderson and Dearman, 1973), minimum ages for deformation and related metamorphism can be estimated. In western Cornwall the initial F_1 deformation (370–350 Ma BP) occurred at the end of the Devonian, whereas elsewhere it represents a mid-Carboniferous event (345–325 Ma BP). The second deformation phase (F_2) overprinted both areas (Sanderson, 1973) during

Geological framework

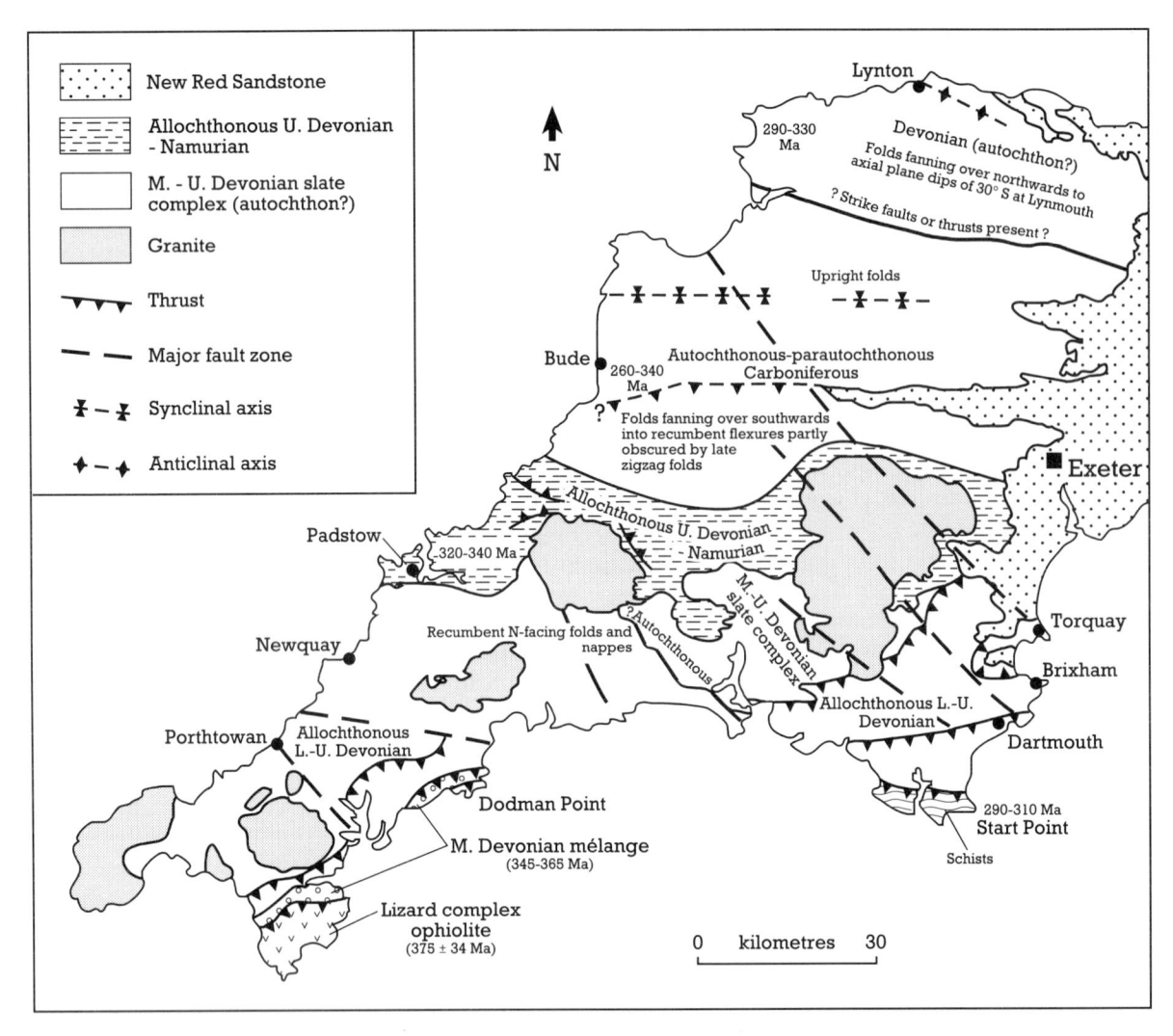

Figure 2.2 Map of south-west England showing major structural features, together with K/Ar age zones (after Dineley, 1986).

the late Carboniferous (315–275 Ma BP). Two subsequent, relatively minor, fold phases (F_{3-4}) are probably related to the emplacement of the batholith.

Volcanic sequences occur at various structural levels within the nappes such that Upper Devonian lavas can be found overlying Lower Carboniferous volcanics of similar eruptive character. For example, detailed structural schemes for central south-west England have been recorded by Isaac *et al.* (1982), Selwood *et al.* (1985) and Selwood and Thomas (1986a, 1986b), and these emphasize the presence of various nappes thrust over parautochthonous/autochthonous 'basement'. Figure 2.3 summarizes the relationship between recognized structural units in this area: the majority of the Upper Palaeozoic volcanics occur within the Port Isaac and Tredorn Nappes (north Cornwall), the Greystone Nappe (central area between Bodmin and Dartmoor) and the Chudleigh Nappe (east Dartmoor). The Greystone Nappe is considered to be an early, synsedimentary, gravity-transported slice emplaced in the late Viséan and subsequently overridden by other 'younger' nappes (for example, Port Isaac Nappe, with mid-Carboniferous ages; Tredorn Nappe, with late Carboniferous ages) derived from further south (Isaac *et al.*, 1982; Selwood and Thomas, 1986b). Basically, tectonic instability was heralded not only by prograding flysch and sedimentary *mélanges*, but also by volcanic activity concentrated within the late Devonian–Dinantian period, both coeval with, and prior to, the advancing deformation front.

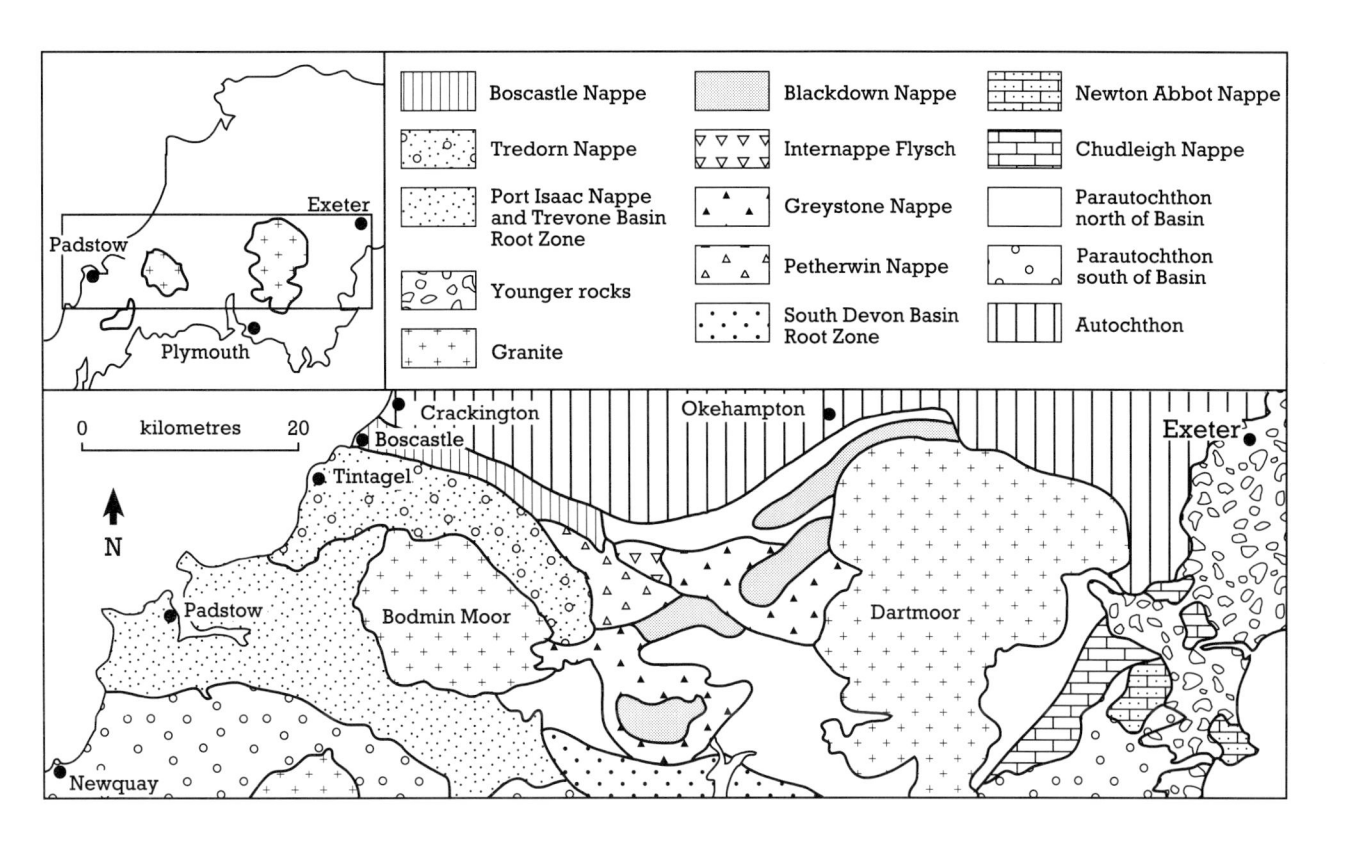

Figure 2.3 Distribution and correlation of structural units in central south-west England (after Selwood and Thomas, 1986a).

Geological framework

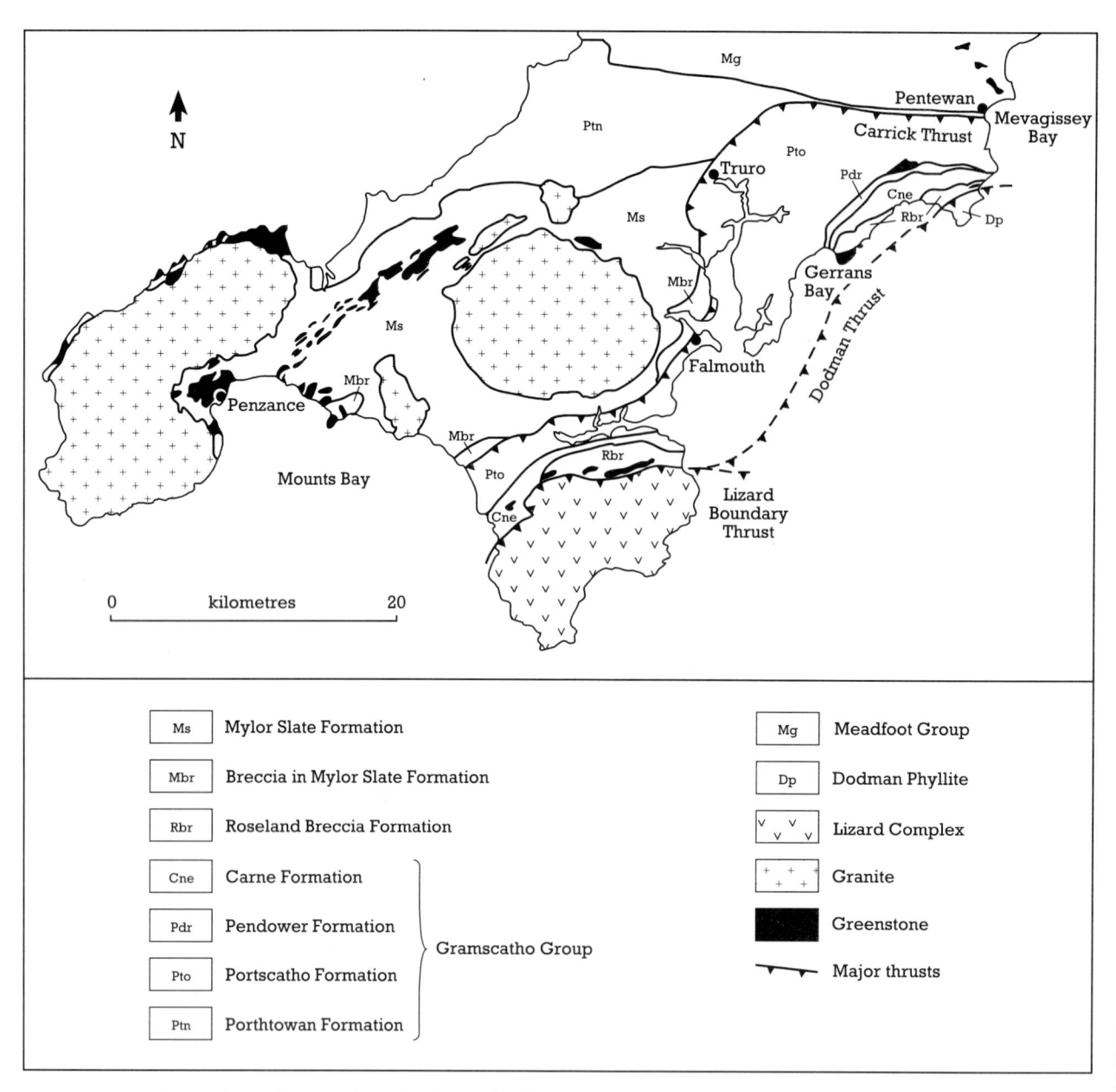

Figure 2.4 Geological map of south Cornwall showing various allochthonous units, including the Lizard Complex, resting on the northern parautochthon of the Porthtowan Formation and Mylor Slate Formation (after Holder and Leveridge, 1986).

We shall now consider in more detail the stratigraphical position and environmental significance of the various magmatic groups outlined in Chapter 1 and how they relate to the geological development of south-west England.

Devonian volcanic activity

Lizard and west Cornwall region

The most significant part of the magmatic story starts in southern Cornwall with the development of the Gramscatho Basin in the late, early Devonian (Barnes and Andrews, 1986). The Gramscatho Group (Figures 2.4 and 2.5) comprises a deep-water turbiditic sandstone–mudstone association and can be divided into two main parts (Holder and Leveridge, 1986):

1. a northern parautochthonous region which is separated by the Carrick Thrust (Leveridge *et al.*, 1984) from

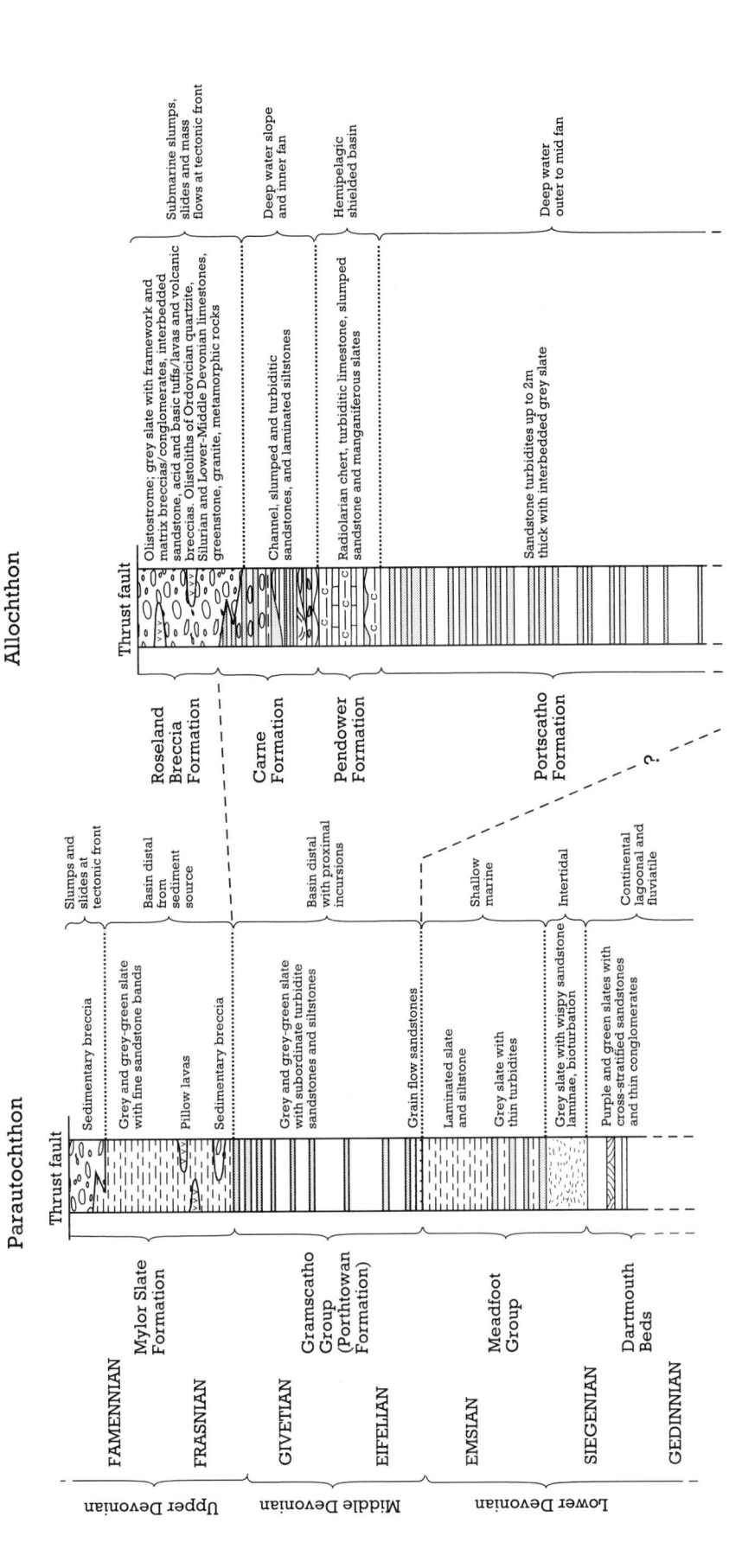

Figure 2.5 Devonian lithostratigraphical sequences in the parautochthon and allochthon units of south Cornwall (after Holder and Leveridge, 1986).

Geological framework

2. an allochthon composed of a series of thrust slices which includes the Lizard Complex at the highest structural level.

The Gramscatho Group is generally considered to be mid- to late Devonian in age (Sadler, 1973; Le Gall *et al.*, 1985); it is tectonically juxtaposed against the early Devonian shallow-water Meadfoot Group along the Perranporth–Pentewan Line (Dearman, 1971).

The importance of the magmatic rocks in this region (Figure 2.2) centres on the interpretation of the Lizard ophiolite as the oceanic basement to the Gramscatho Basin (Barnes and Andrews, 1986; Holder and Leveridge, 1986), and the close association of various, largely mid–late Devonian, basaltic volcanics that have chemical compositions akin to mid-ocean ridge basalts (MORB) (Floyd, 1982a, 1984; Barnes, 1984). The ophiolite and exotic volcanics thus constitute the only good evidence for oceanic crust in this part of the Rhenohercynian belt. The following magmatic suites can be identified:

1. Lizard ophiolite (Strong *et al.*, 1975; Bromley, 1976, 1979; Styles and Kirby, 1980; Kirby, 1979a, 1979b, 1984) and associated hornblende schists.
2. Metabasic clasts within the south Cornish *mélange* zone (Barnes, 1983) – the Roseland Breccia Formation of Holder and Leveridge (1986) (Figures 2.4 and 2.5) – and possibly including the pillow lavas of Mullion Island.
3. *In situ* Middle Devonian submarine volcanics (Tubbs Mill unit) at the base of an allochthonous slice of the Gramscatho Group.
4. Basic, submarine volcanics and associated high-level intrusives within the late Devonian Mylor Slate Formation of west Cornwall (Taylor and Wilson, 1975).
5. An additional basic sequence (undifferentiated Devonian age, possibly early Devonian; Dineley, 1986) are the metavolcanic greenschists of the Start region. Magmatically and tectonically, they could form part of the south Cornish allochthonous terrane and represent another segment of oceanic crust thrust over the early Devonian sediments to the north (Coward and McClay, 1983). There is evidence to suggest that they may well form part of a more extensive 'metamorphic basement' to the sea floor to the south and west of Start Point (Phillips, 1964).

A simplified interpretation of events during the mid- to late Devonian in this region might be as follows. The Gramscatho Basin was generated initially by extension and thinning of continental crust (Matthews, 1977; Leeder, 1982). The attendant magmatism eventually developed oceanic crust that subsequently formed the base to part of the basin (Rattey and Sanderson, 1984; Barnes and Andrews, 1986) and is now represented by the Lizard ophiolite and exotic volcanics within the south Cornish *mélange* (groups 1 and 2 above). The latter have a similar chemistry to the Tubbs Mill volcanics (group 3), which appear to be the earliest (middle Devonian), contemporaneous, submarine lavas now located at the thrust-terminated base of one of the flysch units. The basin was filled from the south by turbiditic sandstones (Gramscatho Group) derived from the erosion of nappes predominantly composed of acidic continental arc materials (Floyd and Leveridge, 1987; Shail and Floyd, 1988). The deepening of the basin is marked by the parautochthonous argillites of the late Devonian Mylor Slate Formation which are probably coeval with the upper part of the northwards-advancing Gramscatho turbidites and *mélanges* (Holder and Leveridge, 1986). Within the Mylor Slate Formation are discrete pillow lava and intrusive horizons (group 4 above) which are chemically distinct from the earlier MORB-like volcanics mentioned above. The filling of the basin is indicated by a proximal heterogeneous sedimentary *mélange* composed of MORB-like volcanic clasts and continental-derived rudaceous debris (Barnes, 1984). Suboceanic slicing and metamorphism of the ocean crust (Vearncombe, 1980), soon after its formation, heralded the onset of obduction, which eventually emplaced the Lizard ophiolite on top of a stack of Gramscatho nappes and the *mélange* towards the end of the Devonian (Styles and Rundle, 1984).

Central south-west England region

To the east of the Perranporth–Pentewan Line, which marks the boundary of the Lizard/west Cornwall region, is the magmatically distinct central region, that for the convenience of this summary covers north Cornwall and Devon. Idealized stratigraphical relationships in Devon are illustrated in Figure 2.6.

The early Devonian (Dartmouth Slate, Staddon Grit and Meadfoot Formations) in south Devon is characterized by shallow-water tidal facies as well

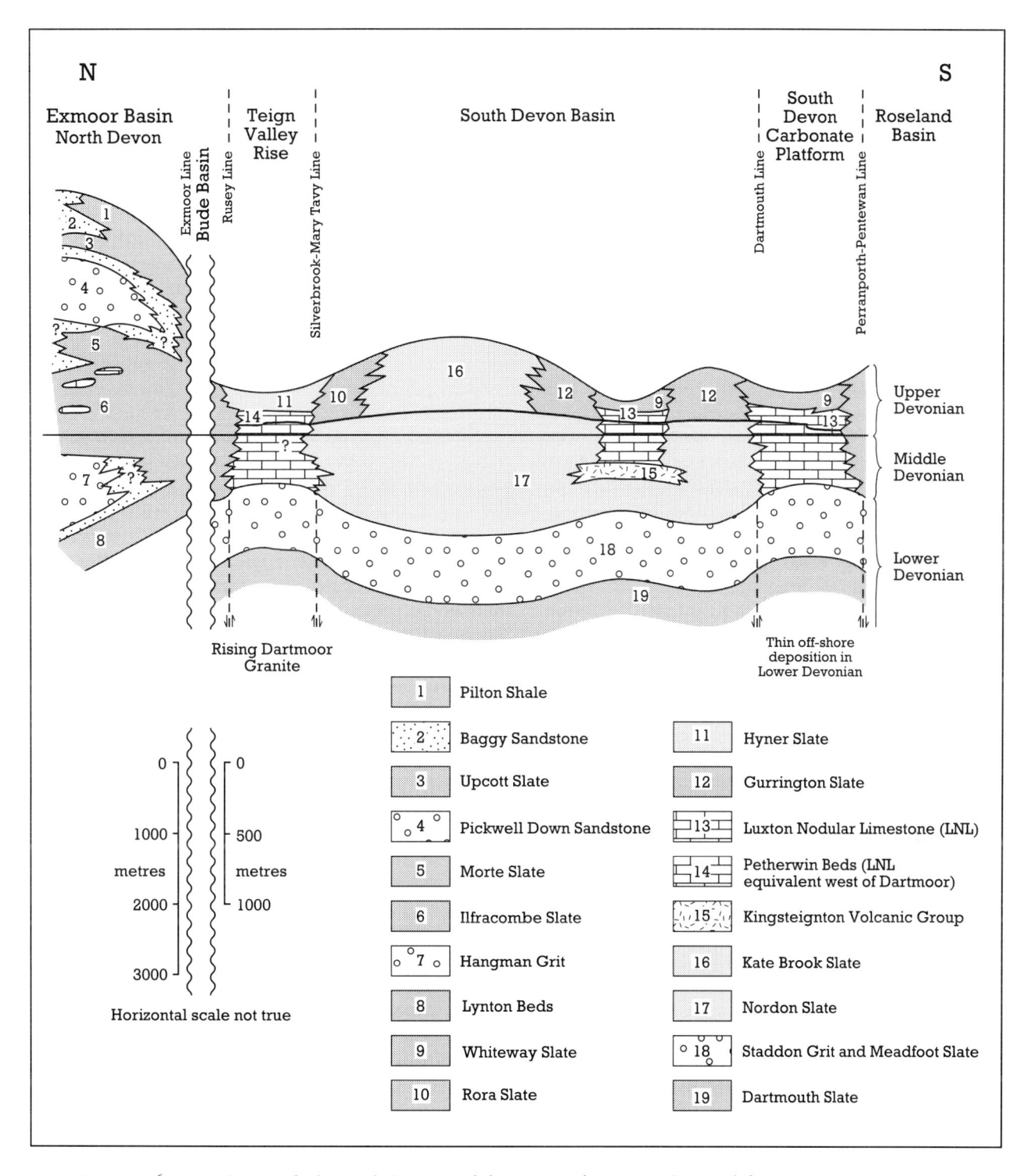

Figure 2.6 Correlation of observed Devonian lithostratigraphy across Devon (after Durrance and Laming, 1982).

as non-marine fluvial and alluvial plain sequences (Dineley, 1966; Richter, 1967; Selwood and Durrance, 1982; Pound, 1983). The earliest record of Devonian volcanism in south-west England is seen within the Dartmouth Slates; this comprises limited rhyolitic and basaltic lavas and intrusives with spatially associated tuffs (Shannon and Annis, 1933; Durrance, 1985a). The rhyolites are typical of an 'active margin environment' (Durrance, 1985a) and may have been produced by the melting of the underlying continental crust. During the middle Devonian, more open marine, shallow-shelf conditions prevailed with the local development of massive carbonate platforms (for example, Plymouth Limestone) which formed an east–west ridge extending from Dartmouth and Plymouth westwards towards Bodmin (Selwood and Durrance, 1982; Burton and Tanner, 1986). Volcanism was again restricted to relatively small volume, but numerous, tuff horizons and minor basaltic, highly vesicular lavas. Acid and basic volcaniclastics (for example, Kingsteignton Volcanics) are best developed in south-east Devon (Totnes, Newton Abbot), where they interdigitate with reef limestones (Ussher, 1912; Middleton, 1960; Richter, 1965). A completely different and interesting igneous association is the Clicker Tor ultramafic body and associated basic lavas and volcaniclastics near Liskeard (Burton and Tanner, 1986) developed within middle Devonian argillites and tuffaceous slates.

In a similar fashion, initially shallow-water sedimentation in north Devon started to the south of the broad Old Red Sandstone coastal plain, and throughout the Devonian was characterized by a migratory sequence of interdigitating marine and non-marine clastic sediments (Webby, 1966; Tunbridge, 1983). No significant magmatic activity is associated with this regime, in contrast to south Devon over the same time period.

From the magmatic viewpoint, significant and widespread volcanic activity did not start until the late Devonian and its products now form a roughly linear belt from the north Cornish coast (Padstow) to east of Dartmoor at Chipley (Figure 1.1). The generally shallow-water conditions over the whole area during the early Devonian were terminated by differential subsidence at the southern margin of the shelf, with the development of a basinal facies during the mid and late Devonian. A schematic diagram showing the general relationship between the various facies during the Devonian and Carboniferous is illustrated in Figure 2.7.

The northern shelf–basin margin is interpreted as a half-graben fracture system (Selwood and Thomas, 1986b) and has a linear belt of volcanic activity closely associated with it. Extensive volcanism thus appears to be associated with the development of the basin, but related to marginal fractures rather than forming an incipient basement. This is probably a similar situation to the submarine lavas within the Mylor Slate Formation of west Cornwall of comparable age and clearly distinct from the Lizard and the *mélange* volcanics. Classic pillow lava sequences (for example, at Pentire Point and Chipley) are seen within the deep-water argillites of the late Devonian basinal facies (Dewey, 1914; Gauss and House, 1972, Middleton, 1960). High-level doleritic intrusives are also common in the Padstow area, and exhibit two petrographically and chemically distinct primary suites (Floyd, 1976; Floyd and Rowbotham, 1982). Rare ultramafic material, found within late Devonian slates, also occurs within the volcanic belt and, spatially, this is closely associated with dolerites and basalts. Such a body is the Polyphant Complex to the north-east of Bodmin Moor (Reid *et al.*, 1911; Chandler *et al.*, 1984), which in the tectonic scheme of Isaac *et al.* (1982) is considered to be part of an ophiolitic *mélange* possibly equivalent to the Gramscatho Basin *mélanges* (Isaac, 1985).

To summarize, a number of magmatic suites can be identified that are broadly related to their eruptive setting and age:

1. Minor early Devonian rhyolites, basic lavas and volcaniclastics that are representative of the earliest expressions of magmatism in the Variscan fold belt (south Devon).
2. Volcaniclastic-dominated sequences and shallow-water basaltic lavas within the middle Devonian, that are often associated with the margins of carbonate reefs and platforms (south and south-east Devon).
3. Deep-water pillow lavas and high-level sills related to the northern, fractured margin of the late Devonian Trevone Basin (north Cornwall).
4. Rare association of ultramafic intrusives with late Devonian basic volcanics and deep-sea sediments (east Bodmin Moor).

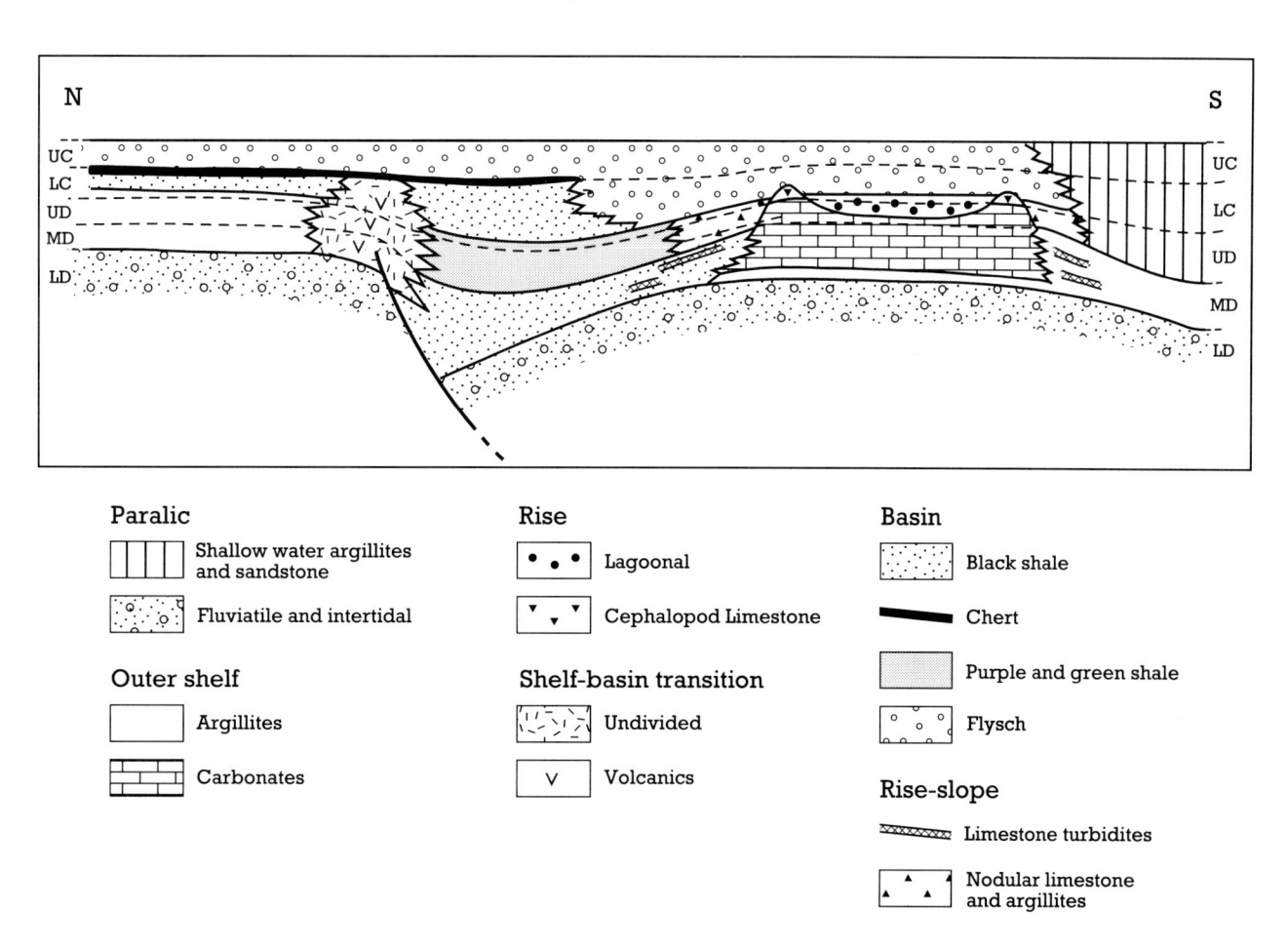

Figure 2.7 Generalized facies relationships throughout the Devonian and Carboniferous of central south-west England (after Selwood and Thomas, 1986b). LD = Lower Devonian; MD = Middle Devonian; UD = Upper Devonian; LC = Lower Carboniferous; UC = Upper Carboniferous.

Carboniferous volcanic activity

Sedimentation in the early Carboniferous consisted, in north Cornwall and north Devon, of deep-water argillites deposited in basins that spread northwards across the shelf margin (Goldring, 1962; Matthews, 1977). However, in south-east Cornwall and south Devon, a shallow-water paralic facies dominated, and the Devonian reefs and carbonate platforms became lagoonal areas (Whiteley, 1984; Selwood and Thomas, 1986b). Due to a rising source region to the south, prograding flysch deposits began to spread northwards across the whole area, eventually filling and shallowing the northern basins by the Namurian (Selwood and Thomas, 1986b). Subsequent sedimentary history in the late Carboniferous reflects the continuation of basin filling by clastic materials until largely non-marine conditions prevailed (Freshney and Taylor, 1972; Higgs, 1984). Magmatic activity was, however, mainly confined to the early Carboniferous in the north of the region, and terminated prior to the onset of the Namurian flysch phase.

Carboniferous volcanicity was thus temporally restricted (Viséan–Tournaisian) and again concentrated along shelf–basin margins. It forms a broad belt from the north Cornish coast (Tintagel) inland to central south-west England and around Dartmoor to the Teign Valley (Figure 1.1). The Carboniferous belt roughly parallels late Devonian activity, but lies further north, reflecting the northward migration of the Carboniferous basins.

On the basis of macrofaunal evidence in the associated sediments, the Tintagel Volcanic Formation is correlated with the Meldon volcanics (north-west Dartmoor), representing volcaniclastic-dominated eruptions dated at near the

Viséan–Tournaisian boundary (Francis, 1970). The Tintagel volcanics are enclosed in pyritous black mudstones suggesting a deep-water basinal eruptive setting, whereas the Meldon volcanics are associated with calcareous argillites, thin limestones and cherts that probably flanked a rise slope. Owing to the extensive within-bed shearing and the late development of white mica, the Tintagel volcanics are now heterogeneous greenschists with identifiable basalt lava only preserved as relicts within low-strain lenses. Geochemical data demonstrate that they are a comagmatic series with alkali-basalt affinities (Robinson and Sexton, 1987; Rice-Birchall and Floyd, 1988).

The central south-west England thrust–nappe terrane between Bodmin Moor and Tavistock includes acid and basic volcaniclastics, vesicular (commonly pillowed) basaltic lavas and thick basic intrusives (dolerites and some gabbros) – the latter being typical of the greenstones of Variscan south-west England. Some of the massive basic bodies were intruded at a high level and may show pillowed tops. The majority of these volcanics are late Viséan to late Tournaisian in age (Stewart, 1981) and confined to the early Greystone Nappe. The volcanics are associated with deep-water, black mudstones, distal turbiditic mudstones, radiolarian cherts and olistostromic material, and they were developed in a small pelagic basin over a 19 Ma period (Chandler and Isaac, 1982). The interest in this pelagic–olistostrome–volcanic association concerns its interpretation as an ocean-floor sequence, with the basalts and dolerites apparently exhibiting chemical affinities with oceanic crust (Chandler and Isaac, 1982). On this basis and the (possible) initial close stratigraphical association of the late Devonian mafic/ultramafic Polyphant Complex, it has been suggested that this was an ophiolite association (Chandler and Isaac, 1982) and that the marginal basin model for the Devonian (for example, Leeder, 1982) could be applicable to the early Carboniferous. The evidence on which this interpretation is based is considered tenuous at best (Selwood and Thomas, 1986b) and the application of the geochemical data is open to question. A completely different eruptive setting is exhibited by apparently isolated volcanic centres, such as that seen at Brent Tor (north of Tavistock) where a reworked apron of hyaloclastite debris bordering a seamount-like edifice occurs.

To the east of Dartmoor in the Teign Valley, relatively minor pillow lavas and acidic volcaniclastics are interbedded with radiolarian cherts, black limestones and mudstones, although much of the activity is characterized by huge, massive, basic intrusives (greenstones). Many of these bodies are dolerite sills or nearly concordant sheets (sometimes 100 m or more thick) showing internal differentiation, with the development of granophyric fractions (Morton and Smith, 1971).

In summary, the volcanics have broadly similar eruptive ages and volcanic assemblages, although a number of suites can be recognized by their tectonic association (Figure 2.3):

1. Tintagel Volcanic Formation basaltic volcaniclastics and minor intrusives within the Tredorn Nappe (north Cornish coast and inland).
2. Volcaniclastics, pillow lavas, and massive intrusives mainly within the Greystone Nappe (between Bodmin Moor and Tavistock).
3. Volcaniclastics, minor lavas and massive intrusives within the Blackdown Nappe (west Dartmoor).
4. Minor volcaniclastics and pillow lavas, with dominant massive intrusives within the autochthon (Teign Valley, north-east Dartmoor).

Carboniferous–Permian plutonic activity

The geography of Cornwall and west Devon is dominated by the granite batholith which constitutes the spine of the peninsula. Like all batholiths, it is made up of many individual intrusions embracing a range of ages and compositions. Although small by global standards and lacking the dioritic and granodioritic components often seen, it compensates for these shortcomings by showing an unrivalled sequence of late- and post-magmatic processes, including mineralization.

The detailed relationship of the batholith with the structural environment outlined above is still somewhat obscure, but from geophysical evidence it is known to extend from a few kilometres east of Dartmoor into the Atlantic west of the Isles of Scilly, a distance of about 250 km. Beyond that, a row of gravity 'lows' extends WSW to the continental margin, and Day and Williams (1970) have interpreted these as being due to granite. If the 'lows' are part of the peninsular granite, then they would extend the length of the batholith to some 500–600 km. Over the Dartmoor–Scilly Isles distance, the batholith varies from about 20 km thick in the east, to half that thickness in the west, and is 40–60 km wide at its base. The Mohorovičić discontinuity beneath south-west England occurs at a depth of approximately 27–30 km (Bott *et al.*, 1958; Bott *et al.*, 1970; Holder and Bott, 1971; Tombs, 1977; Shackleton *et al.*, 1982; Brooks *et al.*, 1983; Brooks *et al.*, 1984). About 95 km WNW of the Isles of Scilly is the small submarine granite outcrop of Haig Fras. This has affinities with the Cornubian batholith and has been interpreted as being separated from it by wrench faulting (Exley, 1966). However, it also contains gneissose material and an alternative suggestion is that it may represent basement that has been thrust northward in a nappe (Goode and Merriman, 1987). A third possibility is that it is related to separate submarine granites of unknown age rumoured to be present closer to Ireland. Day and Williams (1970), for example, consider the Ladabie Bank gravity 'low' (about 40 km north-west of Haig Fras) to be due to granite, but no granite north of Haig Fras was noted in the SWAT seismic survey (BIRPS and ECORS, 1986). Edwards (1984), on the other hand, believes that Haig Fras belongs to a separate batholith parallel to the one exposed on the mainland, and he notes that both have been displaced dextrally.

The most substantial geophysical evidence as to the nature of the base of the batholith comes from Brooks *et al.* (1984) who have found a southwards-dipping seismic reflector at a depth of about 10–15 km. This is the level at which the densities of country rock and granite merge and Brooks *et al.* (1984) interpret this reflector as a thrust. Although this conforms well with the general structural picture, it creates a problem by suggesting a separation of the batholith from its original roots and thus from what would normally be the source of hydrothermal and metallic constituents. Shackleton *et al.* (1982) have included similar thrusts in their interpretation and have proposed that 'the granites were generated farther south and injected northwards as a sheet-like body from which protrusions moved diapirically upwards and by stoping to form the separate exposed granite masses'. They do not specify whether this sheet was magmatic, and their suggestion creates other problems to do with heat, density and vertical and horizontal granite movement, to the extent that granite hot enough to rise diapirically would not also be expected to travel subhorizontally for perhaps 200 km.

The upper surface of the batholith is exposed in a series of seven major outcrops. These are (Figure 1.1), from east to west: the Dartmoor, Bodmin Moor, St Austell, Carnmenellis, Tregonning–Godolphin, Land's End and Isles of Scilly bodies (approximate exposed areas are, respectively, 650 km^2, 220 km^2, 85 km^2, 135 km^2, 13 km^2, 190 km^2 and 30 km^2). Several minor bodies also crop out and include Hemerdon Ball (south of Dartmoor), Hingston Down and Kit Hill (between Dartmoor and Bodmin Moor), Belowda Beacon and Castle-an-Dinas (north of St Austell), Cligga Head, St Agnes Beacon, Carn Brea and Carn Marth (north of Carnmenellis) and St Michael's Mount in Mounts Bay, east of the Land's End Peninsula (Figure 1.1). Each of the major outcrops is itself made up of more than one intrusion, and recent work (Hill and Manning, 1987) suggests that some may contain many more intrusions than hitherto suspected.

The ages of the exposed biotite-bearing granites indicate that the main period of magmatism lay between 290 Ma and 280 Ma BP, and that this was followed by a second phase some 10 Ma later (Darbyshire and Shepherd, 1985; Table 2.1). The source of the early granites has always been regarded as crustal, largely because of the nature of their included xenolithic material, and this view has been reinforced in recent years by determinations of their initial $^{87}Sr/^{86}Sr$ ratios (Table 2.1), Pb and Sr isotopes (Hampton and Taylor, 1983), and general chemistry and mineralogy which show such features as high K_2O/Na_2O and low Fe_2O_3/FeO ratios, high normative corundum, and rare or absent hornblende, sphene and magnetite. These are all characteristics of the 'S-type' granites *sensu* Chappell and White (1974). Most authors suggest partial melting of some such lower-crustal source rock as poorly hydrated garnet granulite (Charoy, 1979; Floyd *et al.*, 1983; Stone and Exley, 1986; Stone, 1988) or cordierite–sillimanite–spinel gneiss 'similar to Brioverian basement' (Charoy, 1986), a suggestion supported by the presence of sillimanite-bearing pelitic xenoliths within the Bodmin Moor and Carnmenellis Granites (Ghosh, 1927; Jefferies, 1985a, 1988). The predominantly pelitic character of the source is emphasized by the high $\delta^{18}O$ content of the granite (10.8–13.2‰; Sheppard, 1977; Jackson *et al.*, 1982) and their high average ammonium content (36 ppm; Hall, 1988).

The partial melting of the crustal component is envisaged as occurring at pressures of 7–8 kbar and temperatures of about 800°C and, by the time that it reached the present level of erosion and exposure, the granite was manifestly hydrated.

Geological framework

Table 2.1 Ages and initial Sr isotopic ratios of granitic rocks from the Cornubian batholith (data from Darbyshire and Shepherd, 1985, 1987)

Intrusive phase	Outcrop and granite type	Rb-Sr age (Ma)	Initial $^{87}Sr/^{86}Sr$ ratio	Comments
Major	Dartmoor (B)	280 ± 1	0.7101 ± 0.0004	-
	Bodmin Moor (B)	287 ± 2	0.7140 ± 0.0002	Mineral age
	St Austell (B)	285 ± 4	0.7095 ± 0.0009	-
	Carnmenellis (B)	290 ± 2	0.7130 ± 0.0020	Mineral age
	Tregonning (E)	280 ± 4	0.71498 ± 0.00381	Highly evolved, lithium-rich
	Land's End (B)	268 ± 2	0.7133 ± 0.0006	Mineralization re-set age
Minor	Hemerdon Ball	304 ± 23	0.70719 ± 0.01025	Heavily mineralized
	Kit Hill	290 ± 7	0.70936 ± 0.00228	-
	Hingston Down	282 ± 8	0.71050 ± 0.00119	-
	Castle-an-Dinas	270 ± 2	0.71358 ± 0.00122	Later intrusion re-set age
	Carn Marth	298 ± 6	0.70693 ± 0.00207	-
Dykes	Meldon 'Aplite'	279 ± 2	0.7098 ± 0.0017	-
	Brannel Elvan	270 ± 9	0.7149 ± 0.0031	Re-analysed
	Wherry Elvan	282 ± 6	0.7120 ± 0.0025	Re-analysed
Mineral veins	South Crofty	269 ± 4	-	-
	Geevor	270 ± 15	0.7122 ± 0.0012	-

It is believed that this state was achieved by the drawing-in of water from the surrounding rocks. That the granite was also cool is evident from the low grade of thermal metamorphism displayed by the enveloping sediments.

In addition to the crustal component, there is an enrichment in such elements as U, Th, F, Cl and Sn, Cu, W, not all of which could easily have been contributed by incorporation of normal crustal material, and which must, therefore, have been derived either from a previously enriched source or by a direct addition from the mantle. The general conclusion (Simpson et al., 1976; Simpson et al., 1979; Lister, 1984; Watson et al., 1984; Thorpe et al., 1986; Thorpe, 1987) is that the latter is more likely, with an addition of radiogenic, halogen and some metallic elements from the mantle. Jefferies (1984, 1988), in a study of radioactive minerals from the Carnmenellis Granite, concluded that these were of magmatic origin, their radio-element content having been derived from fluids circulating in the upper mantle.

Apart from the possible restite origin of some of the xenoliths (Jefferies, 1985a, 1988), the great majority are high-level metapelites or semipelites. Bromley and Holl (1986) have calculated the settling rates of the xenoliths. Most would have been enveloped by the stoping action of the rising magma and sunk within the magma body. They suggest, from their findings, that a seismic reflector found within the batholith at a depth of about 5–6 km. (Brooks et al., 1983) might mark the top of the 'zone of xenolith accumulation'.

Attention has been drawn to the K-rich nature of magmas associated with volcanic rocks in South-west England (Thorpe et al., 1986; Leat et al., 1987; Thorpe, 1987). These authors have deduced from isotopic compositions (for example, $[^{87}Sr/^{86}Sr]_{291}$ = 0.704–0.705; $[^{143}Nd/^{144}Nd]_{291}$ = 0.5123–0.5127) that they originated in a heterogeneous mantle which had previously undergone modification during a subduction event and that LIL-enriched mantle material might also have contributed to the granites. Leat et al. (1987) have proposed, largely on the basis of high concentrations of heat-producing elements, that potassic magmas originating in the mantle became contaminated by pelitic material

Table 2.2 (Opposite) Main evolution and alteration stages of the St Austell Granite (after Bristow et al., in press)

Stage	Process	Age (millions of years) *	Depth (km)	Temperature (°C)	Salinity of fluids	Source of heat	Direction of least stress	Main changes in mineralogy					Associated metalliferous mineralization	Comments
								Feldspar	Quartz	Mica			
I	Emplacement of biotite granite, forming main batholith	290-285	?3	500-600	-	Magmatic	Variscan (E-W)	-	-	-		-	Biotite granite which now forms eastern part of the St Austell granite
II	First phase of post-magmatic alteration and mineralization	285-275	2-3	500-?200	Moderate	Magmatic	Initially E - W, then N - S	Limited greisenization alongside veins	-	-		Sn, W	Early greisenization and mineralization e.g. Castle-an-Dinas (W)
IIIa	Emplacement of evolved lithium-rich granites and biotite granites in western part of St Austell granite	275-270	2-3	500-600	-	Magmatic	N - S	-	-	-		-	Granites belonging to this phase may underlie much of the batholith. Granites hydraulically fractured
IIIb	First part of second phase of post-magmatic alteration and mineralization	275-270	?2	450-380	Moderate	Mainly magmatic, some radiogenic	N - S or NW - SE	Greisenization: converted to quartz, mica and topaz by F-rich fluids, mica of gilbertite type. Tourmalinization: replaced by tourmaline	Repeatedly fractured and fractures annealed by fresh growths of quartz	Some re-crystallization, biotite loses iron which is taken up by tourmaline growth		Sn, W, Cu	Main phase of metalliferous mineralization
IIIc	Emplacement of felsitic elvan dykes	275-270	?2	600-500	Moderate	Magmatic	N - S	-	-	-		Sn, W, Cu	Further input of magmatic heat
IV	First phase of argillic alteration and NW-SE or N-S quartz-hematite veins and faulting	270-260	?1-2	350-200	Moderate to high	Mainly radiogenic, possibly some magmatic or mantle heat	E - W	Na feldspar: altered to smectite/illite assemblage, little kaolinite K feldspar: altered to illite, maybe some smectite	Free silica released by argillation, forms overgrowths on quartz and new iron-stained non-tourmaline bearing lodes (NW - SE and N - S)	Much iron liberated from biotite which is carried out of the granite to form iron lodes. Some mica hydrated to gilbertite		Fe/U/Pb/Zn	Note: Salinity, lack of kaolinite and change in stress direction. Low temperature metalliferous mineralization
Quiescent period?														
V	Second phase of argillic alteration. Main period of kaolinization ?Deep Mesozoic supergene alteration?	260 to present	0.2-1.5	50-150	Low	Radiogenic	Variable E - W or N - S, later becoming vertical	Na feldspar: altered readily to kaolinite K feldspar: altered less readily to kaolinite Smectite: altered readily to kaolinite	Free silica released by argillation, forms overgrowths on quartz and some minor quartz veins	Some iron liberated from biotite, not carried out of granite so colours matrix. In areas of intense kaolinization mica/illite altered to kaolinite		Fe/U (minor)	Note: Fresh water and main episode of kaolin formation. Isostatic uplift may have played a part
VI	Early Tertiary chemical weathering (also Mesozoic?)	25-60	0.0-0.2	20-50	Low	High surface temperature	Vertical	Altered kaolinite, is b-axis disordered in Eocene/Oligocene weathering	Some solution of silica from quartz grains	Some iron liberated from biotite, not carried out of the granite so colours matrix. In areas of intense kaolinization mica/illite altered to kaolinite		-	Tertiary weathering mantle is source of material for ball clays and associated sediments

* Radiometric dates from Bray (1980), and Darbyshire and Shepherd (1985, 1987)

to form a gabbroic, dioritic or minette magma body at the base of the crust, and that this then differentiated to form the main granite magma. The chief difficulties here seem to be the inappropriate isotope ratios of the granite (cf. Table 2.1) and the lack of unequivocal geophysical data; although not ruling out a subgranite body of dioritic composition, the only calculated seismic velocity of 6.9 km s. would only just accommodate a body of such a composition (Holder and Bott, 1971). Of course, if the upper part of the batholith has been thrust away from its roots, as noted above, and suggested by the seismic reflection data of Brooks *et al.* (1984), such a body might exist somewhere to the south.

The second stage of magmatic activity, dated at about 270 Ma, was preceded by the development of a sheeted vein system in fissures opened in the carapace of the earlier granite. This was brought about by the build-up of pressure in a hydrothermal vapour with a low density and moderate salinity of 10–25 equiv. wt.% NaCl (Jackson *et al.*, 1982; Shepherd *et al.*, 1985). Its principal effect in rock terms, however, was the emplacement of a second magma which crystallized to give the albite–zinnwaldite–topaz granite exposed at St Austell, Tregonning and St Michael's Mount. Associated with the Li-mica granite intrusion at St Austell is a Li- and B-enriched variety of the earlier, main biotite granite generated as a metasomatic aureole, and also pockets of fluorite-bearing granite (Manning and Exley, 1984).

At the end of this stage, a series of roughly E–W, fine-grained, granite-porphyry dykes ('elvans') were intruded (Table 2.1). These are believed to have been derived from biotite granite magma, but in a number of cases show evidence of fluidization of solid material (Stone, 1968), indicating that some were not entirely magmatic. All the major plutons contain examples of fine-grained granite, but the origins and ages of these are not clear. Some are magmatic and intrusive, but others may be granitized sediment.

Although an early phase of mineralization, involving Fe–Mn and Fe–Cu, was related to pre-batholith sedimentary deposits (Jackson *et al.*, 1989), the main period of mineralization started at about 270 Ma and resulted from the build-up of hydrothermal fluids which were of high density and salinity (20–40 equiv. wt.% NaCl), and which, on entering the cover, deposited a series of roughly E–W-striking veins containing, for example, Sn, Cu, W and As minerals. The violence of such outbursts, accompanied by sudden decompression, sometimes produced spectacular breccias which are usually cemented by tourmaline. The field evidence as to the age of these breccias is somewhat equivocal, and in the absence of radiometric dating it is not always certain whether they were formed at the end of the first or the second magmatic phase.

Until the main stage of mineralization, argillic alteration was limited to greisening, some bordering mineral veins and some pervading the granites through a well-developed pore system, but temperatures, while fluctuating, probably never fell below about 380°C. Sheppard (1977) has shown that meteoric water was present during this process, emphazising the role played by groundwater from the country rocks in hydrothermal reactions.

Over the next 5–10 million years, as magmatic heat was lost, the temperature was maintained by radiogenic heat from K, Th and U, which are present in unusually high concentrations (Tammemagi and Smith, 1975), but nevertheless it fell slowly to an estimated 350–200°C. Fissures approximately at right angles to the earlier vein systems provided channels at this time for a further stage of mineralization which included Fe, Pb, Zn and U among other elements. The temperatures and salinities of the fluids at this time have been calculated at 200–150°C and 19–27 equiv. wt.% NaCl (Alderton, 1978; Shepherd *et al.*, 1985) and they gave rise to the first true clay minerals, namely smectite and illite, by alteration of the felsic minerals. The granite was thus prepared for subsequent kaolinization (Bristow, 1987; Bristow *et al.* in press).

The last-mentioned process has continued intermittently until the present day as a result of the circulation of water through the granites by convection cells driven by radiogenic heat (Durrance *et al.*, 1982; Durrance and Bristow, 1986; Bristow *et al.* in press). The reality of the circulation of meteoric water and its mixing with older water, made saline by reaction with plagioclase and biotite, has been confirmed by isotopic and selective cation analysis (Edmonds *et al.*, 1984). It has been established that the main radiogenic heat source resides in the accessory minerals apatite, zircon, monazite, xenotime and uraninite, of which the last is much the most significant (Ball and Basham, 1979; Jefferies, 1984, 1988). However, it has been suggested that additional heat derived from the mantle may also be required (Tammemagi and Smith, 1975), so

Post-orogenic volcanic activity

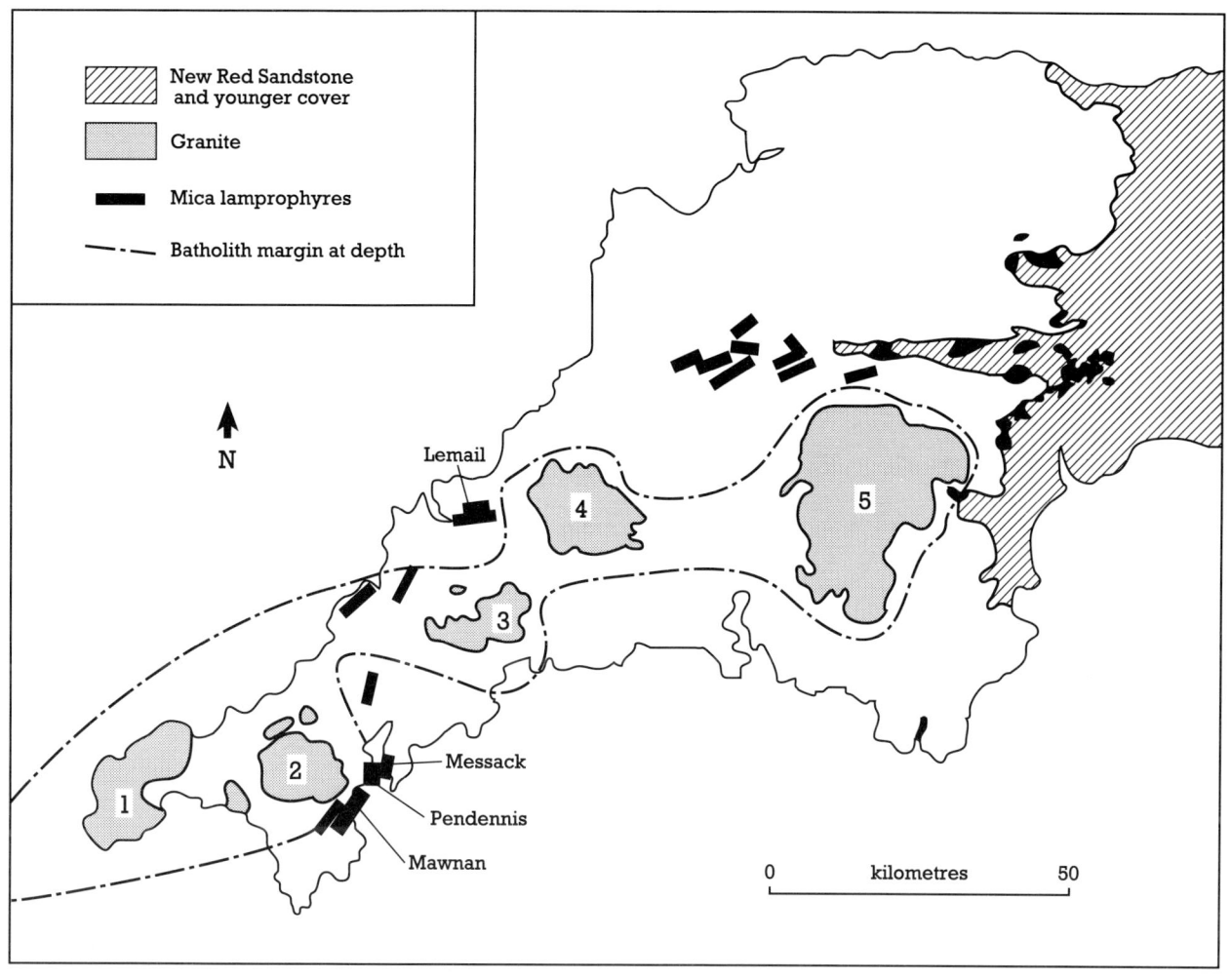

Figure 2.8 Geological outline map of south-west England, showing the distribution of mica lamprophyres relative to the exposed Cornubian granite batholith (grey) and its geophysically determined margin (after Leat *et al.*, 1987). The Exeter Volcanic 'Series' is shown within the New Red Sandstone outcrop and younger Mesozoic rocks (diagonal ruling). 1 = Land's End Granite; 2 = Carnmenellis Granite; 3 = St Austell Granite; 4 = Bodmin Moor Granite; 5 = Dartmoor Granite.

invoking further speculation about the sub-batholith geology.

Sheppard (1977) has made it clear that the circulating water involved in the main process of kaolinization was meteoric, and it now seems likely that a period of Palaeogene subtropical weathering intensified the alteration already started (Bristow, 1988; Bristow *et al.* in press). The main stages of evolution are summarized in Table 2.2.

Post-orogenic volcanic activity

After emplacement of the batholith and isostatic uplift towards the end of the Carboniferous, south-west England underwent rapid denudation during the Stephanian, Permian and Triassic that eventually unroofed the marginal facies of the granite plutons (Dangerfield and Hawkes, 1969). The eroded desert landscape was subsequently buried by aeolian sands and the pebbly sands and silts of flash floods and fluvial channels (Laming, 1966, 1968). Rapid lateral variations are common and in some cases reflect local contemporaneous faulting (Bristow and Scrivener, 1984). Locally derived debris within the basal conglomerates includes granitic material, aureole metasediments and acidic volcanics (Groves, 1931; Hutchins, 1963; Laming, 1966).

Within this 'red bed' continental environment during the late Carboniferous and probably early Permian period, various small-volume volcanic rocks were erupted. In terms of gross composi-

Geological framework

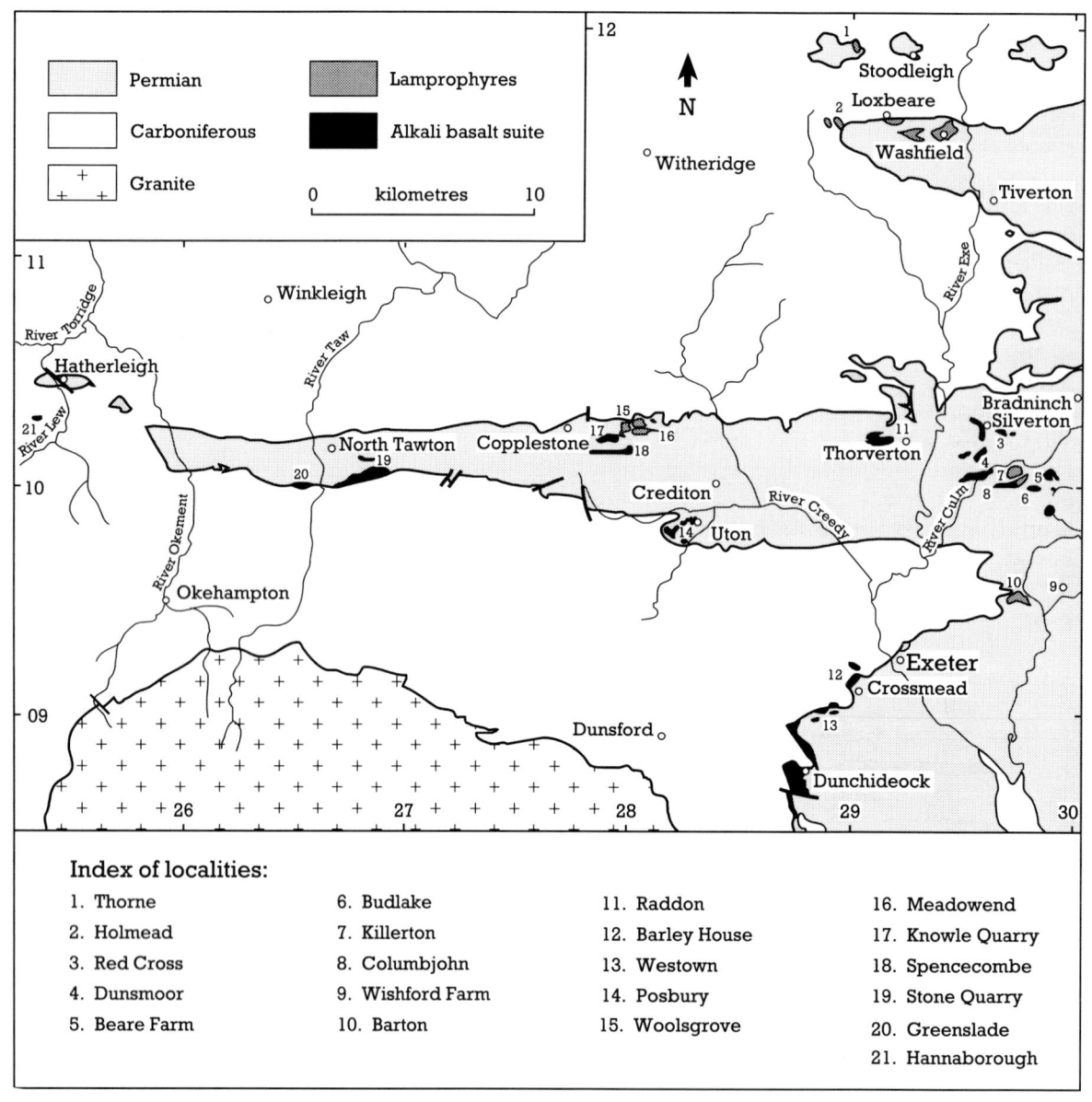

Figure 2.9 Distribution of the two main magmatic groups within the Exeter Volcanic 'Series', mid-Devon (after Edmonds *et al.*, 1969).

tional variation, three main suites can be recognized – lamprophyres (limited occurrence throughout Cornubia), Exeter Volcanic 'Series' (restricted to the Crediton Trough and west of Exeter) and rhyolites (only seen at Kingsand, south of Plymouth).

Lamprophyre suite

Throughout south-west England is seen a set of NE–SW-trending minette-type lamprophyre dykes (Figure 2.8) that cut the Hercynian country-rock fabric and are considered to be post-orogenic in character (Hawkes, 1981; Darbyshire and Shepherd, 1985). These small intrusions are spatially associated with the granite, but apparently lie outside the surface outcrop (Exley *et al.*, 1982) and within the batholithic 'shadow zone' (Leat *et al.*, 1987). The post-orogenic nature of the lamprophyres, coming after major granitic emplacement, appears to be a global characteristic of some types of lamprophyric volcanism (Rock, 1984; Turpin *et al.*, 1988). Hawkes (1981) gave an average age of 291 Ma (whole-rock, K/Ar) for various lamprophyres that may be representative of this suite.

Post-orogenic volcanic activity

Exeter Volcanic 'Series'

Small-volume lava flows and dykes occur within the 'red bed' sequences to the north and west of Exeter (Figure 2.9), and on the basis of limited radiometric dates these are probably late-Stephanian in age. Although often considered to be Permian (Knill, 1982) with ages around 280 Ma (Miller and Mohr, 1964), the originally determined K/Ar dates have been recalculated to 291 Ma (Thorpe *et al.*, 1986), and suggest some of the volcanics must be of very late Carboniferous age. Two main groups of volcanics are recognized – a basaltic suite and a potassic lava suite – that are collectively referred to as the Exeter Volcanic 'Series' (Knill, 1969, 1982; Cosgrove, 1972). The potassic suite is dominated by lamprophyres that are probably related to the minettes mentioned above and found throughout the region (Cosgrove and Hamilton, 1973; Exley *et al.*, 1982; Leat *et al.*, 1987).

The Exeter Volcanic 'Series' was erupted into and over 'red bed' sequences and may show lava–soft-sediment mixing, as well as eroded lava surfaces infilled with sandstone (Knill, 1982). These volcanics are largely restricted to east–west trending grabens (for example, Crediton Trough, Figure 2.9) which developed during uplift after the Variscan Orogeny (Whittaker, 1975) or specifically via the reactivation of a major Variscan basement thrust by post-orogenic tension (Durrance, 1985b).

The significance of these volcanic rocks lies not only in their being representative of post-orogenic volcanism, but also in their relationship to the lamprophyres and the nature of the underlying mantle that supplied the mafic magmas. By far the most interesting feature concerns the potassic lavas; these exhibit a chemical fingerprint characteristic of subduction-related magmas which suggests they were derived from lithosphere subducted during the Variscan (Thorpe *et al.*, 1986; Leat *et al.*, 1987; Thorpe, 1987). Although this does not necessarily imply the presence of an active subduction zone during the Permian (for which there is no direct geological evidence), Cosgrove (1972) originally suggested that the whole 'shoshonitic suite', as he referred to the Exeter Volcanic 'Series', was generated at a plate margin.

Rhyolitic lava group

Remnants of flow-banded porphyritic rhyolitic lavas, apparently lying on deformed Devonian argillites, occur at Kingsand (near Plymouth, Figure 1.1) and also as numerous pebbles within the 'red bed' conglomerates (Laming, 1966). Both occurrences form a comagmatic suite (Cosgrove and Elliott, 1976) and are considered to be representative of widespread subaerial acid volcanism fed by late-stage dykes (Hawkes, 1974). The rhyolites are not apparently related to the granites but have petrographic and chemical affinities with the late, granite-porphyry dykes (Exley *et al.*, 1983) that were emplaced as fluidized systems during the 280–270 Ma interval (Exley and Stone, 1982; Darbyshire and Shepherd, 1985).

Chapter 3

Lizard and Start Complexes (Group A sites)

Lithological and chemical variation

INTRODUCTION

This chapter presents some of the classic sites from the Lizard Complex of south Cornwall, illustrating various aspects of its ophiolitic character in the light of relatively recent work. It is one of the few good examples of preserved Variscan ocean crust in the Rhenohercynian zone of northern Europe; together with the adjacent *mélange*, it illustrates the development and subsequent closure of the Devonian Gramscatho Basin. The metavolcanic greenschists of the Start Complex are also included here as they have chemical characteristics that are comparable with the ophiolite and basaltic clasts in the *mélange* and thus contribute additional evidence for the local preservation of Variscan ocean crust. The localities of all the sites are shown on Figure 3.1.

LIST OF SITES

A1 Lizard Point (SW 695116–SW 706115)
A2 Kennack Sands (SW 734165)
A3 Polbarrow–The Balk (SW 717135–SW 715128)
A4 Kynance Cove (SW 684133)
A5 Coverack Cove–Dolar Point (SW 784187–SW 785181)
A6 Porthoustock Point (SW 810217)
A7 Porthallow Cove–Porthkerris Cove (SW 798232–SW 806226)
A8 Lankidden (SW 756164)
A9 Mullion Island (SW 660175)
A10 Elender Cove–Black Cove, Prawle Point (SX 769353–SX 769356)

LITHOLOGICAL AND CHEMICAL VARIATION

The following lithological units are now recognized within the Lizard Complex (Figure 3.2):

1. Partially serpentinized peridotite (Lizard Serpentinites);
2. Massive and weakly layered gabbros (Crousa and Trelan gabbros);
3. Variably metamorphosed basaltic dykes;
4. Heterogeneous acid/basic intrusive complex (Kennack Gneiss);

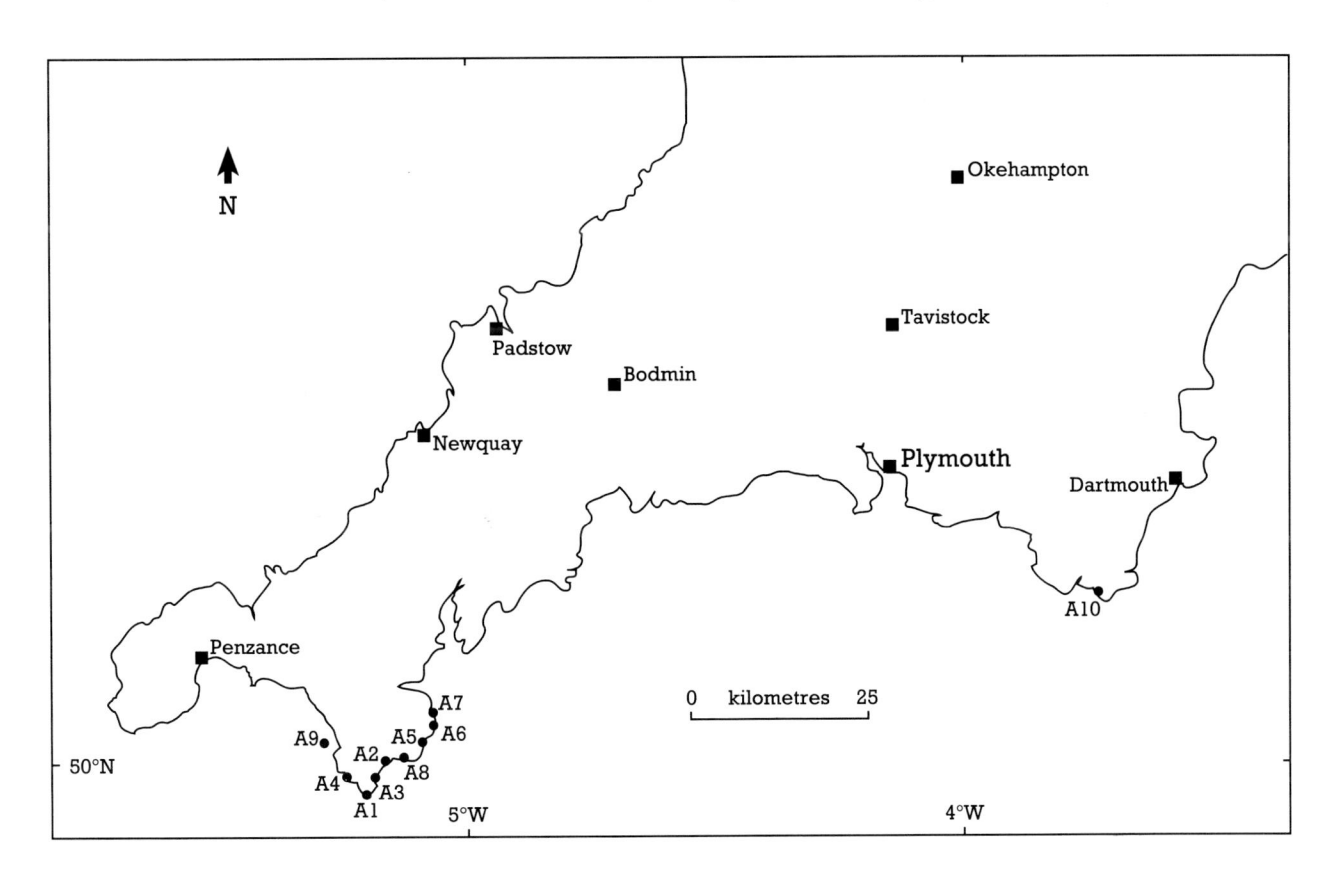

Figure 3.1 Outline map of south-west England, showing the location of Group A sites.

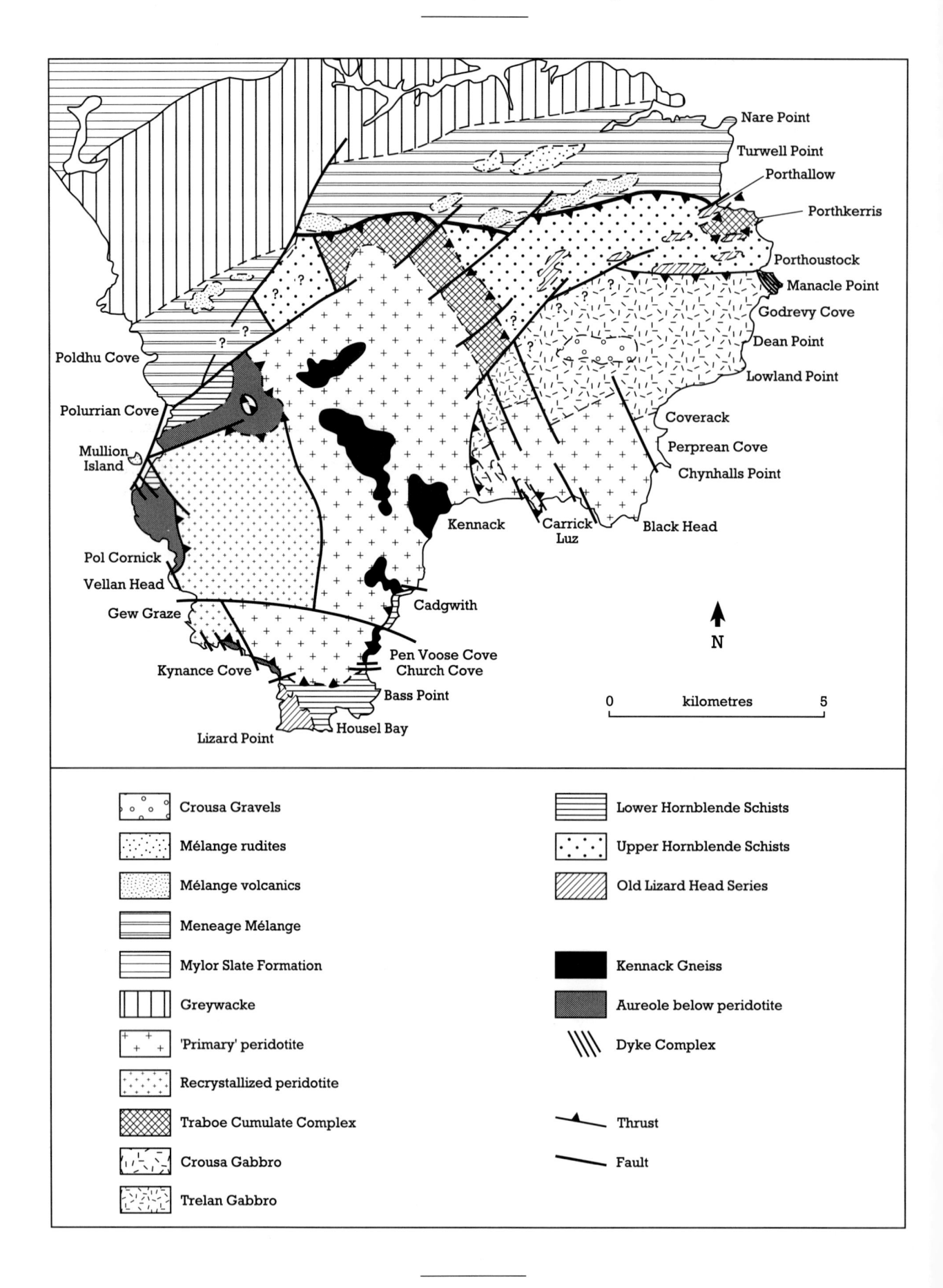

Lizard and Start Complexes (Group A sites)

Legend:

- Crousa Gravels
- Mélange rudites
- Mélange volcanics
- Meneage Mélange
- Mylor Slate Formation
- Greywacke
- 'Primary' peridotite
- Recrystallized peridotite
- Traboe Cumulate Complex
- Crousa Gabbro
- Trelan Gabbro
- Lower Hornblende Schists
- Upper Hornblende Schists
- Old Lizard Head Series
- Kennack Gneiss
- Aureole below peridotite
- Dyke Complex
- Thrust
- Fault

Locations labelled: Nare Point, Turwell Point, Porthallow, Porthkerris, Porthoustock, Manacle Point, Godrevy Cove, Dean Point, Lowland Point, Coverack, Perprean Cove, Chynhalls Point, Black Head, Carrick Luz, Kennack, Cadgwith, Pen Voose Cove, Church Cove, Bass Point, Housel Bay, Lizard Point, Kynance Cove, Gew Graze, Vellan Head, Pol Cornick, Mullion Island, Polurrian Cove, Poldhu Cove

34

5. Traboe Cumulate Complex (Traboe Hornblende Schists);
6. Metabasalt amphibolites (Landewednack Hornblende Schists);
7. Pelitic, semipelitic and hornblendic metasediments (Old Lizard Head 'Series');
8. Intermediate orthogneiss (Man of War Gneiss).

The structure of the Lizard Complex has been a controversial issue for many years. The tremendous difference between the high-grade metamorphic rocks of the Lizard and the Devonian rocks to the north was recognized by the earliest workers (De la Beche, 1830). This gave rise to theories that it was either an upfaulted block of basement or part of a large thrust sheet, possibly associated with rocks of the Start Complex in south Devon (discussed in Flett, 1946). All workers had accepted that the peridotite was essentially a diapir-like intrusion until the work of Sanders (1955) who suggested that it was a thin sheet-like body. Green (1964a, 1964b, 1964c) carried out a very detailed study of the peridotite and concluded that it had formed as a diapiric intrusion of hot mantle, with a well-developed metamorphic aureole.

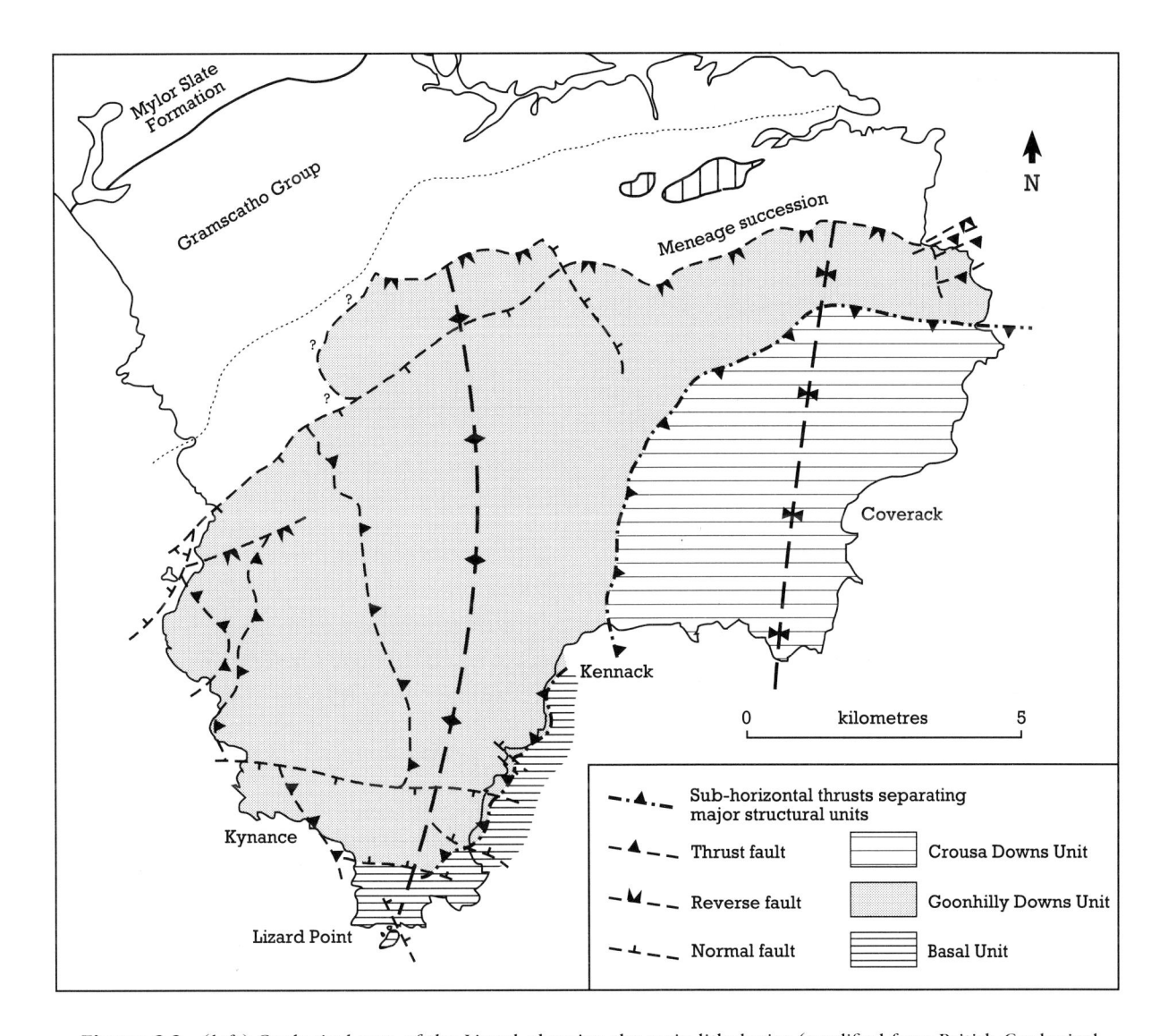

Figure 3.2 (left) Geological map of the Lizard, showing the main lithologies (modified from British Geological Survey Sheet 359; Green, 1964a; Leake *et al.*, 1990); and (above) their division into three tectonic units (after Bromley, 1979).

Lizard and Start Complexes (Group A sites)

In the 1970s ophiolites became a topic of great interest and several workers interpreted the Lizard Complex in this light and, again, suggested that it was a thrust sheet. A borehole by IGS (now BGS) in the centre of the peridotite finally showed that it was only 360 m thick and proved its sheet-like form. Regional seismic studies have shown that the whole of the Lizard Complex is less than 1 km thick and that it is underlain by Devonian sediments. It is now well established that the Lizard Complex is relatively thin thrust sheet that overlies the south Cornish *mélange* at the top of an allochthonous nappe pile.

Bromley (1979) suggested an internal structure for the complex consisting of three thrust sheets, that were from top to bottom:

1. Crousa Downs Unit – an essentially continuous ophiolite stratigraphy along the east coast of the Lizard from mantle peridotite through gabbros to a sheeted dyke complex; the sequence being tectonically truncated at this point.
2. Goonhilly Downs Unit – the main part of the complex; largely peridotite and overlain by the Traboe Hornblende Schists and other amphibolites and metasediments to the north-east.
3. Basal Unit – this occurs in a narrow strip around the south and south-east of the peridotite and includes the Landewednack Hornblende Schists, Old Lizard Head metasediments and the Kennack Gneisses intruded roughly along the contact between this and the overlying unit.

This interpretation has been broadly accepted by most recent workers. However, Leake and Styles (1984) and Leake *et al.* (1990) have questioned the separation of the Crousa Downs Unit from the Goonhilly Unit as they found some continuity of rock sequences across the supposed thrust contact in the Traboe–Trelan area. They have also suggested that the arm of the Goonhilly Downs Unit to the north of the Crousa Downs Unit is part of an imbricate sequence associated with the Lizard boundary. This interpretation is shown on the map (Figure 3.2).

The composition of the main lithological units is reviewed below.

Peridotite

Various schemes have been suggested for subdivision of the Lizard peridotites, the two best known being those of Flett and Hill (1912) and Green (1964a), although none of them is totally adequate to encompass the current level of knowledge. Flett and Hill's scheme had four main types:

1. Coarse lherzolitic type – this was the least-deformed type and was called 'Bastite serpentine' after the prominent orthopyroxene pseudomorphs. It occurred in the central and eastern parts of the body.
2. Tremolite serpentine – a highly deformed recrystallized type that was derived either by metamorphism of the bastite serpentine or an earlier intrusion. It occurred mostly in the western part.
3. Dunite serpentine – this had a significantly different bulk composition and was thought to have originated as dunite and occurred around the margins of the peridotite, particularly in the north around Traboe.
4. Chromite serpentine – was restricted to pods and veins and contained prominent chromite.

Green (1964a) carried out a major study of the peridotite and found similar groups and geographical distributions. From his detailed study of mineral assemblages and textures, he concluded that there was essentially a single intrusion that had undergone several stages of recrystallization and re-equilibration to produce the other types. The primary assemblage was closely analogous to the 'bastite serpentine' of Flett, and consisted of olivine, aluminous orthopyroxene, aluminous clinopyroxene and olive-green aluminous spinel. This had crystallized at 1250–1300°C and 15 kbar, a clear indication that it had originated in the mantle. Large areas had re-equilibrated to a recrystallized anhydrous assemblage, broadly analogous to the tremolite serpentine. This consisted of olivine, low-alumina pyroxenes, chromite and plagioclase and had equilibrated at around 1075°C and 7.5 kbar. The third type was the hydrous, recrystallized assemblage that occurred in narrow zones through the anhydrous types. It consisted of olivine, orthopyroxene, pargasite amphibole and chromite and had equilibrated at 900°C and 5 kbar. Green (1964a) suggested that the dunite serpentine (Flett and Hill, 1912), was a highly serpentinized version of this. The

Lithological and chemical variation

sequence of assemblages with decreasing temperature and pressure gave rise to the model of a diapir rising from the mantle, into the crust and undergoing re-equilibration. It has since been shown that the actual form of the peridotite is sheet-like not diapir-like, but there is no reason to doubt the petrological evidence and mineral assemblages on which the model was based. Styles and Kirby (1980) suggested that the peridotite was possibly the sliced-off top of a suboceanic mantle diapir.

Green (1964a) produced a map that showed the distribution of the mineral assemblages. This is the most detailed map of peridotite variation available to date, but is possibly somewhat misleading. Large areas are shown as consisting of the primary assemblage, but the olive-green spinels essential to the assemblage are seen only in a handful of thin sections, and these come mostly from a relatively small area in the centre of the peridotite, around Gwenter. The aluminous pyroxenes are much more widespread, but the distribution shown on the map is actually based on the recognition of pseudomorphs after spinel.

The volume of rock that has equilibrated during these phases of recrystallization is very small; it is possible to see all three assemblages in one thin section. Only when deformation has been intense and has accelerated reaction have larger volumes re-equilibrated. These features are also partly obscured by later serpentinization. The degree of serpentinization varies considerably from place to place and from one rock type to another. The lherzolitic and amphibole-rich peridotites tend to be less serpentinized and the freshest rocks are around 30% serpentinized. The most serpentinized tend to be the dunitic types and many have no olivine remaining, but still preserve good ghost dunite fabrics. Serpentinization is often more intense close to shear zones and faults. These features make reliable mineral assemblage maps difficult to produce.

Green (1964a) proposed that all the ultrabasic rocks originally had a similar lherzolitic composition; this contrasted with the opinion of Flett and Hill (1912) who suggested there were several, particularly a dunitic type. The dunitic type can readily be distinguished from the others by the lack of orthopyroxene and amphibole and occurs in two distinct modes:

1. large bodies of regional extent that form part of the cumulate complex and will be discussed in a following section (dunite serpentine),

and

2. as veins cutting through the primary type peridotite (chromite serpentine).

The latter is probably formed by the passage of melts or fluids through the upper mantle. More recent work, (Styles and Kirby, 1980; Leake and Styles, 1984) has shown that there is a wide range of bulk compositions that encompasses lherzolitic, harzburgitic and dunitic types, which is indicated by the range of Al_2O_3 and CaO contents (Figure 3.3). There does not seem to be a distinct division between lherzolitic and harzburgitic compositions, more a continuum; for example, both types can be found associated on the beach at Coverack. There are not yet sufficient chemical data to establish if there are regional variations in composition. Figure 3.2 shows just two types, a 'primary' type that is coarse grained and less deformed and a recrystallized type that is finer grained, strongly foliated and intensely recrystallized. This refers to features visible in hand specimen and thin section, and hopefully will be improved when sufficient data are available for meaningful subdivision.

The rare-earth-element geochemistry of Lizard peridotites has been studied by Frey (1969) and Davies (1984). Frey showed that the primary type (spinel lherzolite) had very depleted light REE, typical of residual, mantle-derived, alpine peridotites (Figure 3.4). This is rather unusual as they have major-element chemistry similar to undepleted mantle where a relatively flat REE pattern would be expected. The anhydrous and hydrous recrystallized types were progressively less depleted with close to chondritic abundances and slight light-REE depletion. Davies (1984) also found a plagioclase lherzolite type that had five times chondrite abundances and slight light-REE enrichment. He suggested a model where the spinel lherzolite resulted from repeated extractions of very small fractions of melts which depleted the light REE, but showed little effect on the major-element chemistry. The plagioclase lherzolite and pargasite peridotite were interpreted as being produced by infiltration of melts and associated mantle metasomatism.

Deformation has had a marked affect on most of the peridotites, which often have a porphyroclastic texture. Rothstein (1977), however, has reported evidence of an original cumulate crystal framework preserved between the later planar fabric.

Serpentinization is regarded as a prolonged,

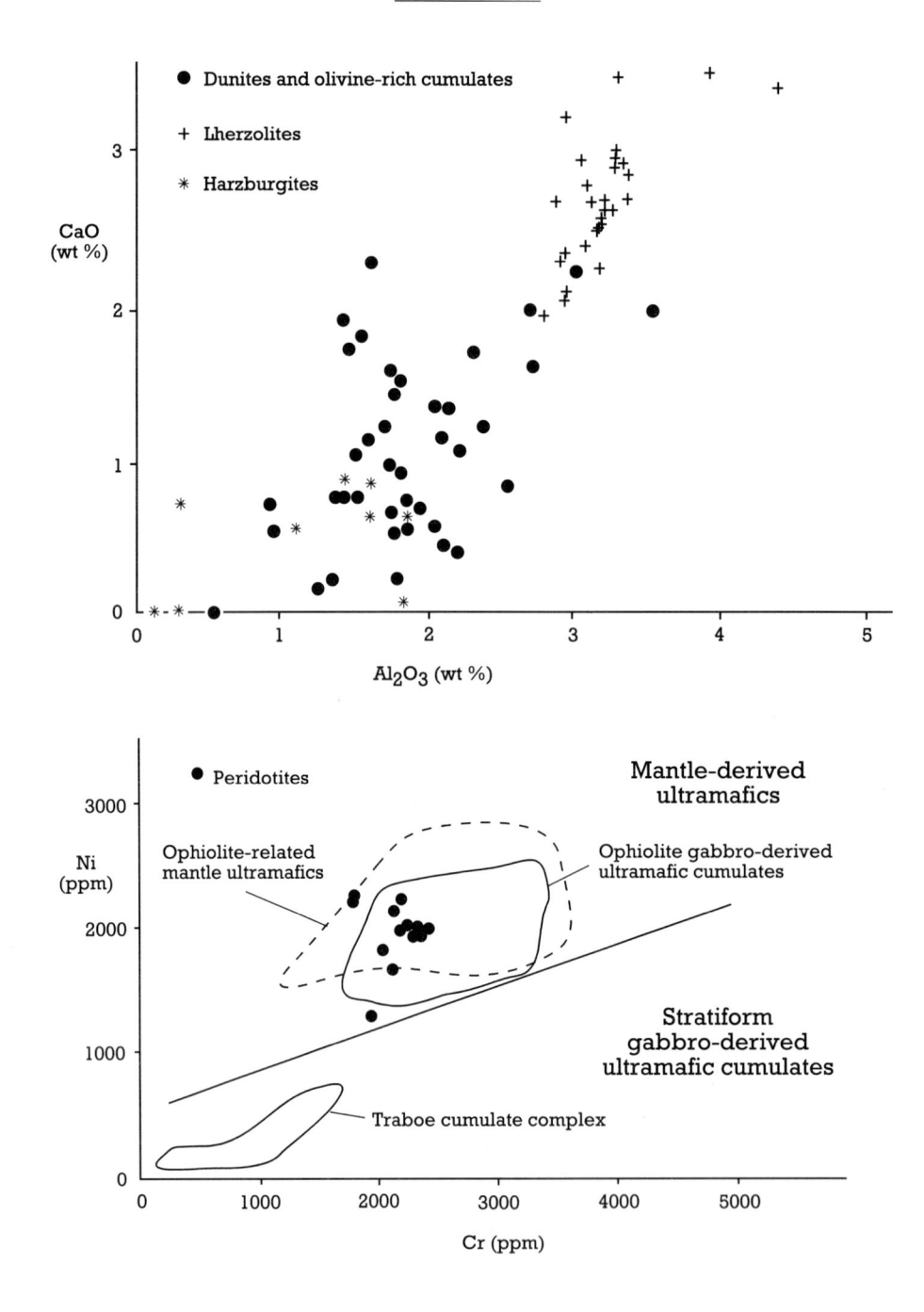

Figure 3.3 Distribution of Al_2O_3–CaO and Ni–Cr in Lizard peridotites, dunites and ultramafic cumulates (data from Parker, 1970; Kirby, 1979a; Leake and Styles, 1984) relative to typical ophiolite- and stratiform-related ultramafics (data from Rivalenti *et al.* (1981) and the literature).

retrograde alteration phenomenon. It probably started at relatively high temperatures, 400–500°C, with the formation of coarse lizardite associated with obduction in Devonian times and probably continues at the present time with the formation of fine chrysotile in veins, etc. due to the presence of groundwater.

Gabbro and dykes

The Crousa gabbro has had a rather turbulent history, with evidence of several closely spaced intrusive phases causing back-veining, autobrecciation and pegmatite formation. In many places there is evidence of deformation at various times,

Lithological and chemical variation

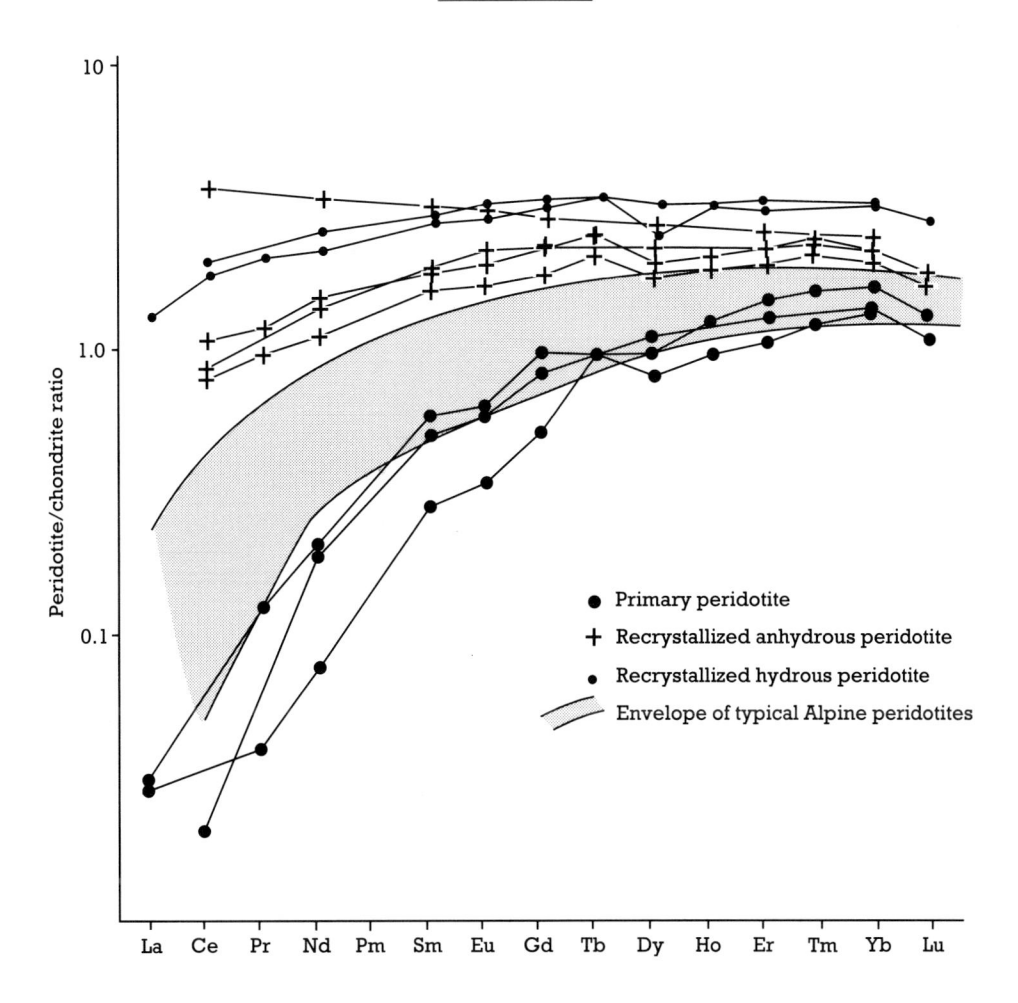

Figure 3.4 Chondrite-normalized REE data for the different assemblages of the Lizard peridotite (from Frey, 1969; Davies, 1984) and typical Alpine peridotites (data from Frey, 1984).

with flaser zones and shears with amphibolite and greenschist-facies mineral assemblages developed. Igneous lamination and layering are developed in a few places (Kirby, 1978), although, in general, apparent layering is due to secondary alteration.

The gabbros consist mainly of plagioclase and augite, with minor ilmenite and magnetite (Bromley, 1979). Olivine is also present, but is confined to the southern part of the outcrop. The top of the intrusive body is to the north, as there is evidence of progressive fractionation in that direction, as well as decreasing Mg/Mg + Fe in rocks and minerals and increasing incompatible elements such as titanium and phosphorus (Kirby, 1979a). There are also fractionated rocks such as hornblende diorites in the north. Leake *et al.* (1990) have described a second type of gabbro from the Trelan area which is a more fractionated type with iron-rich clinopyroxenes, a more sodic plagioclase and often considerable enrichment in Ti and Fe. Overall, the Crousa gabbros have low incompatible-element contents and light REE depletion which indicate derivation from a depleted mantle source and are consistent with an oceanic origin. In this respect they differ from the coarse-grained dolerites and gabbros in the rest of south Cornwall (Floyd, 1984).

In the north, between Godrevy Cove and Porthoustock, the gabbro is cut by numerous basaltic dykes that can occupy between 50 and 80% of outcrop; these are thought to represent the root zone of a sheeted dyke complex (Bromley, 1979). Similar, but much less numerous, dykes also occur in the main gabbro and peridotite further south (lower down in the ophiolite pseudostratigraphy) and some are part of the same sequence. The dykes were intruded throughout the cooling and solidification of the gabbro, as contact relationships show gradations from early, disrupted dykes without chilled margins to cross-cutting vertical dykes. Mineralogically, the dykes are aphyric and sparsely

Lizard and Start Complexes (Group A sites)

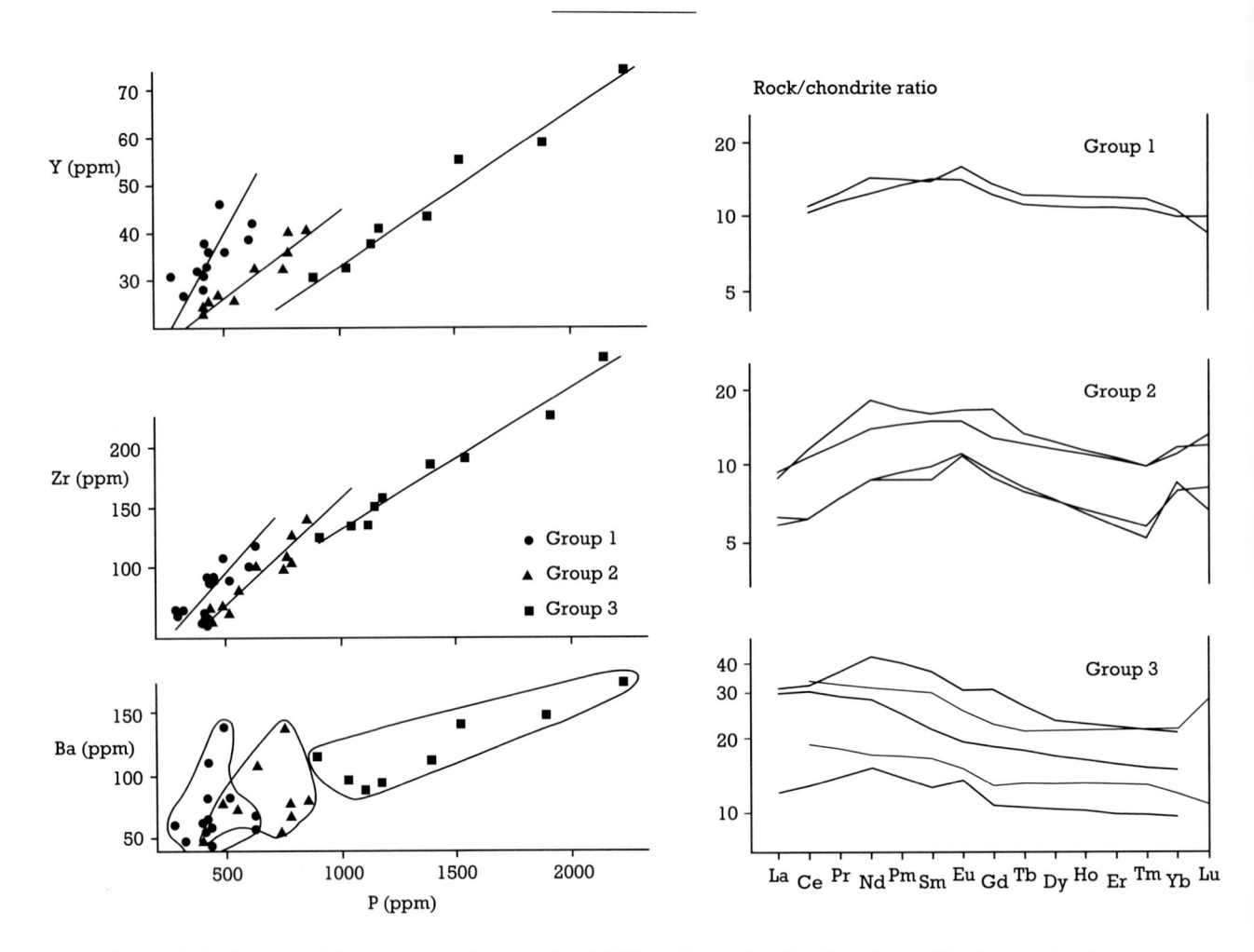

Figure 3.5 Incompatible-element and normalized REE patterns for the Lizard basaltic dykes, showing the distinctions between the three chemical groups (data from Davies, 1984; Kirby, 1984).

olivine- and plagioclase-phyric tholeiites as well as their amphibole-bearing metamorphosed equivalents (Kirby, 1984). They show variable incompatible element ratios and depleted to mildly light REE-enriched patterns (Davies, 1984; Kirby, 1984; Floyd, 1984). Three chemical groups (Figure 3.5) were identified by Kirby (1984), each with its own fractionation sequence involving mainly plagioclase, together with lesser olivine and clinopyroxene. Overall, dyke Groups 1 and 2 (with La/Nb <1, Zr/Nb $c.$ 30 and depleted light REE patterns) are chemically similar to oceanic basalts (Kirby, 1984), whereas Group 3 (considered to represent the earliest dyke phase) is more enriched than either group 1 or 2 and was probably derived from a 'recently' light REE-enriched or metasomatized source, similar to that of the recrystallized peridotites (Davies, 1984).

Kennack Gneiss

The Kennack Gneiss is a series of interbanded acid and basic gneissose rocks that occur along the south-east coast of the Lizard, roughly at the base of the Goonhilly Downs structural unit. They have been controversial since the time of Bonney (1877b) and interpretations of their origin fall into two main groups:

1. migmatization of Old Lizard Head metasediments and Hornblende Schist by a hot overriding sheet of peridotite (Sanders, 1955; Kirby, 1979b; Vearncombe, 1980; Malpas and Langdon, 1987);
2. composite intrusions of mafic and felsic magmas, probably along shear zones, followed by deformation (Flett and Hill, 1912; Green, 1964c; Bromley, 1979; Sandeman, 1988).

A consensus of opinion has yet to emerge, but the most recent and most detailed study by Sandeman (1988) favoured the composite intrusive model.

Hornblende Schists

The Hornblende Schists were divided by Flett and Hill (1912), into two types, the Landewednack type and the Traboe type. The Landewednack type is generally associated with sediments, has epidotic layers and masses, a well-developed, regular, flat-lying foliation and a rather homogeneous appearance. The Traboe type is usually associated with the 'serpentine' (peridotite), has a steep foliation, is often coarser grained and has a very variable mineralogical character. Flett and Hill (1912) thought that the Landewednack type was derived from basalt lavas and sills, and the Traboe type from gabbros that predated the intrusion of the peridotite.

The distribution of the two types of hornblende schist has been more or less accepted by all subsequent workers. The Landewednack type occurs largely around the southern tip of the Lizard and the Traboe type around Traboe, Mullion and Predannack. Green (1964b) suggested a radically different hypothesis for their origin. He proposed that all the hornblende schists had originated as basaltic lavas which had undergone amphibolite-facies regional metamorphism. Those close to the peridotite (Traboe type) had suffered an additional contact metamorphism during intrusion of the hot peridotite to form brown-hornblende granulites, whereas those immediately adjacent to the peridotite developed two-pyroxene granulites. He recognized several mineral assemblages formed at increasing temperatures and, on the basis of a few bulk rock analyses, suggested that the two types had similar compositions.

This interpretation has been rejected by most subsequent workers. On the basis of extensive field observations, Bromley (1979), suggested that the Traboe type were metamorphosed gabbros and basic dykes and not formed by the effects of metamorphism due to intrusion of the peridotite. Kirby (1979b), in addition to his field observations, produced numerous chemical analyses, which showed that although there was some overlap in composition, the Traboe type had a much wider range and could not be derived from the Landewednack type. He also thought that they were metagabbros. Styles and Kirby (1980) suggested that there were two types of Traboe Hornblende Schist; most of it metagabbro, but a small proportion (very close to the basal thrust of the peridotite along the south-east coast near Cadgwith and at the base of the peridotite in the Predannack borehole) was derived from 'contact' metamorphism of Landewednack type. Leake and Styles (1984) described three boreholes from the Traboe area and confirmed previous work that the Traboe type were metagabbros and apart from the few exceptions mentioned previously, could not be derived from the Landewednack type. They also showed that a series of metapyroxenites and dunites were intimately associated with the metagabbros, together forming the 'Traboe cumulate complex' (Figure 3.6). This complex overlies the main peridotite slab and is similar to cumulates found at the mantle–crust transition zone in many other ophiolites. This is a very important member of the Lizard ophiolite stratigraphy, as previously it was thought that the cumulate zone at the base of the crustal sequence was missing; the only possible rocks of this type being the small development of troctolite and associated rocks seen at Coverack. The only good surface outcrop of this cumulate complex is at Porthkerris.

The chemical composition of the Landewednack Hornblende Schists is very similar to ocean-floor basalt (Floyd *et al.*, 1976; Kirby, 1979b) and this is good evidence to suggest that they are part of an ophiolite complex. The Traboe Complex has more primitive chemical compositions but, as they are plutonic rocks that have formed by crystal fractionation, they cannot be directly compared with the Landewednack type which are dominantly metabasaltic lavas.

Mélange metabasalts

The south Cornish *mélange* (Barnes and Andrews, 1981) includes Mullion Island and the Meneage and Roseland areas (Figure 3.2); characteristically it contains many exotic clasts and huge blocks of basaltic parentage within a Middle–Upper Devonian argillite matrix. The larger blocks (Mullion Island, Nare Head) are composed of pillow lavas and pillow breccias, whereas smaller clasts include 'greenstones', aphyric and plagioclase-phyric basalts, dolerites and rare gabbros, together with

Lizard and Start Complexes (Group A sites)

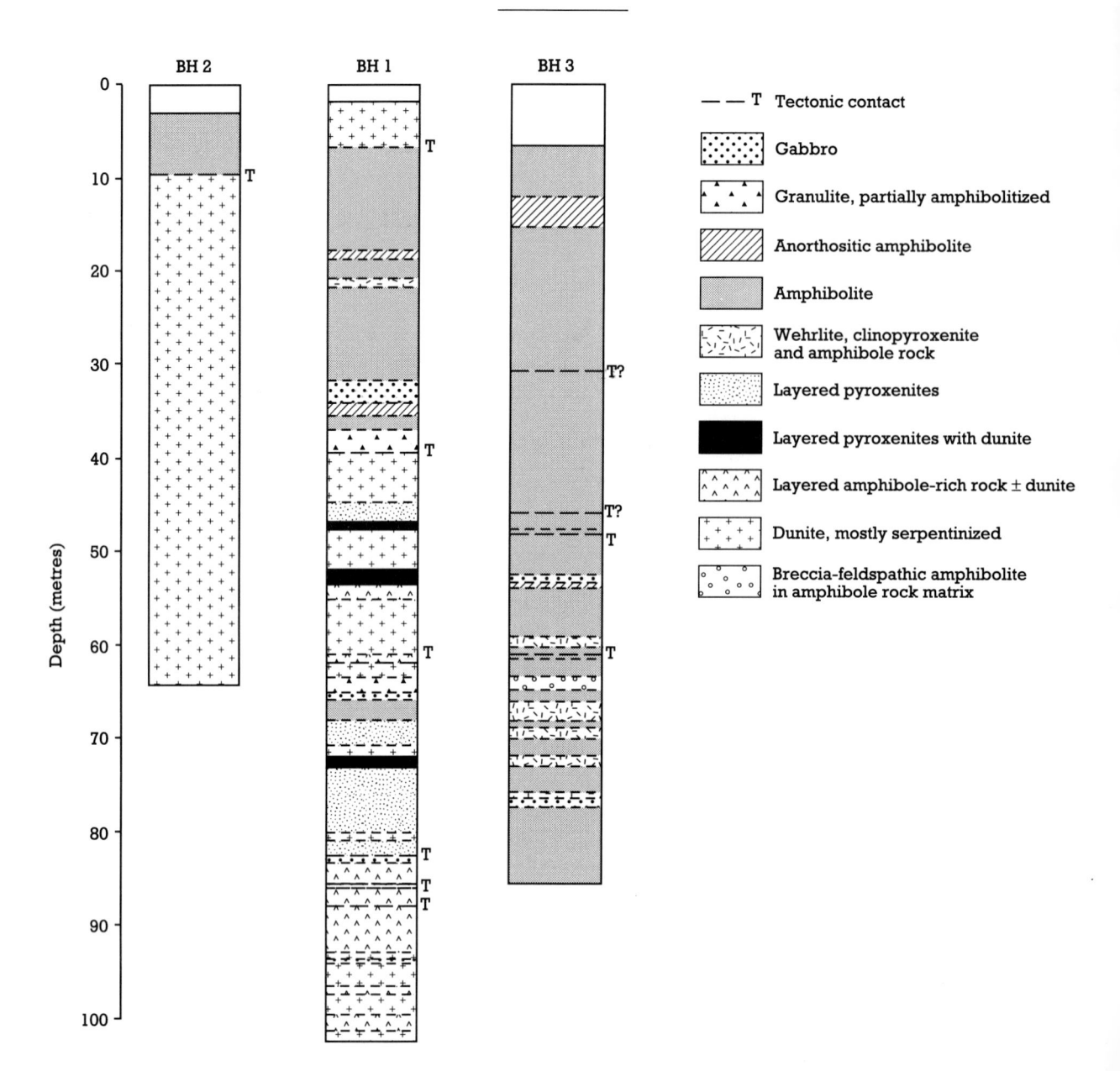

Figure 3.6 Lithological borehole logs for the Traboe ultramafic–mafic cumulate complex at Traboe, Lizard area (data from Leake and Styles, 1984).

their variably metamorphosed equivalents, including epidote–plagioclase–hornblende amphibolites (Barnes, 1984). Virtually all the basic rocks have undergone some degree of low-grade metamorphism, generally in the pumpellyite facies (Barnes and Andrews, 1981; Floyd, 1983). The Mullion Island pillow lavas show relict quench textures and primary calcic augite in a secondary groundmass (Floyd and Rowbotham, 1979).

Floyd (1984) demonstrated that the south Cornish metabasalts (including those in the *mélange*) were mainly tholeiitic and exhibited distinct incompatible-element ratios (Ce/Y, Nb/Y, Y/Zr) when compared with the rest of the basic volcanics in south-west England. In general, all the metabasalts and amphibolites within the south Cornish *mélange* have a range of chemical compositions from normal type to variably enriched, transitional-type MORB (Floyd, 1982a, 1984; Barnes, 1984). They are dissimilar to the ophiolitic basic rocks of the normal-type MORB composition in displaying slightly enriched incompatible-element patterns (Figure 3.7). It is concluded that the volcanic debris within the

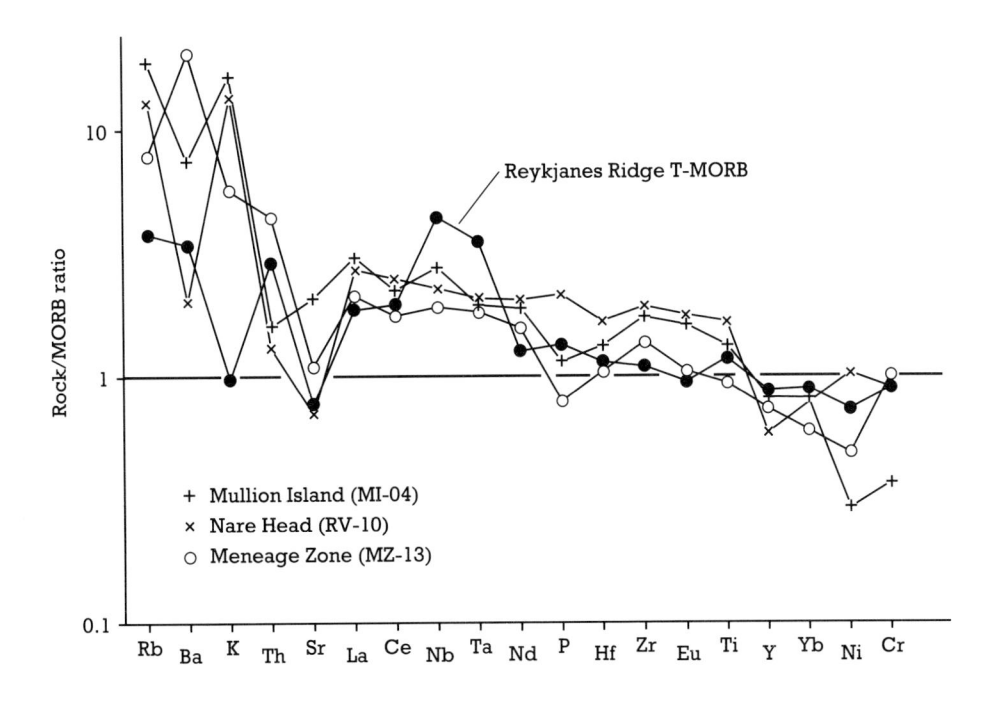

Figure 3.7 MORB-normalized multi-element patterns for selected *mélange* metabasalts compared with an example of transitional-type MORB from the Reykjanes Ridge.

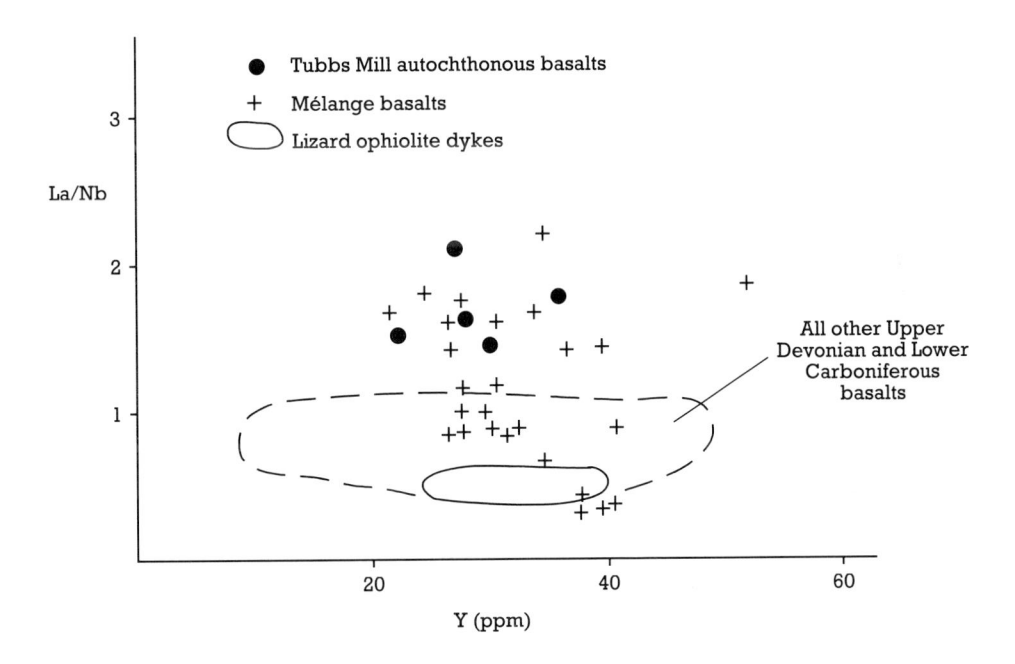

Figure 3.8 Diagram showing the variation in the La/Nb ratio for the Tubbs Mill pillow lavas and some of the *mélange* metabasaltic clasts relative to the Lizard dykes and Upper Devonian–Lower Carboniferous basic volcanics from south-west England.

mélange cannot be satisfactorily matched chemically with the ophiolitic dykes, in that the clasts do not represent the eroded volcanic carapace to the Lizard ophiolite. A stratigraphically *in situ* pillow lava sequence is found at the base of the Portscatho Formation (Tubbs Mill volcanics) in the allochthonous Gramscatho Group to the north of the *mélange* zone (Holder and Leveridge, 1986). These lavas have similar enriched chemical characteristics to the basic clasts and blocks within the *mélange*, and this suggests that the latter could have been derived from an ocean crust of similar composition, rather than Lizard-type ocean crust.

The Tubbs Mill pillow lavas and some of the *mélange* metabasalts in Roseland, however, show higher La/Nb (>1.5; Figure 3.8) and Th/Ta (>0.5) ratios than typical transitional-type MORB, together with chondrite-normalized negative Nb and Ta anomalies, which might reflect the influence of a subduction (back-arc?) environment. However, these minor chemical deviations could also have been generated by continental-crust contamination which, if the Gramscatho Basin was initially generated by the rifting of ensialic crust, might be expected for early segments of Devonian oceanic crust.

Start Complex greenschists

The greenschists occupy a unique position in southern Cornwall and have not been clearly identified with any other unit of basic composition. They are composed of fine-grained, variably foliated and laminated, quartz–albite–chlorite–epidote–amphibole schists, and they are generally considered to be a series of metavolcanic basic tuffs (Tilley, 1923) interbedded with pelitic and semipelitic metasediments. Little extensive chemical work has been done on the greenschists, although our unpublished data indicate that they were derived from a relatively uniform basaltic series with a tholeiitic character. The most interesting chemical feature is that these rocks are the closest to normal-type MORB in southwest England. In terms of the compositional variation exhibited by volcanic rocks in southwest England, therefore they have chemical characteristics that link them with the Lizard dykes and south Cornish *mélange* metabasalts. On this basis they could represent a volcanic segment of Variscan ocean crust docked against the Devonian parautochthon to the north.

A1 LIZARD POINT (SW 695116–SW706115)

Highlights

The best sections of the Old Lizard Head 'Series' metasediments and associated metabasalts, the Landewednack Hornblende Schists, occur here. Also exposures and boulders of the Man of War Gneiss are of importance.

Introduction

The earliest detailed description of the micaceous and hornblendic schists, that occur around the southern tip of the Lizard Peninsula was by Bonney (1883), who regarded them all as metamorphosed sediments. Somervail (1884) suggested that the hornblendic rocks were magmatic and this has been accepted by all subsequent workers. Flett and Hill (1912) give a full review of all earlier work, together with their own detailed descriptions of the range of metasediments. They regarded the hornblendic rocks as a series of metamorphosed basaltic lavas, tuffs and intercalated sediments and introduced the name Landewednack Hornblende Schists, derived from the hamlet one kilometre to the north. They pointed out that the sediments and metabasalts were interbedded and components of a coherent rock group but there was no 'way up' evidence to establish which group was the older. They described the mineralogy of the schists, which indicated high grades of metamorphism with the development of garnet and sillimanite. The metamorphic assemblages were discussed in greater detail by Tilley (1937) who also described unusual cordierite–anthophyllite rocks from Pistil Ogo, about 400 m north of Lizard Point.

Boulders of a dioritic gneiss with a very characteristic banded appearance are abundant on the beaches along the section; they can be seen *in situ* on the southern tip of Vellan Drang at very low tides. This is the Man of War Gneiss, named after the largest of the skerries and is the least accessible and least known of all the Lizard rock types.

Description

The Old Lizard Head metasediments and the Landewednack Hornblende Schists are intimately associated and in many places are interbedded. To the east of Polbream Cove the rocks are almost entirely amphibolite, whereas to the west they are dominantly metasediments. The large cliffs below the lighthouse form a thick, gently dipping, bedded sequence of very uniform amphibolites. Closer inspection shows that the bedding is also a metamorphic foliation and that the sequence has been isoclinally folded. This is shown by a strong lineation of the hornblende prisms and the presence of small intrafolial folds. The intensity of deformation and metamorphic recrystallization has destroyed most of the original igneous structures in these rocks. However, within the amphibolites and the schists there are a few coarser-grained, dyke-like bodies. Within the amphibolites are thin epidotic layers, formed during a very early stage, prior to deformation, which makes the folds more obvious. Most writers have suggested that the epidote was derived from calcic sediments. Also present are epidotic pods and cross-cutting veins that have clearly formed later. The majority of the rocks here give the impression that they are derived from a series of lava flows or sills, but, in a few places, horizons are present that have a less-uniform character and a fine-grained, fragmental nature that possibly could be derived from tuffs or volcaniclastic sediments.

To the west of Polbream Cove the sequence is dominantly of mica schists, although some layers of amphibolite and green hornblendic schists are also present. The hornblendic schists are transitional in lithology between the amphibolites and the sediments and are probably derived from tuffaceous sediments with a mixture of clastic and basic volcanic detritus. The interbedding and common deformational history of the sediments and amphibolites is good evidence for them being a largely extrusive sequence. The schists vary somewhat, but are dominantly muscovite rich, although quartz-rich varieties are also present. They tend to be fine-grained and rather phyllitic in appearance, although much of this is due to late shearing and belies the original high metamorphic grade shown by the wide occurrence of garnet, fine sillimanite and possibly kyanite.

At Pistil Ogo, associated with the greenschists, are the unusual cordierite–anthophyllite schists described by Tilley (1937) that have a characteristic blotchy appearance due to the clots of cordierite.

In a few places around Polpeor and Pistil Ogo, there are small pods of quartzofeldspathic material that could possibly be the first stages of partial melting. In the cliffs close to Lizard Point there is a much more substantial sheet of granitic rock which is several metres in thickness.

The beaches along this section are littered with boulders of a banded dioritic gneiss with a marked segregation of the mafic and felsic minerals. This is the Man of War Gneiss which forms many of the offshore skerries, and the southern part of Vellan Drang, the series of reefs which are accessible at very low tides. The reefs and skerries are almost entirely encrusted with barnacles, etc. and it is very difficult to see any contact relations. However, what little can be observed confirms the early report of Fox and Somervail (1888) that it is a *lit-par-lit* intrusion into the Old Lizard Head 'Series'.

Interpretation

This is the only site on the Lizard where the metasedimentary sequence can be clearly seen. The only other coastal section is a short section at Porthallow where it has been affected by intense shearing and alteration. Small, inland exposures are very poor and deeply weathered. The sedimentary origin of the rocks has been recognized since the earliest work (De la Beche, 1839), even though none of the original sedimentary features are preserved. The close relationship with the amphibolites is also clearly demonstrated and their common structural history and transitional rock types establish the former as an extrusive sequence.

The chemical composition of basalts is a good indicator of the tectonic regime of their origin and several workers have analysed rocks from this section and localities nearby (Floyd, 1976; Kirby, 1979b; Styles and Kirby, 1980). Trace-element discriminant diagrams presented by these authors give a strong indication that these amphibolites have the chemical signature of ocean-floor basalts. The recent study of REE geochemistry by Sandeman (1988) shows a generally flat pattern with about 20× chondritic abundances typical of mid-ocean ridge basalt (MORB). The recognition that the Landewednack Hornblende Schists have MORB affinities is important, as it connects this lower tectonic unit with the rest of the ophiolite complex. Their chemistry shows that they are not comagmatic with the other ophiolitic rocks, but this is not surprising, bearing in mind that the Lizard has features suggesting a slow-spreading centre and that small, transitory magma chambers would be expected rather than a single large one. This chemical type contrasts with most of the other basaltic rocks of south-west England, but is similar to the greenschists of the Start Complex.

The chemical data from the Man of War Gneisses (Sandeman, 1988) show that the contents of large-ion-lithophile elements, such as Sr, K, Rb and Ba, are high and that Ti is low compared with MORB. This indicates that they have similarities to arc-related rocks, which shows that they are not related to the Landewednack Hornblende Schists, but have a greater similarity to later magmatic episodes such as the Kennack Gneiss.

This site is very important in Lizard geology as it is the best section of the metamorphosed equivalents of the sea-floor sediments and basalts of the ophiolite complex. This upper part of the typical ophiolite sequence is never seen in its correct position in the Lizard ophiolite stratigraphy; it is only preserved in slivers attached to the base of the main ophiolite slab during obduction.

Conclusions

This site consists of metamorphosed sediments and magmatic rocks of the Lizard Complex. This includes the Old Lizard Head 'Series' (metamorphosed fine-grained sediments) and the Landewednack Hornblende Schists (metamorphosed basaltic lavas) which are intruded by the Man of War Gneiss (metamorphosed dioritic rocks). The Lizard Complex has been interpreted as a piece of ancient, early Devonian, oceanic crust (ophiolite) that was thrust to the north during the late Devonian. Studies of the chemistry of the 'basalts', show clear affinities with basalt erupted on the present-day ocean floor, at mid-ocean ridges. The later 'diorites' have a chemistry which is similar to that seen in the igneous rocks formed in modern island arcs at the margins of the oceans. This is the best site to examine sediments and volcanics which formed part of the floor of the Variscan ocean before it was subsequently destroyed by continent–continent collision.

A2 KENNACK SANDS (SW 734165)

Highlights

This is the type locality for the Kennack Gneiss Group, and it shows good exposures of 'primary' peridotite.

Introduction

The Kennack Gneiss Group is a series of interbanded mafic and felsic gneisses with associated gabbros, basaltic dykes and granites, that occur along the east coast of the Lizard, from Kennack Sands south to Church Cove, and in some of the valleys inland from there. They have been one of the most enigmatic and controversial rock groups of the Lizard Complex, and in this review they are covered by two sites, this locality and one at Polbarrow. To avoid repetition, the introduction for both sites is given here and the discussion at the end of the Polbarrow site description.

Early workers, and indeed most subsequent authors, have found this to be a most perplexing group of rocks. De la Beche (1839) recognized the intrusive nature of the banded gneisses into the peridotite, both here and at Parn Voose site (Polbarrow), but many other localities now attributed to the Kennack Gneiss were originally mapped as hornblende schist. Bonney (1877a) assigned the Gneisses to his 'granulitic group', and suggested that they were highly metamorphosed sediments, with the banding being an expression of the original sedimentary bedding. In contrast to De la Beche (1839), he proposed that the peridotite was intruded into the granulitic group. He noted the similarity of the gneissic rocks to those in north-west Scotland, which led him to suggest that these and the other metamorphic rocks of the Lizard were Archaean in age.

Teall (1887) described the felsic portion of the gneiss as intruding into the mafic and, therefore, that the rocks were igneous and not sedimentary. The banding was due to a 'rolling out' of this intrusive material under pressure, a kind of fluxion structure. Somervail (1888) described the intrusive nature of the felsic portion of the gneiss into the peridotite. Bonney and McMahon (1891) accepted that the banding of the gneisses might be due to fluxioning of mafic rock intruded by felsic rocks, but still maintained that they were older and that the peridotite was intruded into them. Lowe (1901, 1902) stoutly maintained that the 'granulitic group' was the youngest of the area, drawing attention to the conformable nature of the fluxion banding, the numerous inclusions and peripheral alteration of peridotite blocks in the gneiss.

This general view was accepted by Flett and Hill (1912) in the Lizard Memoir. Flett described the field relations in great detail showing that the felsic fraction intrudes the mafic fraction, and that both intrude the peridotite. The banding and many of the structures observed were due to the fluxioning together of composite intrusions of felsic and mafic magmas while they were still hot, together with some reaction to produce rocks of intermediate composition. He also recognized a group of granitic rocks, not accompanied by mafic rocks, that he called the Kennack Granite. Bonney (1914), however, was still reluctant to accept that the 'granulitic group' was younger than the peridotite. Flett's views were restated with little alteration in the second edition of the Lizard Memoir (Flett, 1946).

More recently, Sanders (1955) has restudied the Kennack Gneiss and the structures along the south-east coast of the Lizard. He made the then radical suggestion that the form of the peridotite was a thin sheet rather than the plug-like intrusion accepted by all previous workers. He also suggested that the Kennack Gneisses were a thin development immediately beneath the peridotite, formed by migmatization of the Landewednack Hornblende Schists.

Green (1964c), in his study of the Lizard, largely devoted himself to the peridotite, but also discussed the Kennack Gneiss. He generally accepted Flett's views suggesting *lit-par-lit* injection of felsic magmas into mafic sills, but thought that there was no need for actual mixing of the magmas.

Strong *et al.* (1975), in a brief paper proposing an ophiolitic nature for the Lizard, preferred Sanders' (1955) view that the gneisses were migmatites. Bromley (1979) discussed the migmatitic and intrusive models and favoured an intrusive origin by deformation of a felsic netvein complex. Kirby (1979a) carried out a detailed field and chemical study, and came to the conclusion that the gneisses were probably formed by the migmatization of Old Lizard Head metasediments and Landewednack Hornblende Schists. Styles and Kirby (1980) described a borehole at Kennack where a considerable thickness of gneisses were intersected (150 m). They stressed the heterogeneity of the gneiss group and, with a few words of caution, supported the conclusion of Kirby (1979a).

Styles and Rundle (1984) carried out a Rb/Sr isotopic study of a granite vein from the Kennack borehole. They obtained an age of 369 ± 12 Ma (mid-Devonian), which finally put to rest any lingering thoughts that the gneisses might be Precambrian and, as the initial ratio was low (0.70424 ± 0.00009), they were unlikely to have originated from melting of old crustal material. They suggested that the source consisted of a high proportion of mantle or juvenile crustal material.

Subsequently, Barnes and Andrews (1986) have briefly described the field relations of the Kennack Gneisses, and suggested the simultaneous existence of felsic and mafic magmas. They reinterpreted Kirby's (1979a, 1979b) chemical data, and have suggested that the low incompatible-element contents of the felsic portion do not fit with an origin by melting of amphibolite, but by melting of depleted continental crust material. This mixing of such a magma with oceanic-basalt material possibly occurred in an intracratonic setting where ocean crust was being formed in a rift. However, Malpas and Langdon (1987) presented new major- and trace-element data to support an origin by melting of a mixture of Old Lizard Head metasediments and Landewednack Hornblende Schists, a model similar to that proposed by Kirby (1979b).

The most recent and most detailed study of all is that by Sandeman (1988), who produced maps at a scale of 1:1500 and described many field relations in great detail.

Sandeman (1988) points out the relationships between the different components and concludes that two magmas must have been present at the same time. The felsic was slightly later than the mafic and both physical intermingling and chemical mixing between the two magmas took place. His detailed chemical studies show that the magmas did not have any close relationship prior to their mixing during intrusion. The mafic magma has the chemical characteristics of a calc-alkaline basalt, probably associated with a volcanic arc. After mixing with the granitic magma the composite 'magma package' was intruded along shear zones and faults.

Lizard and Start Complexes (Group A sites)

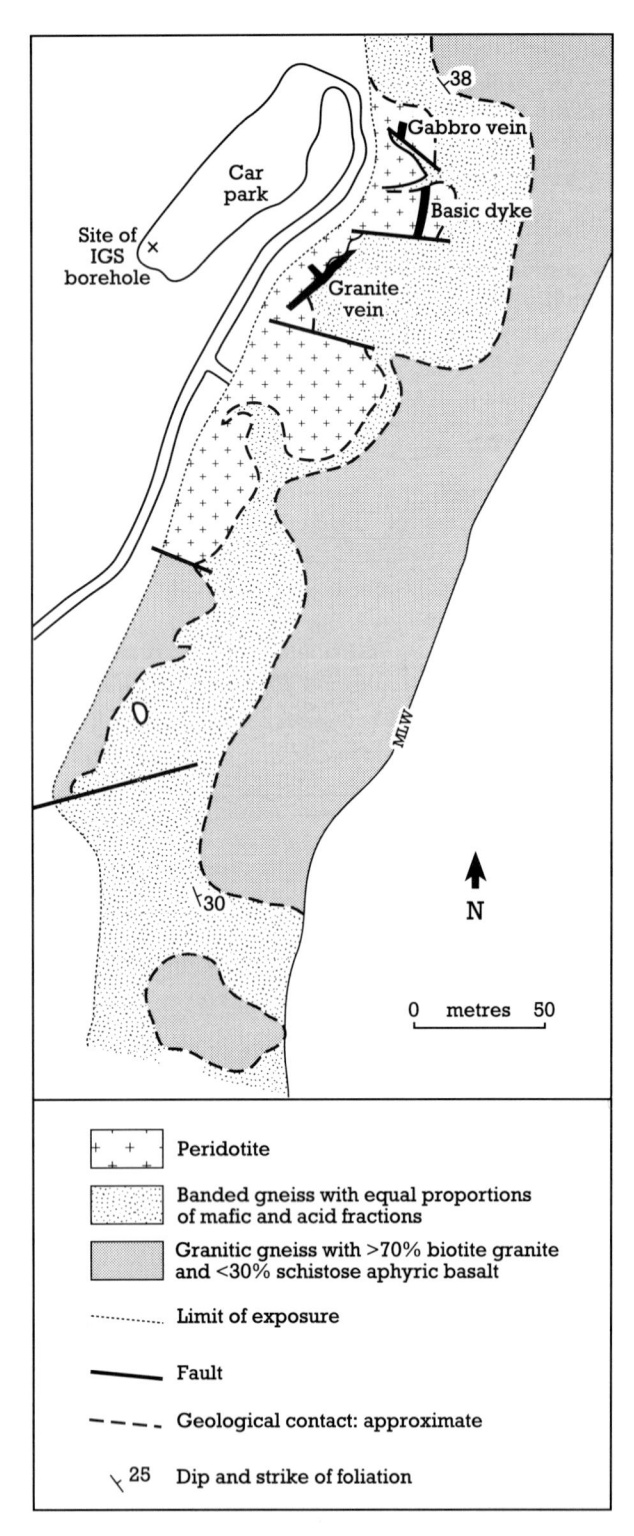

Figure 3.9 Geological sketch map of the Kennack Sands site (A2).

Description

The outcrops along the cliffs and foreshore to the south of the stream at Kennack Sands are some of the best to show the relationships between the various components of the Kennack Gneiss Group and other rock types including peridotite, basaltic dykes and gabbro veins. A sketch map shows the main features (Figure 3.9).

The outcrop in the low cliff a few yards south of the stream shows some of the classic features of the Kennack Gneiss; it was figured by Flett and Hill (1912). Here a vein of banded gneiss intrudes the peridotite as a steep, dyke-like body, with faulting along the northern contact. The felsic portion of the gneiss is slightly younger than the mafic which it cross-cuts locally, but they appear to have been intruded essentially together. The banded gneiss vein truncates a gabbro pegmatite vein to the north and a basaltic dyke to the south, demonstrating that it is the latest intrusive phase in the vicinity.

The banded gneisses in the outcrops close to the cliff are generally of the finely banded variety with subequal proportions of the felsic and mafic components, in bands usually a few centimetres thick (Figure 3.10). In many places, it can be seen that the felsic portion is the younger, particularly in the more mafic varieties where they are cut by numerous thin felsic veins. The degree of deformation is always high, and at least two phases of high-temperature folding can be seen.

The large rocks lying further from the cliffs are composed largely of granite gneiss sheets, several metres in thickness with much thinner interlayers of mafic gneiss. On the seaward side of one of the first large rocks encountered, is a vein of gabbroic to dioritic composition with abundant subrounded xenoliths of gabbro and numerous plagioclase xenocrysts. The large plagioclases, so clearly xenocrysts in this rock, are probably of a similar origin to the 'phenocrysts' in the more homogeneous, fine-grained dioritic rocks nearby.

At Thorny Cliff there are distinctive hybrid rocks. They consist of a monzodioritic host containing blocks of a fine mafic rock. The mafic rocks have a marked lobate outline and appear to have undergone partial digestion by the more felsic host. This is good evidence that here has been assimilation of the mafic rocks to form hybrids of intermediate composition.

The peridotite exposed along the cliffs at Kennack is relatively fresh, often showing only

Figure 3.10 Banded gneiss, Kennack Sands. (Photo: M.T. Styles.)

around 50–60% serpentinization. The rock is very dark green, with prominent, large, bronzy orthopyroxenes up to 10 mm in size and small, dark-green clinopyroxenes. There is a steep foliation trending roughly north–south and a compositional banding due to variation in the proportion of pyroxene, with thin pyroxenite bands in a few places. The contacts between peridotite and the felsic portion of the gneiss are characterized by marked alteration with the formation of various secondary minerals, including talc and chlorite. However, the most striking alteration is to an asbestiform magnesian amphibole.

The features seen along this beach section clearly show that the felsic portion of the gneisses is younger than the peridotite and also the pegmatitic gabbro and basic dykes.

Interpretation and conclusions – see Polbarrow – The Balk below

A3 POLBARROW–THE BALK (SW 717135–715128)

Highlights

This is the best locality for showing complex intrusive relationships between the mafic and felsic components of the Kennack Gneiss Group, and the thrust contact between the Lizard peridotite and the Landewednack Hornblende Schists.

Introduction

This is a beach and cliff section 500 m in length, much of it is accessible only for a short period of time around low tide (Figure 3.11). The site shows two of the important geological relationships of the Lizard Complex. First, that between the Landewednack Hornblende Schists and the

Lizard and Start Complexes (Group A sites)

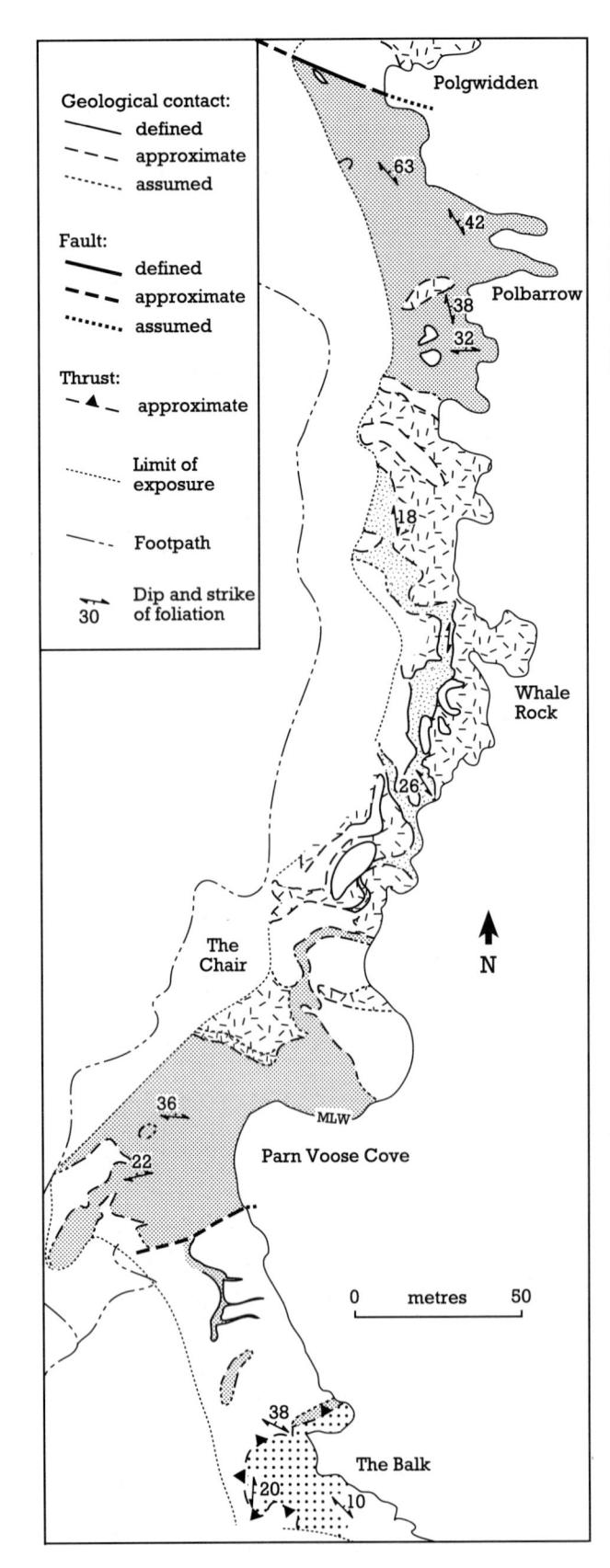

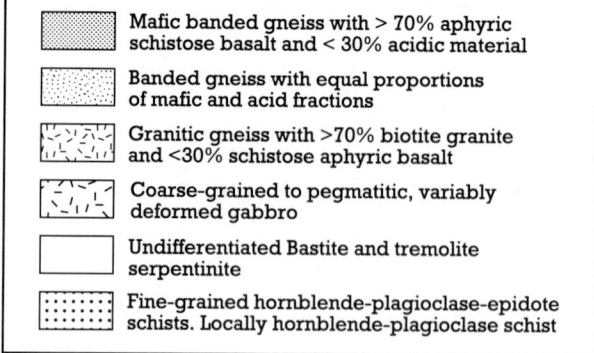

Figure 3.11 Geological sketch map of the Polbarrow–The Balk site (A3) (after Sandeman, 1988).

overlying Lizard peridotite in the southern part of the Complex, and, secondly, between the gabbros and felsic and mafic components of the Kennack Gneiss. The relationships seen from Parn Voose along the coast, past Whale Rock to Polbarrow, show a deformed intrusive complex of gabbro and banded gneiss, and are most important in an assessment of the origin of the Kennack Gneiss Group. A detailed account of the background research on the Kennack Gneiss Group was given previously with the Kennack Sands site description.

Description

The contact between the Lizard peridotite and the Landewednack Hornblende Schist is seen at The Balk, along the road leading into the old quarry from Church Cove and around the lower part of the old quarry face. The Landewednack Hornblende Schist is a dark, slightly banded, schistose amphibolite with a flat-lying foliation. Close to the contact with the overlying peridotite, the amphibolites are noticeably more gneissose. Some of this is possibly the effect of metamorphism, locally induced by the overriding peridotite, similar to that described by Green (1964c) at a similar contact north of Cadgwith. There are also some deformed pods of gabbro at this horizon. The actual contact with the peridotite is highly sheared, with a fairly gentle dip to the north-west; locally there is calcite and altered sulphide mineralization along the thrust plane. The thrust is offset by small, late, upright faults in several places. The peridotite also shows the effects of intense shearing and, in the quarry and on the beach below, blocks can be found with augen of orthopyroxene in a highly sheared matrix of streaky serpentinite, a peridotite mylonite. This section clearly shows that the peridotite was thrust over the Landewednack Hornblende Schists at relatively high temperatures; it is not an intrusive contact.

There is little outcrop between the Balk Quarry and Parn Voose Cove where the rocks are very different. The south side of the cove appears to be along a fault as, along the south wall, the peridotite is at sea-level, but at the back of the cove it is at the top of the cliff, some 50 m higher. A few outcrops and most of the boulders in the cove are dominantly mafic gneiss with thin veins of felsic material such as that figured by Flett (1946, p. 104). The gneiss has been highly folded and bears a striking resemblance to migmatites from high-grade metamorphic terranes. Much of the rock around the back of the cove is a medium- to coarse-grained gabbro (Figure 3.12), locally sheared and cut by flaser zones. Along the western side of the cove, xenoliths of peridotite and gabbro are included in banded gneiss.

Further north, between Parn Voose and Polbarrow, along the rocky ledges (only accessible at low tide), the relationships with the banded gneisses are most clearly seen. The gabbro body at the back of Parn Voose is quite extensive and forms much of the rock for some 500 m, but in many places it is intruded by thick sheets of banded gneiss. The intrusive relations are very clear, and there are numerous xenoliths of peridotite and gabbro within the gneiss. The gabbro intrudes into the peridotite and these are then in turn intruded by the banded gneiss. Often at the margins of the felsic gneiss, there is a rock of intermediate composition which has lobate contacts with the mafic rock and appears to have reacted with it. In some places, offshoots of felsic gneiss finger out into the gabbro, and in others, lenses of gabbro are streaked out within the foliation of the gneiss. The two rock types have clearly been deformed together at high temperatures. Many of the features of the contacts between the felsic gneiss and the gabbro suggest that the gabbro was hot and relatively soft at the time of injection of the gneiss. This was pointed out by Flett and Hill (1912) and Kirby (1979b). The peridotite appears to have acted as rigid blocks during the plastic deformation of the other rocks showing that, at the time of deformation, it was crystalline peridotite not serpentinite which

Figure 3.12 Acid-veined gabbro at Parn Voose. (Photo: M.T. Styles.)

would have had much weaker mechanical properties.

Interpretation (Kennack Sands and Polbarrow–The Balk)

The introduction to both Kennack Sands and Polbarrow showed that there has been a long-running controversy over the origins of the Kennack Gneiss Group. Proposals fall into two main groups:

1. migmatization of various mixtures of pelitic metasediments and hornblende schists, and
2. late composite intrusions of felsic and mafic magmas.

The outcrops within these sites are the most important for assessment of the origin of the rocks as they encompass the widest range of lithologies of this diverse group.

The finely banded, highly deformed rocks seen at Kennack and Parn Voose bear a striking resemblance to migmatitic rocks from high-grade terranes. At localities such as Polpeor local 'sweat outs' of granitic material suggest that local melting might have occurred in places where the temperature was a little higher. Similarly, at Porthkerris, local melting of the amphibolite seems to have taken place, so the circumstantial evidence for anatexis is present. However, even at Kennack the evidence (Sandeman, 1988) shows that the very banded rocks are restricted to the cliffs and innermost foreshore rocks and that overall there is a zonal arrangement. This zonal pattern was most clearly seen in some of the smaller outcrops, to the east of Kennack and at Little Cove Poltesco, but can also be demonstrated at Kennack. On the outer rocks is a granitic core, with few mafic rocks, followed farther inshore by a mixed, interbanded zone and finally a dominantly mafic zone with little granitic rock, close to the cliff. Such a regular pattern found at several localities is difficult to explain by a migmatitic origin, although some original interlayering of sediment and amphibolite as proposed by Styles and Kirby (1980) might be plausible.

The field evidence from Kennack and Parn Voose is somewhat equivocal, but could be used to support a migmatitic origin. The proportions of felsic and mafic rocks are roughly correct, but the rocks have been so extensively deformed that any original contact effects or cross-cutting intrusive relations have been smeared out. The field relations around Whale Rock leave much less doubt as to the origin. It is the opinion of Flett and Hill (1912), Green (1964c), Sandeman (1988) and the author that the relationships seen here where the rocks are least deformed, are the key to the whole group. Here there are clear composite intrusions, with mafic magmas closely followed by felsic magmas and abundant evidence of hybridization processes. The location of the intrusions seems to be controlled by structural features, particularly flat-lying thrusts and, to a lesser extent, minor offshoots along steep faults. At Polbarrow and Parn Voose it may be the 'basal' thrust of the peridotite, but at Kennack a thrust within the lower part of the peridotite is also possible.

A borehole was drilled in 1980 by the Institute of Geological Sciences (now British Geological Survey) in the car park at Kennack (Figure 3.9), about 100 m north-west of the site described here (Institute of Geological Sciences, 1982). This penetrated some 150 m of Kennack Gneiss with interlayers of peridotite (Styles and Kirby, 1980). Such a thickness was much greater than expected, as most previous models had suggested that the gneiss was essentially a thin development at the base of the peridotite sheet. It would be difficult to generate such a thickness of gneiss by melting, as there would be a problem in achieving sufficient heat; sandwiching the protolith material between slabs of peridotite is one of the few possibilities. There is no real problem about the thickness if the origin was intrusive.

Kirby (1979b) and Malpas and Langdon (1987) put forward chemical data which they interpreted as supporting a migmatitic origin. The chemical compositions of parent rocks, felsic anatectites and mafic restites seemed to fall on linear trends, as would be expected if they were related by partial melting. The composition of granitic fractions was also similar to minimum melt compositions. The chemical data of Sandeman (1988), with a much larger number of samples, do not clearly show these important linear trends. He has suggested that only the felsic fractions show linear trends, and that the mafic rocks show either scattered or possibly curved distribution. The trends suggested by Malpas and Langdon (1987) may be a fortuitous accident of their small, and hence unreliable, data set. A summary of this new data (Sandeman *et al.*, in prep) is given below.

The mafic magmas that formed the mafic portion of the gneiss are quite distinct from the Landewdnack Hornblende Schists and most of the ophiolite dykes that have oceanic affinities. They are light-REE-enriched and have affinities with volcanic arcs and show evidence of crystal fractionation. The mafic magma chambers were intruded by felsic magmas, leading to the magma mingling which, on intrusion into the peridotite, etc., gave the very banded gneisses and some mixing and hybridization to produce intermediate magma types. In most cases, these phenomena can only be assumed from the chemical characteristics, and only at localities such as Whale Rock can they actually be seen. The subsequent intense deformation of possibly hot, incompetent rocks accentuated the degree of banding and gneissic appearance of these mixed rocks.

It is interesting to note that the two most detailed studies by Flett some 80 years ago, based entirely on field relations, and by Sandeman with the aid of modern geochemistry, have come to essentially the same conclusion. The Kennack Gneisses are a series of composite mafic and felsic intrusions that subsequently have been extensively deformed.

Conclusions (Kennack Sands and Polbarrow–The Balk)

These are important sites showing the juxtaposition by a major thrust plane of the Lizard peridotite and the Landewednack Schist, two constituent major parts of the Lizard Complex. A wide range of igneous rocks including gabbros, and acidic and basic gneisses, are also exhibited. Although the association of metamorphic and igneous rocks is complex, a sequence of events may be determined that helps to interpret the magmatic structure of the oceanic crust and upper mantle.

Peridotite occurs as blocks in the gneisses. Gabbro intrudes the peridotite and is in turn intruded by gneiss, the last dated to around 370 million years before the present (early to mid-Devonian). The intimate mixing (and, in places, hybridization) between basic and acid gneisses has been the subject of much debate. It is now thought to have been the product of the mixing of two very different but contemporaneous magmas. These sites tell us much about the development of the Lizard Complex as a slice of ancient ocean floor that was finally thrust over continental crust during late Devonian times.

A4 KYNANCE COVE (SW 684133)

Highlights

Primary and recrystallized types of peridotite are well exposed here, and the relationships between these rocks and granite veins, basic dykes and hornblende schist are clearly seen at a locality that is readily accessible.

Introduction

Kynance Cove is one of the few places on the west coast of the Lizard where it is possible to descend the large cliffs. Here, the two main types of peridotite, the primary and recrystallized types are juxtaposed by a fault. In the gap between Asparagus Island and the mainland, and in the cliffs going northwards to Lawarnick Cove, are a variety of granitic, amphibolitic and banded gneisses within the peridotite. This area has not been discussed in many publications, and only Flett and Hill (1912) have described it in detail and published a sketch map. There have been no major controversies about the interpretation of the rocks in this area.

Description

The east side of Kynance Cove has a large wall formed of the primary-type peridotite, though the rock is probably more easily studied on the paths leading down to the cove. It is a coarse, partially serpentinized peridotite with prominent, bronzy orthopyroxenes and a distinct coarse steep foliation. When at its freshest it is a dark, green-black colour, but many samples from this area have been slightly hematized to give a dull red-brown. In the rocks at the base of the cliff there are several basaltic dykes cutting the peridotite. A large fault runs roughly parallel to this western side of the cove and, in the central and western parts the peridotite, is of the recrystallized type. This has a distinct, closely spaced, metamorphic foliation with prominent augening around orthopyroxene porphyroblasts. Amphiboles are common in this type, and most rocks are reddish or brownish due to oxidation of the iron minerals.

Granitic sheets several metres in thickness are present between Asparagus Island and the mainland and also in the cliff to the north towards Lawarnick Cove. They are quartz–feldspar rocks with a little muscovite, and little or no foliation is developed. They do, however, show some boudinage structure. These features distinguish them from the acid fraction of the Kennack Gneiss. Similar granite sheets are also found at several other localities along the west coast of the Lizard. There is often a marked reaction zone at the contact with the peridotite, forming zones of talc and chlorite representing the migration of chemical components across the contact. Silica has moved into the peridotite to form talc, and magnesium migrated into the granite to form chlorite.

Within the rocks in the centre of the gap between the base of the cliffs and Asparagus Island, is a block of amphibolite. The contact relations are not clear, but it appears to be of the more banded Traboe type rather than the massive Landewednack-type amphibolite. This and the presence of banded gneiss a little to the north in Lawarnick Cove, suggest that this is close to the lower contact of the peridotite.

Interpretation

Kynance Cove shows the two main peridotite types, the primary and recrystallized types, in close proximity. However, this is not a transitional contact that records the sequence of events of how one is formed from the other. The two rock types with their contrasting mineral assemblages and previously formed structural fabrics, have been juxtaposed by late, steep faults. The intrusive relations and contact alteration of the late granite sheets are well-displayed.

The presence of the Traboe type amphibolites both here and in George's Cove, 2 km to the north, is of regional structural interest. Their presence and the bodies of Kennack Gneiss here and further south at Pentreath suggest that this level is close to the base of the peridotite sheet similar to the east coast. The base of the peridotite is clearly offset in numerous places by late faults, such as that at the east side of Kynance Cove. It is generally close to sea-level both here and at George's Cove, but is downfaulted to somewhat deeper levels in the intervening area. Previous discussions of the structure of the peridotite, covered in the introduction to the chapter, and particularly Green (1964c), suggested that, although the contact on the east coast might be flat lying, that on the west coast was steep, and consistent with a diapir-like form. The evidence outlined above, however, points to a structure similar to the east coast, although less well exposed, and hence does not support a diapir-like form.

Conclusions

Here outcrops of both primary and recrystallized Lizard peridotite are seen to be cut by later granitic veins and basaltic dykes. The peridotite represents a slice of upper mantle, thrust over younger rocks of the then continental margin, during the late Devonian. It forms part of the ultramafic base to the Lizard ophiolite, being formed at high temperatures and pressures in the mantle prior to subsequent recrystallization and serpentinization. The site at Kynance Cove therefore presents graphic evidence for the juxtaposition of Lizard rocks with an oceanic origin carried on the back of a major thrust over the top of continental crust.

A5 COVERACK COVE–DOLOR POINT (SW 784187–SW 785181)

Highlights

This is the best section through the contact between the peridotite and gabbro units; this contact is thought to be the Mohorovičić discontinuity between the mantle and oceanic crust of the Lizard Complex.

Introduction

This has been a classic area of Lizard geology since the intrusive relations of the various rock types were described from here by Flett and Hill (1912). These authors demonstrated the intrusive sequence of peridotite, troctolite, gabbro and basic dykes, and produced several field sketches and a map to illustrate this. Green (1964a) showed that the peridotite was a mantle rock and thought that the gabbro was a later ring-intrusion following emplacement and uplift of the peridotite. The area became of great significance after ophiolite models for the Lizard had been proposed (Bromley, 1979; Kirby, 1979b; Styles and Kirby, 1980), as this was then considered to be the transition zone from the crust to the mantle, and thus represent the Moho. This was addressed particularly by Kirby (1979b), who studied the zone in detail and drew attention to the troctolites and associated rocks that are probably representative of a thin sequence of transition zone cumulates. Davies (1984) carried out a Sm/Nd isotope study of a gabbro sample from Coverack and produced a mineral–rock isochron that showed a crystallization age of 375 ± 34 Ma (mid-Devonian).

Description

The section at Coverack is a broad sweep of beach that is covered at high tide, but numerous, low, rounded, rocky outcrops are seen at low water (Figure 3.13). The magmatic stratigraphy consists of three main divisions, an ultrabasic division to the south, a central interbanded division and a northern gabbro division (Figure 3.13).

The ultrabasic division consists largely of the primary type peridotite, the 'bastite serpentine' of Flett and Hill (1912). The peridotites exposed here, particularly around Dolor Point, are some of the freshest on the Lizard, being only around 30% serpentinized in some cases. They are dark-green, coarse-grained rocks that consist largely of olivine with conspicuous orthopyroxene. Clinopyroxene is also present; its abundance varies from 5–10%. Thus there are both harzburgites and lherzolites, but they form part of a continuous spectrum and are not significantly different rock types. Many samples contain a brown chrome spinel and a few exhibit a small amount of plagioclase. The peridotites have a coarse subvertical foliation that trends roughly north-south. This is a high-temperature porphyroclastic texture that here, as in many other ophiolites, is ascribed to solid-state flow that occurred in the upper mantle beneath the spreading centre. Dunites occur as pods several metres in size around the west side of the harbour and also as a series of veins on the south side of Dolor Point. The dunites are conspicuous by their lack of orthopyroxene and reddish colour as the olivine is usually totally serpentinized. The origin of the dunites will be discussed in detail under 'Lankidden' below, but essentially they are thought to have been formed by the passage of picritic melts through the mantle.

At the northern part of this division, some 50 m north of the harbour, is the famous Coverack troctolite. This is a very distinctive rock composed of white plagioclase and red serpentinized olivine. Rarely, the olivine has survived alteration and then the rock is dark green and black and actually looks more like the gabbros than the 'typical' troctolite. Thin-section examination shows that many of the rocks are not true troctolites, as they contain more than 10% clinopyroxene and are, in fact, olivine gabbros, but they are still distinctly different from the main gabbro. The proportion of feldspar to mafics also varies substantially from 20 to 80% feldspar, although most have roughly equal proportions. The troctolite forms two sheet-like bodies, but it and the peridotite are cut by numerous sheets of pegmatitic gabbro and fine basic dykes. The intrusive sequence of these rocks is superbly displayed in the classic locality, close to the sea wall beneath the graveyard, described and figured by Flett and Hill (1912). Here it can be seen that the troctolite intrudes the peridotite (Figures 3.14 and 3.15), and is itself cross-cut by the gabbro sheets, and all three are intruded by the basic dykes.

The central division, some 200 m wide, consists of peridotite cut by numerous sheets of gabbro pegmatite. The thickness of these sheets varies from a few centimetres up to tens of metres. The thicker gabbro sheets tend to form the lower ground between the more prominent peridotite rocks. There are many small bodies of dunite within this area and, in the centre of the zone roughly in front of the store, is a much larger dunite body around 20 m in diameter (Figure 3.13). Many of these gabbro sheets are somewhat deformed and foliated, and, in several places, discrete high-temperature shear zones are seen with a streaky mylonitic fabric. Basaltic dykes are quite abundant, and at one location a dyke cuts through the shear zone, showing that magmatism continued after the onset of deformation.

The northern division is composed of gabbro with a few xenoliths of very altered peridotite and a few basic dykes. The peridotite blocks are all very altered being serpentinized and carbonated. Where the foliation can be seen, it varies from block to block, which contrasts with the very consistent N–S trend in the peridotites further south and implies that these blocks are detached xenoliths. The gabbro, generally referred to as

Lizard and Start Complexes (Group A sites)

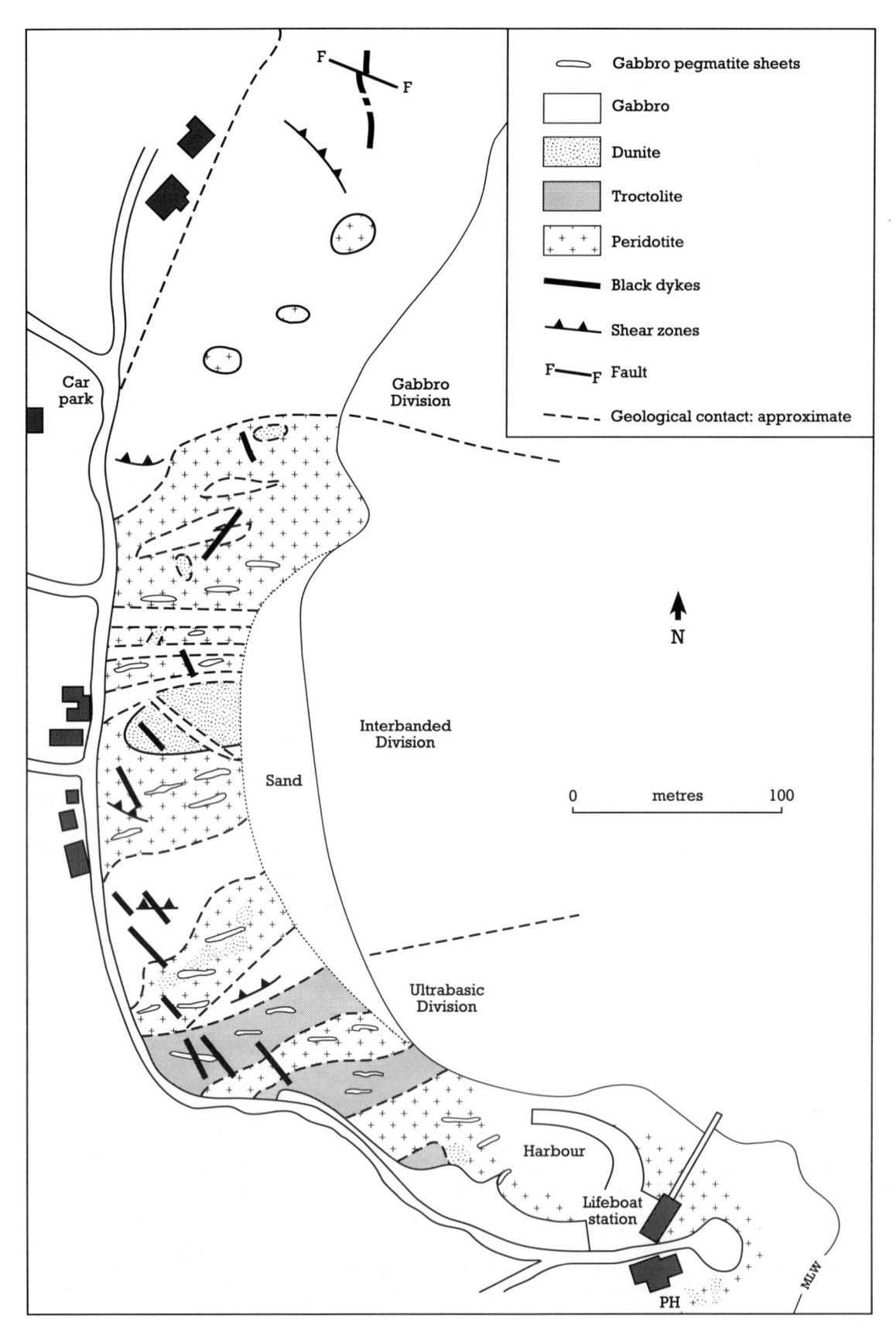

Figure 3.13 Geological sketch map of the Coverack site (A5).

the Crousa Gabbro, is a coarse-grained rock composed largely of plagioclase and augite with olivine, minor ilmenite, brown hornblende and rare biotite as primary constituents. When fresh, the gabbro is very dark coloured, almost black, as the feldspar is very dark purple in colour and thus can easily be mistaken for an ultrabasic rock. Many of the rocks have undergone some hydrothermal alteration which saussuritized the feldspar, turning it white and giving the rocks a typical mottled gabbro appearance. In many rocks, augite is partially or extensively altered to green hornblende. There are shear zones within the gabbro which have a NW–SE trend. At the northern end of the section, in the low cliffs and small outcrops sticking out through the beach boulders, two distinct generations of dykes can be seen. Early dykes have been pulled apart and broken up in the gabbro, whereas the late dykes have sharp planar margins, although even these are affected by late faults. There is a particularly good example at the northern end of the beach with distinct phenocryst-free, chilled margins. This dyke contains fresh olivine which is rare in dykes on the Lizard.

Interpretation

The principal interest of the site at Coverack is the excellent exposure of the transition from mantle to crust in an ophiolite sequence. It is not, however, a typical ophiolite succession, as the substantial cumulate sequence of dunites and pyroxenites separating the mantle peridotites from the gabbros, is missing. Such a sequence only seems to be present here as a very minor constituent represented by the dunites and troctolites. At this junction, but inland to the west, there is the Traboe cumulate complex which could represent a much thicker development (Leake and Styles, 1984). It does not reach the coast, but seems to be cut out laterally by the intrusion of the Crousa Gabbro, which is a later intrusion. The section at Coverack is therefore atypical for an ophiolite, due to the intrusion of later gabbro; possibly off-axis magmatism occurred as suggested by Badham and Kirby (1976). The chemistry of the gabbro was studied by Kirby (1979b), who showed that it was progressively fractionated when traced from south to north, with such features as Fe/Mg ratio and Ti and P

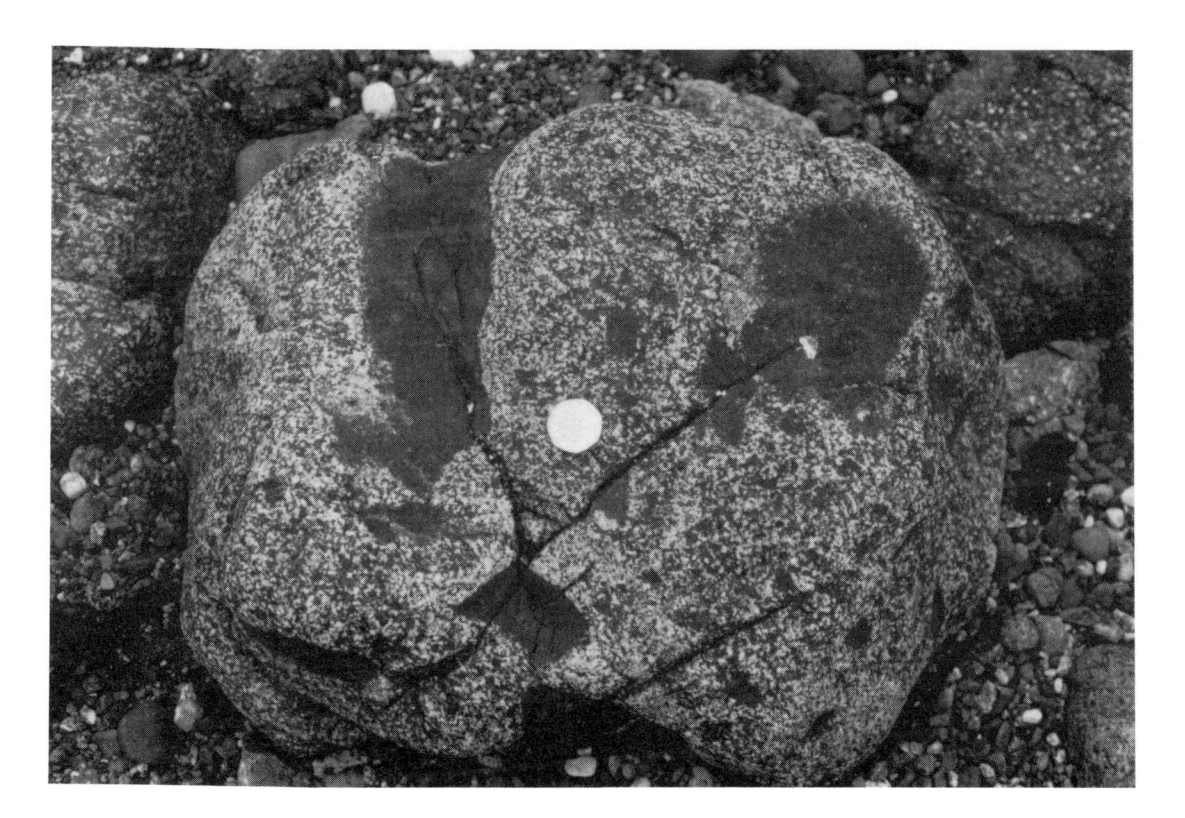

Figure 3.14 Xenoliths of peridotite enclosed within troctolite, Coverack Beach. (Photo: M.T. Styles.)

Figure 3.15 Troctolite veining peridotite, Coverack. (Photo: M.T. Styles.)

contents increasing in that direction. The overall chemical features were consistent with it forming from a magma with MORB-type chemistry.

The basic dykes are an important part of the geological framework at Coverack, but as they form the main feature of the following Porthoustock site, they will be discussed in detail there. The chemistry of the dykes suggests that they are not derived from the same magma as the gabbro, that is, they are not cogenetic. The trace-element ratios of such elements as P, Zr, Y and Ti, V are distinctly different and show they are not derived from the same source. However, this is not surprising: there is field evidence that some dykes cut shear zones through the gabbro, showing a clear time gap between the two intrusive phases.

Overall, the site at Coverack is one of the most important on the Lizard; it shows clearly the intrusive relations and hence time-sequence of many of the main rock types. The difference in fabrics between the peridotites and gabbros shows they were deformed at different times, in totally different regimes. The peridotites have a pervasive porphyroclastic texture that is typical of that formed by solid state 'creep' in the mantle at very high temperatures, perhaps 1000°C. The gabbros, in contrast, have deformation in discrete shear zones with amphibolite-facies metamorphic assemblages showing temperatures of 500–600°C, formed after the gabbros had crystallized and cooled in the ocean crust. The section as a whole records the transition from mantle to crustal rocks in an oceanic-type sequence, a fossil Moho.

Conclusions

The dominant basic and ultrabasic rocks at this site represent a sequence or section through ancient oceanic lower crust and upper mantle respectively, and is thus equated with the oceanic crust–mantle boundary – the Mohorovičić Discontinuity. The gabbro, with a radiometric age of around 370 million years before the present, represents the consolidation of a deep crustal magma chamber emplaced below a spreading ridge; a very different environment to many of the basic rocks in the rest of south-west England. Basaltic dykes cut both each other and the gabbros and peridotite, enabling a sequence of events to be determined for this part of the Lizard Complex.

A6 PORTHOUSTOCK POINT (SW 810217)

Highlights

This site displays the best development of the dolerite dyke swarm, which is interpreted as the sheeted-dyke complex of the Lizard oceanic crust.

Introduction

Dolerite dykes are present in small numbers cutting most rock types of the Lizard Complex, although within the Landewednack Hornblende Schists and Old Lizard Head metasediments they are clearly deformed and not necessarily of the same generation as those discussed here. At Coverack (discussed above), they form a small percentage of the outcrop but, going north through the gabbro, the proportion of dykes gradually increases. Around Leggan Cove there is a dramatic increase in the abundance to around 25%, and in the West of England Quarry at Porthoustock they form about 50% of the outcrop. Locally near the old pier and Porthoustock Point, they form around 80% of the rock with only thin gabbro screens separating the dykes. Flett and Hill (1912) described and figured this great abundance of dykes and commented on the fact that, although many dykes looked very fresh in the field, with sharp straight sides and chilled margins, thin-section studies showed them to be totally altered to actinolite, albite, etc. Dolerites with fresh olivine and pyroxene are rare in the Lizard dykes.

Bromley (1973, 1979) drew attention to the great number of dykes; from his field observations he divided them into three groups and suggested an order of intrusion. The first formed were olivine-phyric, and in some places exhibited a granular texture. They are often disrupted and back-veined by gabbro. The second type were plagioclase-phyric, had chilled or phenocryst-free margins and were occasionally veined by hornblende diorite. The youngest set of dykes are fine-grained, aphyric plagioclase hornblende rocks. He saw no evidence that they had originally contained olivine and pyroxene and suggested that they might be primary amphibolites. This last type is by far the most abundant at Porthoustock.

Kirby (1979, 1984) studied the dykes in detail both in the field and chemically. He also found three groups broadly similar to those of Bromley (1979), but thought that the dykes with the amphibolite mineral assemblages must be older than those with olivine. This is the opposite sequence to that suggested by Bromley (1979). Chemical data showed that the later (olivine) dolerites were the most primitive, with MORB-like characteristics, and that the earlier (aphyric) ones were more fractionated with a slightly calc-alkaline nature (Kirby, 1984).

Davies (1984), as part of his geochemical study of the Lizard, analysed dykes from Porthoustock. He divided the dykes into early and late types along the same lines as Kirby (1984), producing high-quality REE data. He showed that the early dykes were light-REE-enriched and could have been formed by melting of plagioclase lherzolite, similar to that analysed from Coverack. The later dykes were MORB-like, light-REE-depleted and could be formed by melting of pargasite harzburgite. He suggested a model for the generation of the Lizard, similar to the Red Sea, with early rifting, mantle metasomatism and production of light-REE-enriched magmas followed by later MORB-like magmatism.

Sandeman (1988) analysed the dykes as an adjunct to his study of the Kennack Gneiss, and showed the same chemical types as Kirby (1984) and Davies (1984), but suggested, like Bromley (1979), that the aphyric dykes were late, not early. He showed that their light-REE-enriched chemistry was like the basic fraction of the Kennack Gneiss, and suggested that all the rocks with more calc-alkaline affinities are associated with a late, arc-related phase of magmatism, rather than an early rifting phase.

Description

The best places to see the intrusive relations between the various types of dykes and the gabbro are the beach sections around Leggan Cove and Manacle Point, and in the now-disused West of England Quarry. The southern side of the quarry, where the face cuts across the strike of the dykes, is the most informative. The early disrupted dykes and many features of the gabbro, such as shear zones, are best seen in Leggan Cove. At the northern end of the cove is one of the small, late, hornblende diorite bodies that are thought to be differentiates of the gabbro magma. These diorites can be seen veining the dykes in several places.

The sections in the quarry are dominated by the abundance of dykes that form around 50% of the rock exposed. They are mostly of the dark-green aphyric type and are generally around 1 m in thickness. The sides of the dykes are usually fairly straight, and chilling is often seen at the margins. Dykes are seen to intrude other dykes of the same type in many places, but cross-cutting relations between dissimilar dykes are rare. On the upper level of the south wall of the quarry, small veins of a very leucocratic plagioclase diorite invade the aphyric dykes. This is probably the plagiogranite of Davies (1984). The whole sequence is cut by late, steep faults that have calcite veining along them.

Lizard and Start Complexes (Group A sites)

Figure 3.16 Sheeted, basic dykes at Porthoustock Point. The dykes locally form about 80% of the outcrop with only thin gabbroic screens separating them. (Photo: M.T. Styles.)

Along the beach, just east of the old pier on the north side of Porthoustock Point, the abundance of dykes is at its greatest, forming around 80% of the outcrop (Figure 3.16). They are separated by thin screens of coarse gabbro and this is thought to be part of a sheeted dyke complex. The abundance of dykes increases northwards at Porthoustock, but unfortunately the sequence is truncated and the rocks on the north side of the Cove are very different. It is generally held that a concealed, large fault must run along the valley and out through the cove. The rocks exposed along the north side of the cove are highly deformed amphibolites, with numerous mylonitic zones. Generally, they appear to have been formed from quite coarse-grained precursors with a substantial proportion of finer rocks. Whether this is actually a highly sheared version of the dyke complex is open to speculation, but chemical data to test such a hypothesis are not yet available.

Interpretation

The dyke complex at Porthoustock is a major feature of the Lizard geology that has been recognized since the earliest geological surveys (De la Beche, 1839). The recognition by Bromley (1979) that this is a sheeted dyke complex is important as this is a significant feature in the categorization of the Lizard as an ophiolite. This is also the best example of an ophiolite–sheeted dyke complex in the whole of the UK.

Various workers have studied the dykes and there is agreement that several types of dykes are present, but exactly how they should be divided and what the time relations are is not yet clear. Cross-cutting relations between the different types seem to be few, partly because the different types have their main developments in different places. The aphyric types are most abundant in the north around Porthoustock, whereas the porphyritic types are most abundant in the south, around Coverack. There is a certain logic to extensively altered dykes being earlier than much fresher ones, but as they are in different places and in different structural and possibly hydrothermal regimes, this cannot be accepted without corroboratory evidence. If, for example, the source of fluids causing alteration was high-level hydrothermal systems such as are present in many modern ocean ridges, then the rocks highest in the sequence would be the most altered.

The chemical data that are available show clearly that there are at least two distinct types of dykes: olivine- and plagioclase-phyric MORB-types with light-REE-depleted patterns and the aphyric, light-REE-enriched types. The MORB-like dykes are closely similar to the gabbro, but the aphyric dykes are more evolved and not directly related magmatically. This is in accordance with the field relations, as the aphyric dykes do not root down into the gabbro – as the dykes in several ophiolite complexes do where the dykes are essentially comagmatic with the gabbros. Both Davies (1984) and Sandeman (1988) have produced essentially plausible models to account for the chemical variation. Davies (1984) suggested an analogy with the Red Sea, where initial rifting was associated with calc-alkaline magmatism and was followed by later MORB magmatism. Sandeman (1988), in contrast, suggests that the early MORB magmatism was followed by arc-related magmatism. Both the dykes and the Kennack Gneiss, known to be one of the later rock types, were associated with this later phase. This latter model has the attraction of simplicity, as there is a simple evolution from MORB-like to arc-related magmatism, whereas Davies' model would presumably have, in addition to the MORB magmatism, both an earlier rifting-related phase (to form the aphyric dykes) and a later arc-related phase (to account for the Kennack Gneiss).

A definitive work on the dykes, with the detailed field observation and chemical study to back it up, has yet to be carried out. The uncertainty that remains from current work cannot be resolved with the present data. This is, however, a very important site for the history of the Lizard Complex and the excellent features of igneous geology that can be seen here.

Conclusions

At this locality are seen rocks formed at a deep level in a portion of ancient ocean crust. The particular interests are the basalt and dolerite dykes, the site of which represents the channelways for melt feeding the sea-floor lavas above. One of the processes of growth and expansion of present-day oceanic crust is the repeated injection of magma in the form of near-vertical dykes below the spreading centre. At Porthoustock, many such dykes occur, cutting through the older gabbro, and are so closely packed that they make up the majority of rock on the foreshore. Although the chemistry of the dykes is variable, overall they are akin to various types of basalt formed in modern ocean ridges. This site plays a key part in the interpretation of the Lizard Complex, affording definite evidence for the presence and processes that contributed to the growth of early Devonian-aged oceanic crust to the south of what is now South-west England.

A7 PORTHALLOW COVE – PORTHKERRIS COVE (SW 798232–SW806226)

Highlights

The boundary fault of the Lizard Complex is uniquely exposed here; within an imbricate zone, the highly deformed mafic and ultramafic rocks of the Traboe Cumulate Complex are best exposed at this locality.

Introduction

The boundary between the igneous rocks of the Lizard Complex and Devonian sediments to the north is of major geological importance. It has been a point of controversy since the earliest research, and the first Lizard Memoir (Flett and Hill, 1912) left it as an open question, due to disagreement between the two authors. The position of the boundary can easily be located within a few tens of metres, but whether it is a major thrust separating Archaean rocks from Ordovician slates, or a faulted contact bringing together rocks of a similar age (Ordovician?) with different metamorphic grade, was disputed at an early stage. The second version of the Lizard Memoir (Flett, 1946) considered new evidence and favoured a major thrust contact. Controversy broadly along these lines has continued for many years between those who considered the Lizard Complex a thrust mass, possibly part of a much larger Lizard–Start thrust sheet (Hendricks, 1939;

Styles and Kirby, 1980), and those who thought it an upfaulted block of basement (Bromley, 1979; Matthews, 1981). Barnes and Andrews (1984) showed that there was no noticeable increase in metamorphic grade in the Devonian Meneage Formation immediately beneath the Lizard Complex. During the final emplacement of the ophiolite it was no longer above the ambient regional temperature of around 250–350°C.

Description

There are no exposures of the Lizard Complex–Devonian sediment boundary inland, and it can only be seen on the west coast at Pollurian and to the east at Porthallow. The contact at Pollurian is a very distinct, late, steep fault with fault breccia, gouge and slickensides. It separates amphibolites from the Devonian mudstones of the Meneage Formation. The boundary at Porthallow is much less clear, and it seems that the controversy cannot be resolved solely on the basis of the field relations at these localities.

To the south of the Lizard boundary at Porthallow, is a thin zone of Old Lizard Head metasediments and then, south of this for some two kilometres, a spectacular section of the Traboe-type hornblende schists. At Porthkerris Cove there is Traboe-type hornblende schist to the north and Landewednack-type to the south. The main differences between the two are that the Landewednack type has been supposed to be derived from basaltic lavas and tuffs and the Traboe type from either contact metamorphism of the Landewednack type (Green, 1964b), or from metagabbros (Flett and Hill, 1912; Bromley, 1979; Styles and Kirby, 1980). Leake and Styles (1984) proposed that the Traboe type, as seen at Porthkerris, was part of a Traboe cumulate complex which also included dunites, pyroxenites and anorthosites (Figure 3.17). This occurred

Figure 3.17 Interlayered basic and ultrabasic cumulate rocks, Traboe-type schists, Porthkerris. (Photo: M.T. Styles.)

Porthallow Cove–Porthkerris Cove

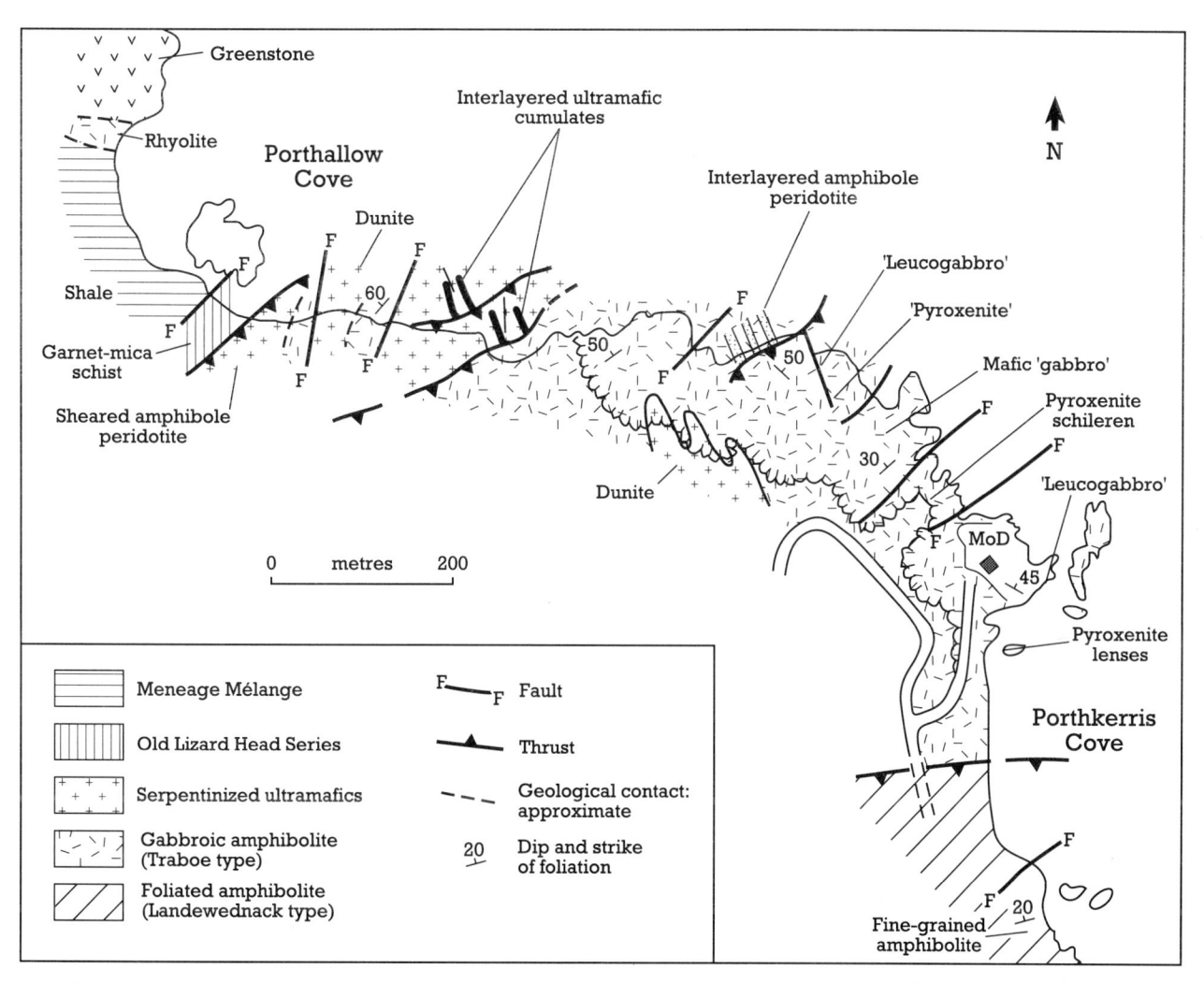

Figure 3.18 Geological sketch map of the Porthallow Cove–Porthkerris Cove site (A7).

extensively inland, but the only good surface outcrop of the cumulate complex was at Porthkerris. This site is an excellent one at which to see the relationships between the different types of hornblende schists.

This section will be described starting at Porthallow and proceeding southwards to Porthkerris Cove, a continuous section of nearly 2 km where a tremendous variety of rocks is encountered (Figure 3.18). On the north side of the cove at Porthallow, are greenstones with pillow-like structures and altered rhyolites within the turbidite mudstones of the Roseland Breccia Formation (Holder and Leveridge, 1986). On the south side of the cove the first rocks encountered are fine-grained, cleaved mudrocks with a gentle southerly dip. They do not contain visible mica. The first few inlets along the cliff are along deeply eroded, steep faults dipping to the south, and the second one brings together the previously mentioned mudrocks with the fine-grained mica schists of the Old Lizard Head 'Series'. Thin-section studies show that these rocks have recrystallized mica and small garnets and thus are quite distinct from the adjacent sediments. This is indeed the contact between the Lizard Complex and the Devonian sediments, a late fault. The outcrops are very weathered, but the schists have a strong, flat-lying, schistosity and abundant evidence of intense shearing, including mylonitic fabrics seen in thin section. On the beach are large boulders of a pink quartzofeldspathic rock that have fallen from the quarry, the Porthallow Granite Gneiss of Flett and Hill (1912). This rock contains garnets in addition to the quartz, feldspar and mica and in the author's opinion are more likely to be metamorphosed quartzofeldspathic sediments than granite. There are also

Lizard and Start Complexes (Group A sites)

Figure 3.19 Folded pyroxenite layers in gabbroic rock, Porthkerris. (Photo: M.T. Styles.)

hornblendic schists within the sequence, although these are now extensively altered to chlorite. Overall, the rocks along this section bear a strong resemblance to those seen around Lizard Point, although here they are much more sheared and retrogressed.

To the east the Old Lizard Head metasediments are overlain by a very altered, carbonated, amphibole-bearing peridotite. The contact between the two is a thrust. The schists are epidotized and chloritized, and near the contact are mylonitized but there is no evidence of high-temperature alteration and it is assumed that the peridotites were 'cold' at the time of thrusting.

From here southwards, for about 150 m, are seen interlayered dunite serpentinite and gabbroic granulites, the contacts probably being tectonic rather than intrusive. The dunite serpentinite is a fine-grained, green-black rock, with small chlorite pseudomorphs after spinel. It is traversed by a network of joints that are the site of later chrysotile formation. The orthopyroxene-bearing, 'bastite serpentine', so familiar elsewhere in the Lizard peridotite, is absent. The amphibolites are fine grained and have a banded appearance but a granular texture, and in thin section contain small clinopyroxenes. These are Traboe-type amphibolites and it appears they have formed from 'layered' gabbroic rocks. On the beach are numerous boulders which have fallen from the quarry, which are superb examples of the highly deformed gabbroic granulite. They show very clearly the effects of high-temperature plastic deformation and shearing (Figure 3.19). Several phases of folding can be seen in the 'layered' gabbros, and classic fold interference patterns are developed.

For about 150 m to the east of another fault, the sequence is largely dunite serpentinite with several interlayers of a pale-grey rock. The observable field relations suggest that this is an early, possibly primary interlayering, but that some of the repetition is due to folding. Thin sections show that these are not basic rocks as might be expected, but are composed of Mg hornblende and chlorite, and probably derived from some kind of aluminous ultrabasic rock, possibly a spinel pyroxenite. This part of the section ends at another thrust, and the remainder of the section can be seen in the quarry at Porthkerris.

At the western end of the quarry are fine-banded amphibolites, and on the seaward slabs at Pol Gwarra are several metre-thick layers of amphibole peridotite. On the small promontory between Pol Gwarra and Pedn Tierre, is a body of leucogabbro, which appears to retain a layering-like feature, even though it is highly deformed. In this area, shear zones abound: many have gabbro pegmatites associated with them and it is possible that the shearing may have induced partial melting in the very hot, basic rocks. To the east around Pedn Tierre, is an area of very dark amphibolites and amphibolitized pyroxenites. To the east of a fault in an inlet there is a fine, granular amphibolite with *schlieren* of pale-green amphibole that contain relicts of rare clinopyroxene. These 'pyroxenite' *schlieren* are usually lensoid and highly deformed but whether they originally formed layers cannot now be established. Very good examples of shear zones that have subsequently been folded can be seen immediately north of the Ministry of Defence buildings (Figure 3.18). The rocks to the south and east of the buildings are largely leucogabbroic amphibolites. There are scattered pyroxenite lenses and many superb examples of high-temperature deformation phenomena.

There are no exposures across Porthkerris Cove and on the south side, the lithology of the rocks is very different. Here are massive Landewednack-type amphibolites with a flat-lying foliation, and in a few places isoclinal folds can be seen. The very regular nature of the folds and foliation, contrasts strongly with the almost chaotic relations in the Traboe-type amphibolites to the north. The contact between the two types cannot be seen, but in the overgrown roadway at the back of the cove there are mylonitized amphibolites, of rather indeterminate nature, which suggest that the contact is possibly a thrust fault.

Most of the rocks along this section have a northerly dip to the foliation, but the 'stratigraphy' in the Traboe rocks youngs to the south. Allowing for some minor movements with later faulting, the sequence from north to south is dunite, pyroxenite, gabbro with pyroxenite, leucogabbro. This implies that the sequence is overturned relative to normal ophiolite 'stratigraphy'. The structures seen in the quarry at Porthkerris suggest a large fold plunging north, that the steep foliations in the wall of the quarry are the steep limb and the floor of the quarry is on a short flatter limb. This folding gives rise to the apparent diapiric shape referred to by Green (1964c) and Flett and Hill (1912).

Interpretation

The section between Porthallow and Porthkerris is well exposed and has received much attention from geologists, as outlined in the introduction. It is an excellent location to study two of the main controversies of the Lizard Complex, the nature of the Lizard boundary and the relations between the Traboe and Landewednack Hornblende Schists.

The contact between the Old Lizard Head schists and the Roseland Formation mudstones is a late, steep fault; as it is on the west coast at Pollurian. However, the schists immediately south of the fault have a gently dipping mylonitic fabric that is truncated by the fault, which suggests that thrusting was important at an earlier stage. The observable field relations show that the actual contact is a steep fault, but can this necessarily be extrapolated to signify the nature of the Lizard boundary in a wider context? Flett (1946) pointed out that the boundary has a sinuous trace across the north of the peninsula, which would be more compatible with a gently dipping structure rather than a simple steep fault. Recent geophysical evidence shows that the Lizard Complex is less than a kilometre in thickness and that it is underlain by Devonian sedimentary rocks (Brooks *et al.*, 1984; Rollin, 1986) as previously suggested by Styles and Kirby (1980). This tends to favour a thrust sheet, and, allied with dating of Lizard rocks showing Devonian ages, (Davies, 1984; Styles and Rundle, 1984) makes models involving upfaulted basement blocks untenable.

The question of the two types of amphibolites may now be examined here. Green (1964c) maintained that the effects seen at Porthkerris were due to the metamorphism of essentially homogeneous Landewednack-type amphibolites by the intrusion of peridotite: seen at the top of the quarry, and in a fault block on the beach at Porthallow. The aureole where these effects occurred was some 350 m wide, with an outer limit through the bay at Porthkerris and unaffected amphibolites at the south side of the bay. The metamorphic assemblages developed in the 'aureole' were clinopyroxene, and, in some places, two-pyroxene granulites. These can be found at Porthkerris, even though most of the rocks are extensively retrogressed.

The most recent authors have questioned this interpretation, and have pointed out the wide range of rock types from dunite to anorthosite (Bromley, 1979; Styles and Kirby, 1980) and also the wide range of chemical compositions of the

Traboe-type (Kirby, 1979b; Leake and Styles, 1984), which are markedly different from the Landewednack-type amphibolite. For this huge variety of bulk compositions to be produced from a homogeneous parent, would require chemical mobility of large amounts of material on a scale of tens of metres, and require a very fortuitous final distribution. In the light of the observed chemical inhomogeneities on a millimetric scale preserved in many rocks, this is extremely unlikely. It is much more likely that the variation in bulk compositions is essentially primary and that this is a fragment of a cumulate complex very similar to that in the Traboe area described by Leake and Styles (1984). It is important to note that none of the peridotites in this area are the 'typical' lherzolite or harzburgite mantle peridotites seen in most of the Lizard: they are mostly dunites and minor amphibole peridotites which are typical of the Traboe cumulate complex. The features seen here therefore cannot be taken as showing the relations between mantle and crustal rocks and extrapolated to the rest of the ophiolite.

The rocks in this site are in a series of thin thrust slices that form an imbricate zone at the base of the ophiolite nappe close to the Lizard boundary. Within these thrust slices are the only good exposures of the Traboe Cumulate Complex – a series of rocks formed by the high-temperature deformation of cumulates from magma chambers in the lower oceanic crust. The composition of coexisting ortho- and clino-pyroxenes is controlled by the temperature at which they formed. Hence analysis of coexisting pyroxenes can be used to calculate their temperature of formation by geothermometry. Pyroxene geothermometry using the method of Wells (1977), gives temperatures in the range 900–1050°C, for granulites from the Traboe cumulate complex in the Traboe area (Styles, unpublished data). This is much hotter than normal regional metamorphism, and probably indicates the breakup of hot, newly formed crust, rather than prograde metamorphism of colder rocks.

Conclusions

This locality is situated at the faulted boundary between hornblende schists of the Lizard Complex which are juxtaposed with low-grade 'normal' Devonian sedimentary rocks typical of most of south-west England. The Hornblende Schists here are part of the metamorphosed mafic–ultramafic unit within the Lizard Complex and representative of a segment of ancient ocean floor thrust subsequently into the adjacent continental margin. The thrust slice is now terminated by a high-angle steep fault that separates it from the continental sediments of the rest of Cornwall. The site provides evidence for major Earth movements that brought together rocks formed in very different environmental situations.

A8 LANKIDDEN (SW 756164)

Highlights

An important high-temperature shearing event that occurred shortly after the emplacement of the Lizard gabbro is best developed at this locality.

Introduction

The cliffs around the prominent gabbro headland of Lankidden have many spectacular geological exposures but they have received only passing interest from most previous workers. Teall (1886, 1888) described the 'beautiful augen structure in the gabbro' and Flett and Hill (1912) mentioned that there were numerous gabbro dykes to either side of 'the great spectacular dyke of Carrick Luz' (the rocks at the southern tip of Lankidden). Green (1964c) also noted these features and suggested movement of the dyke walls during intrusion had produced them. Kirby (1979b) interpreted it as a feeder to the main Crousa gabbro magma chamber. Styles and Kirby (1980) suggested that the thrusts seen here represented the major thrust zone that separated an upper, eastern tectonic unit from a lower western unit. Bromley (1979) had, however, maintained that the thrust was further east at Poldowurian just east of Kennack. Recent work further north, inland by Leake *et al.* (1990), suggests that there is little evidence for a major thrust or the existence of the two major tectonic units, and that the thrusts, although spectacular at Lankidden, are only of local importance.

Description

The cliff-bounded headland of Lankidden, with the site of a Bronze Age castle at the southern tip,

Lankidden

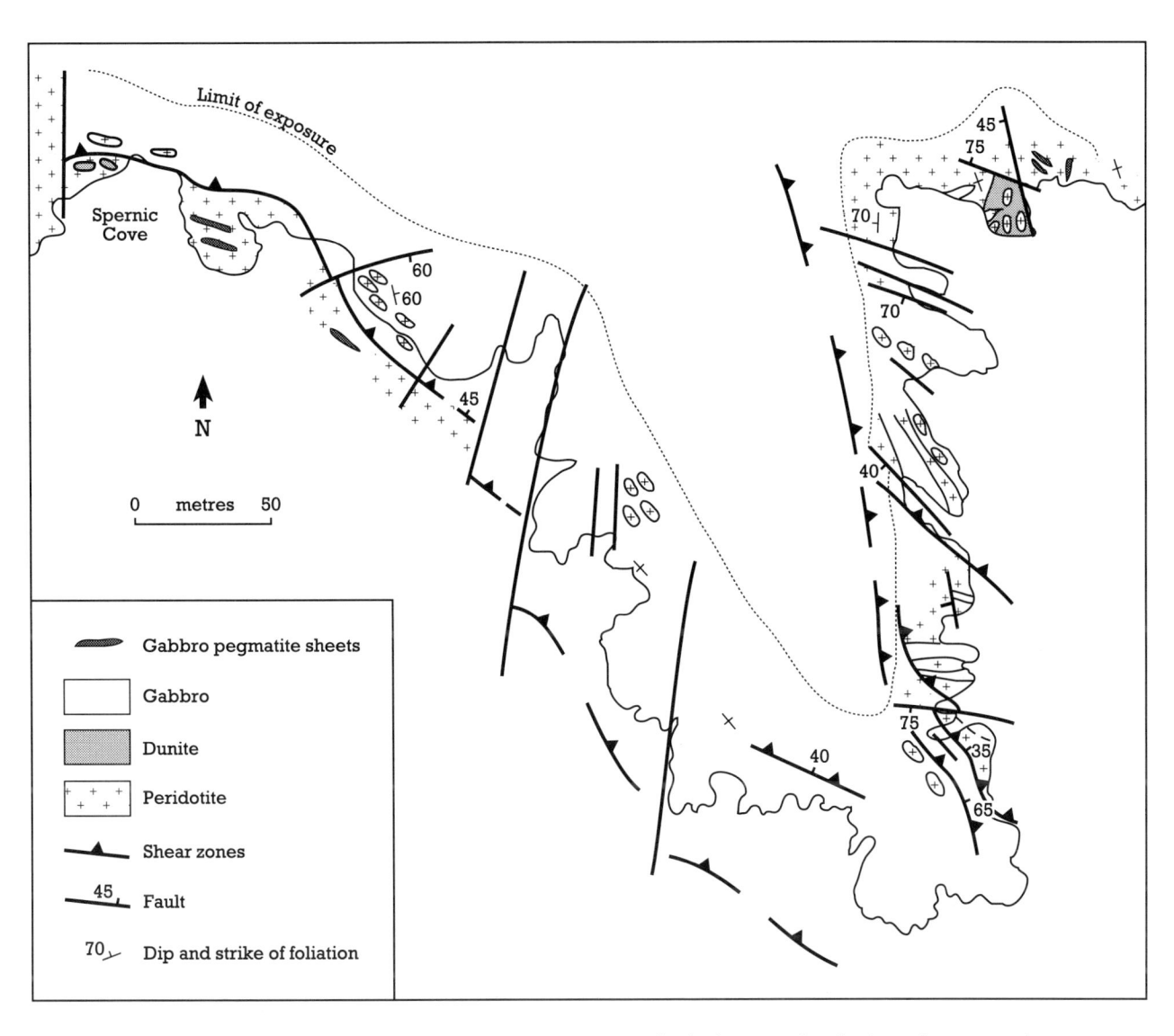

Figure 3.20 Geological sketch map of the Lankidden site (A8) showing distribution of outcrops between landward exposures and low-water reefs.

is an area of outstanding beauty. The headland is largely formed of a sheet or dyke of coarse gabbro around 100 m in thickness and dipping around 45° to the east. In most places the contacts with the peridotite are highly sheared; they can be seen in Spernic Cove to the west and Lankidden Cove to the east (Figure 3.20).

The peridotite is the coarse, primary type, with prominent orthopyroxene crystals, up to 5 mm in size, in a matrix of partially serpentinized olivine. A coarse, rough foliation is present that is essentially vertical and trends north–south.

In Spernic Cove, the western contact between the peridotite is not a sharp one and there are numerous inclusions of peridotite within the gabbro and thin gabbro veins in the peridotite below the main contact (Figure 3.21). The foliation in the flaser gabbro dips at around 45° to the east. This sheared contact is truncated by one of the many late faults at the small cove on the west side of Lankidden; it is presumably offset out to sea as it is not seen further south on the headland.

The outcrops further south on this west side, show a superb range of features produced by the high-temperature deformation of a coarse-grained gabbro. At one end of the range are the coarse, augen-bearing, flaser gabbros (Figure 3.22), with augen of clinopyroxene up to several centimetres in size, in a fine schistose matrix of recrystallized plagioclase and pyroxene. Thin sections show that the augen of clinopyroxene are now partly

Figure 3.21 Lenses of peridotite enclosed within later gabbro, Carrick Luz, Lankidden. (Photo: M.T. Styles.)

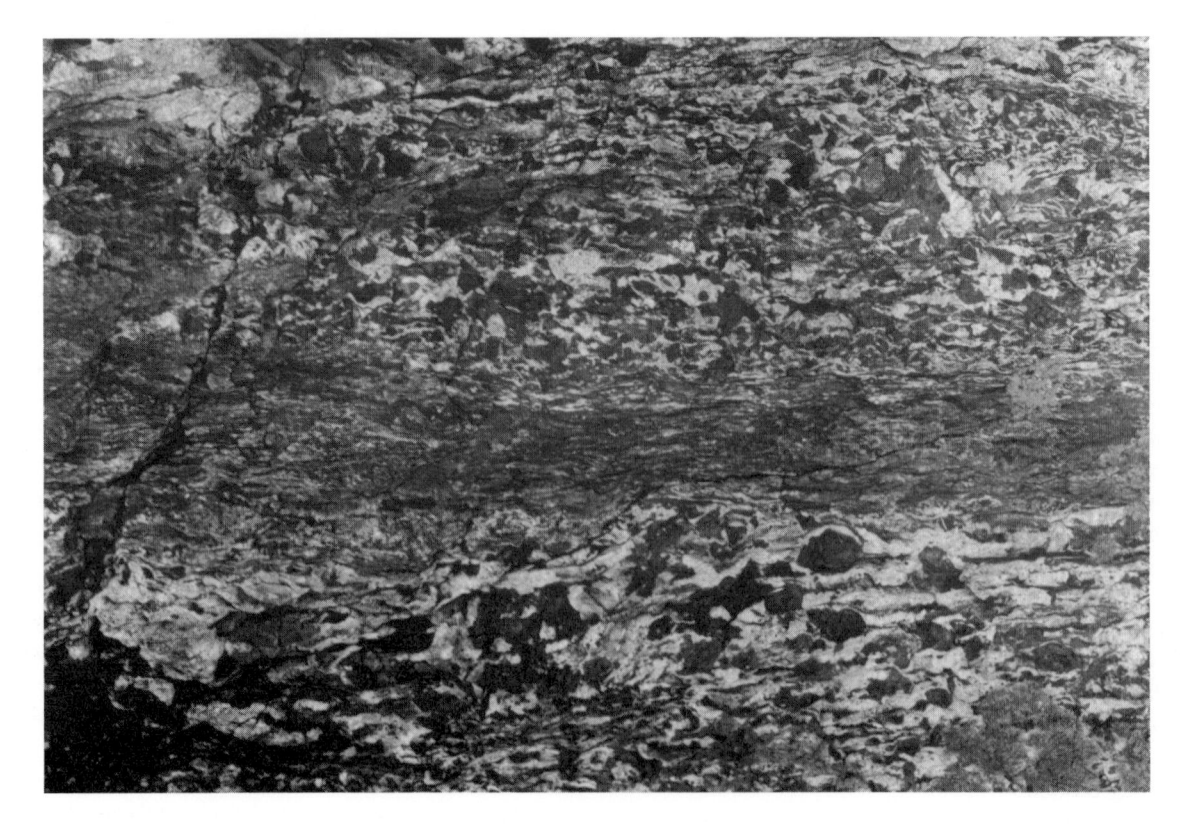

Figure 3.22 Flaser gabbro, Carrick Luz, Lankidden. (Photo: M.T. Styles.)

Lankidden

Figure 3.23 Shear zones developed in gabbro, Carrick Luz, Lankidden. (Photo: M.T. Styles.)

altered to hornblende. Much plagioclase is now saussuritized, although some fresh crystals remain in rocks that have large feldspar porphyroblasts. The presence of pyroxene shows that temperatures must have been at least 600°C during deformation. Where the rocks have been deformed to a greater degree, a considerable reduction of grain size has taken place, and fine 'gabbro schists' give way to streaky mylonites in the zones of most intense shearing (Figure 3.23). All these features can be seen within a few metres of each other, at many places along the west side of Lankidden, but are perhaps best exemplified in the small gully, near the southern tip. Late, basic dykes can be seen cutting through the shear zones.

Along the eastern side of the headland, access is more difficult and relationships more complex. This appears to be the intensely gabbro-veined 'hanging'-wall of the main intrusive sheet, and there are numerous small shears and gabbro veins. Both the peridotite and gabbro are intensely hematized, and this can make identification of rock types difficult.

The small rock promontory in the centre of Lankidden Cove, to the east of the headland, has the best examples on the Lizard of the dunite veins that cut through the peridotite. In many places, the dunites form single veins or pods, but here they form a series of criss-crossing anastomosing veins up to 3 m in thickness. The veins are composed of serpentinized olivine and, in many places, small stringers of chromite and *schlieren* of partly digested peridotite. The margins of the veins are not sharp, but diffuse and gradational over a distance of 1–2 cm, with the prominent orthopyroxenes of the peridotite decreasing in abundance into the vein. These features suggest that this is a high-temperature phenomenon with wall-rock reaction involved.

Interpretation

The cliffs along the west side of Lankidden show with great clarity the effects of high-temperature shearing deformation of various intensities in a coarse-grained gabbro, producing intense banding and augening. This was recognized long ago by Teall (1888) and is the best example in southwest England, and possibly the whole of the UK.

The dunite veins are of considerable interest,

even though their origin has yet to be definitely established. Flett and Hill (1912) suggested they were formed from a separate 'chromite serpentine' magma, distinct from the bastite serpentine (primary peridotite). Green (1964a), however, thought they were just the result of extreme serpentinization of the primary peridotite. It is the author's view that they are a high-temperature phenomenon, as outlined above, and that they were possibly the pathways of picritic melts ascending through the hot upper mantle. As the melts passed up through the mantle, they crystallized olivine, and the heat of crystallization – and possibly some fluid from the magma – was sufficient to cause local partial melting of the wall rock which was already close to its solidus. This caused the melting of pyroxenes: the lowest-melting fraction in the peridotite, and produced the diffuse margins that may now be observed. Whatever their origin, these dunites are a significant feature of the Lizard peridotite, and this is the best exposure.

The outcrops around Lankidden give superb examples of many phenomena of interest to both igneous petrologists and structural geologists. They are of relevance to studies of Lizard geology and of geological phenomena in general. Such is the clarity of these features, that their interpretation has changed little during the last 100 years.

Conclusions

The outcrops here are composed of sheared gabbros and altered peridotites of the Lizard Complex. The gabbros are demonstrably younger than the peridotite, and contain included fragments of the latter. The sheet-like gabbro body has suffered deformation at high temperatures, such that it is heavily sheared, with the development of intense banded texture with augen (German, meaning literally 'eyes') – large remnant crystals set in a finer crushed matrix. Dunite (an olivine–rich peridotite) occurs as veins and represents mantle-derived ultramafic melts which solidified within the peridotite host rock.

A9 MULLION ISLAND (SW 660175)

Highlights

The best-developed pillow lavas (submarine basalts of mid-ocean ridge type) in the Lizard area can be dated here by their association with fossiliferous cherts of Devonian age.

Introduction

Mullion Island is situated about 0.5 km southwest of Mullion Cove, off the west coast of the Lizard. It is one of the classic areas of Variscan greenstone and, although depicted on early nineteenth century maps of the region, little was known about the rocks and their association with the Lizard mainland. The morphology of the volcanics and their recognition as pillow lavas, together with their association with cherty sediments, were initially described by Fox (1893) and Fox and Teall (1893). Whereas basic volcanics and minor intrusives occur in the choatic Meneage Zone to the north of the Lizard Complex, Mullion Island was the best example of pillow lavas in the area, and possibly represented part of the same tectonic province faulted off from the Lizard (Flett, 1946). Modern ideas extend this view, whereby the lavas represent either a megaclast or a thrust slice incorporated within the tectonosedimentary *mélange* of southern Cornwall (Barnes and Andrews, 1981). Mullion Island thus forms part of the Roseland Breccia Formation (Holder and Leveridge, 1986) that incorporates all the *mélange* rocks within a thrust slice directly below the Lizard thrust segment.

A Frasnian age for the Mullion Island pillow lavas was reported by Hendricks *et al.* (1971) on the basis of conodonts extracted from the associated siliceous limestones. This is the same as the age of the *mélange* matrix of the Roseland Breccia Formation, which contains many basaltic clasts of generally similar chemistry to the Mullion Island pillow lavas (Floyd, 1984; Barnes, 1984).

Description

The site comprises the whole of the island and exposes about 30–35 m stratigraphical thickness of pillow lava (Figure 3.24). On the basis of the draping of pillows over each other and their interrelationships, the sequence is the correct 'way up'. The lava tubes are orientated approximate east–west, although in view of their likely disorientation during incorporation into the *mélange*, this cannot be used to infer the

Mullion Island

Figure 3.24 A spectacular development of pillow lavas, of Frasnian age, on Mullion Island. (Photo: P.A. Floyd.)

direction of lava flow. Most of the lavas are fractured and a number of subhorizontal shear or movement zones cut through the sequence. At a high point on the island is a c. 100-cm-thick sedimentary sequence of alternating siliceous argillite and intermittent, irregular, pinkish radiolarian chert (recrystallized), conformable with and overlying the pillows.

The pillow lavas are elongate, draped tubes, with nearly circular or ovoid cross-sections ranging from 0.15 to 1.2 m in diameter. Although some pillows have virtually no vesicles, most exhibit concentric zones of small vesicles 1–2 mm in diameter that hardly diminish in size towards the rim portions. Vesicle infillings are generally chlorite with a little quartz or calcite. A few pillows have a large, central, unfilled vacuole, left after the tube had been drained of lava. Interpillow material may not always be present, but is generally either cherty argillite or a grey carbonate. As well as the pillow lavas, thin, intrusive sill-like bodies may also be seen forming part of the low, sea-swept platforms on the south of the island.

The lavas look relatively 'fresh' compared with many similar exposures in south-west England with the interiors being bluish, although the fine-grained and originally glassy margins are pale green and heavily altered and fractured. Microscopic examination indicates that the lavas are aphyric tholeiites with variable, fine-grained quench fabrics that grade from the rims to medium-grained subophitic textures within the cores of large tubes. Primary minerals – plagioclase, brownish (sometimes zoned) clinopyroxene, ilmenite – are relicts set in a secondary mineral matrix (Figure 3.25) indicative of the pumpellyite facies of regional metamorphism (Floyd and Rowbotham, 1979). It is the development of the secondary phases (chlorite, pumpellyite, amphibole) that give the outer portions of the tubes their greenish tinge.

Interpretation

The significance of this site within the magmatic framework of south-west England mainly concerns the importance of the pillow lavas in the tectonic development of southern Cornwall. Although there are other good exposures of Upper Devonian pillow lavas, the Mullion Island basalts are distinctive in that they are:

Lizard and Start Complexes (Group A sites)

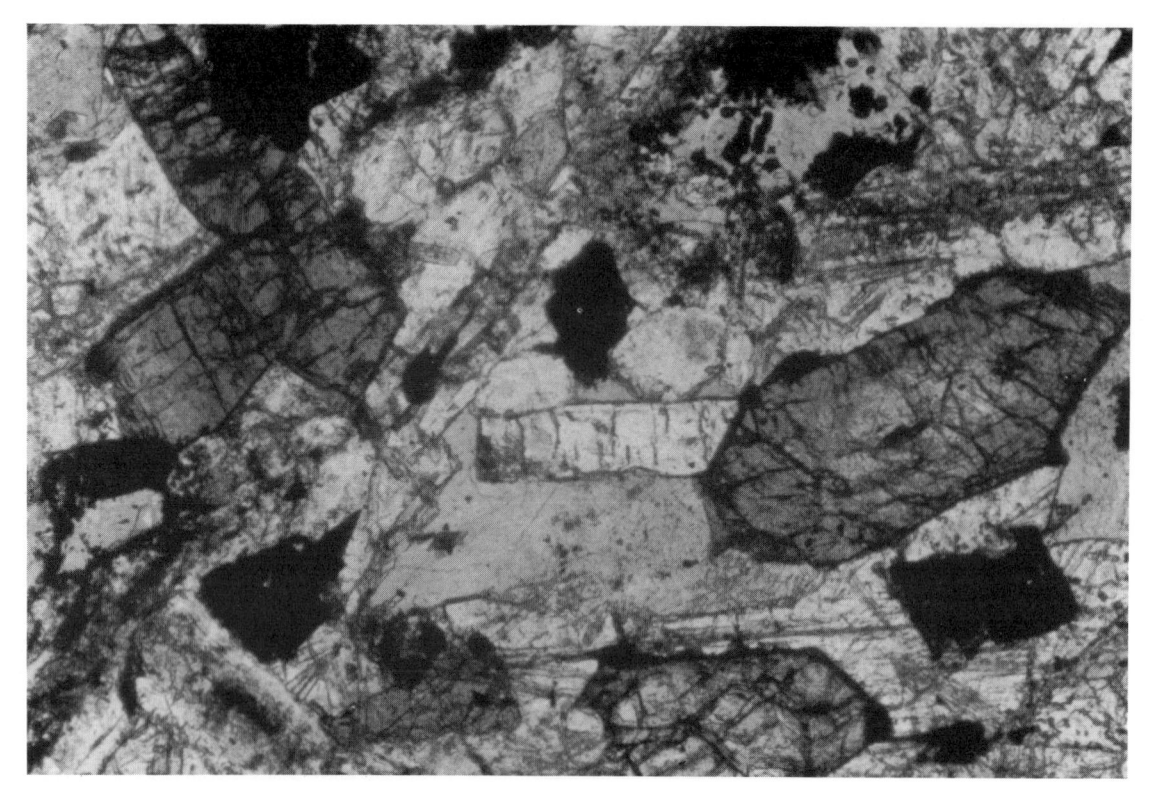

Figure 3.25 Photomicrograph of pillow lava from Mullion Island. Primary plagioclase, zoned clinopyroxene and ilmenite are set in a secondary pumpellyite-facies mineral matrix. (Photo: P.A. Floyd.)

1. relatively fresh, retaining primary quench-textured clinopyroxene with enhanced contents of Al, Ti and Na (Floyd and Rowbotham, 1979),
2. tholeiites with a mildly enriched MORB-like chemistry (Floyd, 1984), and
3. were subsequently metamorphosed in the pumpellyite facies.

These features distinguish them from the highly altered (no primary phases present), alkali-basalt pillow lavas of north Cornwall and the incompatible-element-enriched tholeiites in the Mylor Formation of the Penwith Peninsula to the north. In terms of relict and secondary mineralogy, and their particular chemical composition, they have an affinity with the other MORB-like clasts within the *mélange* of the Roseland Breccia Formation of southern Cornwall. In the context of the tectonic model for this area during the late Devonian, they are therefore representative of the volcanic portion of the tectonically disrupted oceanic crust of the Gramscatho Basin. Along with the Lizard ophiolite, they are among the remaining segments of the oceanic floor of the Rhenohercynian zone of northern Europe.

Apart from the relatively rare preservation of primary minerals in south-west England pillow lavas, the clinopyroxenes are distinctive in having abundances of minor components (Al, Ti and Na) that are more typical of alkali basalts than tholeiites. This feature emphasizes the importance of crystallization history (the relative order of phase growth) in governing the composition of pyroxenes in basalts (Floyd and Rowbotham, 1979). For MOR-type basalts the Mullion Island lavas have relatively high contents of U (>1 ppm) which is mainly resident in altered (originally glassy) matrix (Williams and Floyd, 1981) and was largely absorbed during early sea-water alteration.

Conclusions

This site represents an isolated relict of the volcanic portion of an ancient oceanic floor with basalt lava flows and minor sediments. The lavas were extruded on the sea-floor in the form of bulbous, tube-like 'pillows' and exhibit textures indicative of rapid quenching by the cold seawater. The basalts have compositions similar to some modern-day lavas formed at mid-ocean ridges and, together with the Lizard Complex, provide evidence for the existence of an ocean during the Devonian period, to the south of a continental area.

A10 ELENDER COVE–BLACK COVE, PRAWLE POINT (SX 769353–SX 769356)

Highlights

This is the best section through the highly deformed and metamorphosed oceanic basalts of the Start Complex.

Introduction

This locality lies on the west side of Prawle Point, the southernmost landward extremity of the so-called Start Complex (Figure 3.26). The southernmost peninsula of south Devon makes up this complex, which exhibits a range of variably schistose rocks of different aspect to the Lower Devonian argillites and phyllites to the north. The actual age of the complex is not known, although it is generally assumed to be Devonian (Dineley, 1986), with a structural history not dissimilar to that in south Cornwall generally (Marshall, 1962; Hobson, 1977). Many tectonic syntheses have suggested that the Start boundary fault is a continuation of the important Perranporth–Pentewan Line (Sanderson and Dearman, 1973; Sadler, 1974; Matthews, 1977) and that it is not linked to the Lizard–Dodman Thrust (Hendricks, 1939). In this context, the Start Complex junction has been interpreted as either:

1. originally a low-angle thrust that emplaced the Start Complex over the Devonian to the NNW (Coward and McClay, 1983);
2. a basement fault forming a terrane boundary to southern pull-apart, ocean-crust-floored, basins (Holdsworth, 1989).

Thus, apart from the special lithological characters of the magmatic rocks at this site, the whole complex has an important tectonic place in the early Variscan geology of South-west England.

The Start Complex exhibits two main groups of schists (Ussher, 1904; Tilley, 1923): metasedimentary micaceous greyschists and metavolcanic greenschists, together with minor variants of mixed sedimentary and volcanic character. The greenschists exhibit two simple end-member mineral assemblages: chlorite–epidote–albite and hornblende–epidote–albite. Mineralogical and chemical data (Tilley, 1923) indicate that initially the greenschists constituted a series of basaltic lavas which were subsequently highly tectonized and metamorphosed to a low grade. Relative to the low-grade greenschist-facies sediments to the north, the Start Complex schists belong to the same intermediate P–T facies series, but were produced at slightly higher pressures (Robinson, 1981). Illite crystallinity and phyllosilicate cell parameters link the Start Complex metamorphic regime to that of south Cornwall generally, rather than the lower-pressure environment to the north of the Perranporth–Pentewan Line (Primmer, 1983b).

Early, major-element, chemical data established the basaltic character of the metavolcanic greenschists (Tilley, 1923), whereas trace-element data (Floyd, unpublished) indicates that they constitute a series of essentially undifferentiated tholeiites with a MORB chemical signature. This feature is clearly important in elucidating the tectonic evolution of south Cornwall during the Devonian, and it indicates that the Start greenschists originally constituted a volcanic segment of the Variscan ocean floor.

Description

The site comprises the steep cliffs and coves on the west side of Prawle Point; it can be reached via the coastal path. Typical lithological (and small-scale structural) features are well displayed in the section, with low-strain areas exhibiting greenschists in their least-deformed state.

The greenschists are fine to medium grained; they are characteristically schistose with fine banding produced by variations in mineralogy and grain size. Petrographically, they comprise associations of amphibole, chlorite, epidote, clinozoisite, albite and sphene, with accessory calcite, pyrite, quartz, white mica and iron oxides. Rapid changes in lithology are common, with intimately interbanded chlorite-rich, epidote-rich and white mica-rich assemblages. Pyrite is abundant in some layers. Quartz is frequently present in the form of fine, banding-parallel metamorphic segregations

Figure 3.26 The rocky cliffs of Elender Cove expose metavolcanic greenschists of the Start Complex. Elender Cove, near Prawle Point, Devon. (Photo: David Noton Photography.)

or veinlets. Albite may form large (few millimetres-sized) porphyroblasts, which are emphasized by the weathering and stand out in relief on the rock surface. Small, resistant 'nodules' composed mainly of epidote and quartz are also characteristic of some layers, and these testify to element migration during metamorphism prior to the superposition of the enclosing deformed tectonic fabric. Tilley (1923), however, described similar epidote nodules as the metamorphosed products of infilled lava vesicles. It is unlikely that infill material would retain its nodular shape after the deformation suffered by the volcanics: it would be sheared out into lenticular bodies, as seen in less-deformed lava sequences.

Interpretation

The original nature of the volcanics is problematic, and this is difficult to pronounce on this at this site, or indeed anywhere else in the Start Complex. Tilley (1923) suggested that the greenschists represented a series of basalt lavas and sills, although typical lava features, such as curved pillow-lava surfaces, have not be observed. The delicate nature of the fine laminations and rapid changes in lithology, especially in low-strain areas, strongly suggests that much of the sequence was composed of basaltic volcaniclastics (tuffs) rather than lavas. However, elsewhere the sequence shows greater variability in terms of gross banding, which could reflect original differences between lava flows, sills and tuffaceous material, all now heavily sheared to a degree of schistose uniformity.

On the basis of their lithological and chemical characteristics, together with tectonic considerations, the Start greenschists are related to the south Cornish nappe–thrust belt and compositionally have their counterpart in the basic rocks of the Lizard Complex and *mélange*. One of the most significant features of this site is that the greenschists are tholeiitic, with a depleted normal-type MORB chemistry, much more primitive than any ocean-floor basalt recognized in south Cornwall. In this respect they have affinities with some of the Lizard dykes and the Landewednack hornblende schists with depleted incompatible-element contents. Thus, one of the main reasons for the inclusion of a Start greenschist site is the tectonic significance of the MORB character of the metavolcanics. This suggests that they represent another segment of Devonian ocean crust at the western end of the Rhenohercynian Basin along with the Lizard ophiolite and MORB-like basaltic clasts within the *mélange*. Like the allochthonous units of south Cornwall with their remnants of ocean floor, the Start Complex was subsequently docked adjacent to the magmatically distinct Lower Devonian autochthon to the north.

Conclusions

Here occur highly deformed and altered rocks which were once basic rocks of volcanic origin. Their chemistry indicates that they are basalts with a composition similar to those currently forming at mid-ocean ridges. Rocks formed on an ancient ocean floor have little in common with basalts in the area to the north of Start or in Devonian volcanic sequences extensive in southwest England. They do equate, however, with basaltic rocks found in south Cornwall (including the Lizard Complex), being similarly formed as ancient ocean floor. Both areas, therefore, represent exotic terranes thrust and welded on to the margins of a northern continental plate in late-Devonian times.

Chapter 4

Pre-orogenic volcanics (Group B sites)

List of sites

INTRODUCTION

The sites covered in this chapter are listed below, and arranged in broad stratigraphic–tectonic groups from west Cornwall to east Devon. The majority are examples of extrusive and intrusive basaltic rocks covering the late Devonian–early Carboniferous period of maximum volcanic activity in south-west England. The actual intrusion age of the large sills is not known, although many in western Cornwall appear to be pre-main-phase deformation, that is, pre-late Devonian (Taylor and Wilson, 1975). Other bodies showing evidence of high-level intrusion, soft-sediment deformation and spatially associated lavas, have an age corresponding with the enclosing sediments. In central south-west England, where faunal evidence of depositional age is good, lavas interbedded with dated sediments are typically Frasnian (for example, Pentire Point pillow lavas), Famennian (for example, Chipley pillow lavas) or Dinantian (for example, Tintagel Volcanic Formation). Lavas within the poorly fossiliferous Mylor Slate and Gramscatho Formations of western and southern Cornwall are probably late Devonian (Turner *et al.*, 1979; Wilkinson and Knight, 1989).

Both the magmatic and superimposed contact-metamorphic features of late Devonian extrusives and closely associated intrusives within the Land's End Granite aureole are described, as these sites often exhibit classic contact effects of high-level granites.

The locations of all sites are shown in Figure 4.1.

LIST OF SITES

Intrusives in parautochthonous late Devonian, west Cornwall:

B1 Porthleven (SW 628254–SW 634250)
B2 Cudden Point–Prussia Cove (SW 548275–SW 555278)

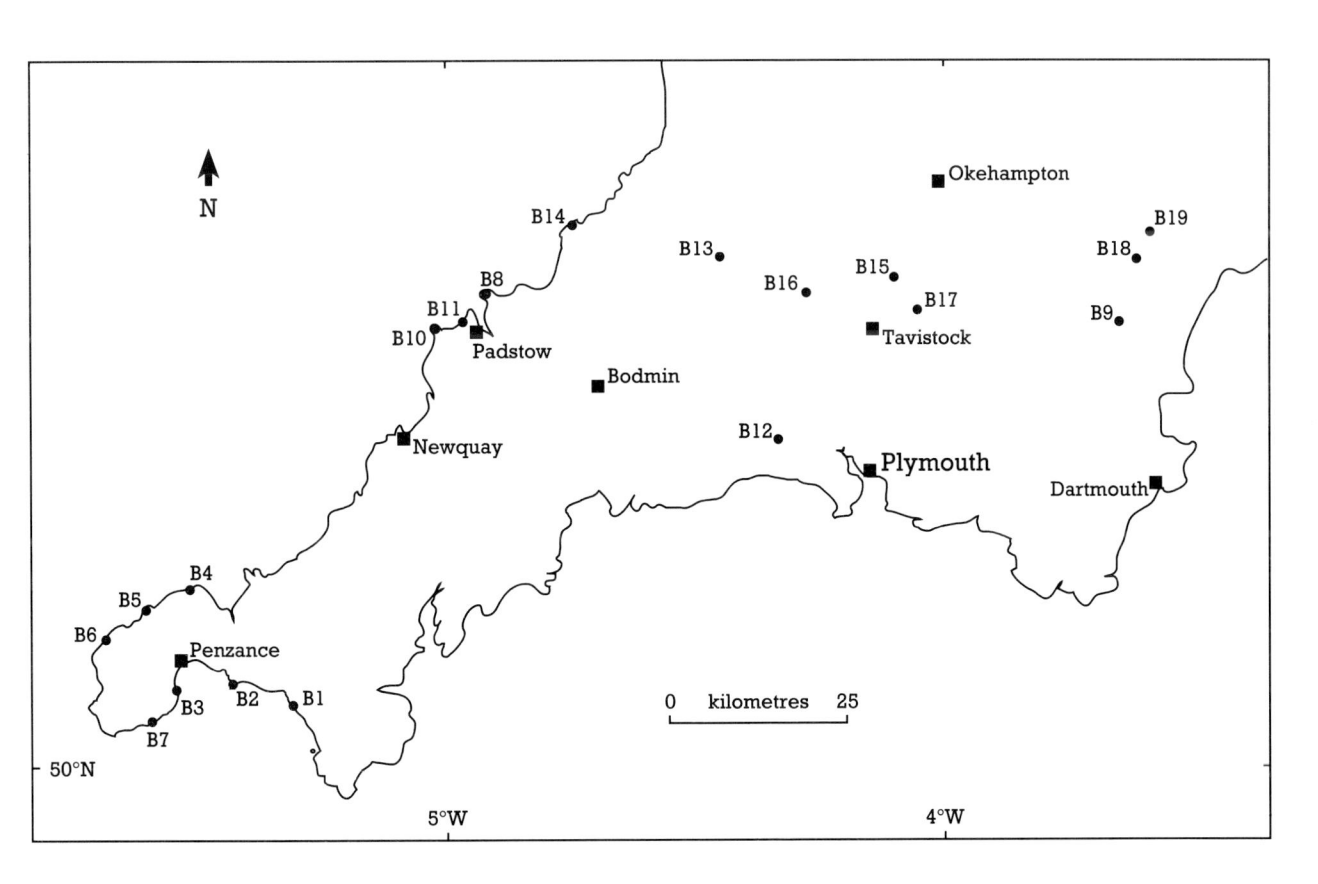

Figure 4.1 Outline map of south-west England showing the location of Group B sites.

Pre-orogenic volcanics (Group B sites)

Contact-metamorphosed extrusives and intrusives in parautochthonous late Devonian, west Cornwall:

B3 Penlee Point (SW 474269)
B4 Carrick Du–Clodgy Point (SW 507414–SW 512410)
B5 Gurnard's Head (SW 432387)
B6 Botallack Head–Porth Ledden (SW 362339–SW 355322)
B7 Tater-du (SW 440230)

Extrusives and intrusives in allochthonous late Devonian, north Cornwall and Devon:

B8 Pentire Point–Rumps Point (SW 923805–SW 935812)
B9 Chipley Quarries (SX 807712)
B10 Dinas Head–Trevose Head (SW 847761–SW 850766)
B11 Trevone Bay (SW 890762)
B12 Clicker Tor Quarry (SX 285614)
B13 Polyphant (SX 262822)

Extrusives and intrusives in allochthonous early Carboniferous, north Cornwall and Devon:

B14 Tintagel Head–Bossiney Haven (SX 047892–SX 066895)
B15 Brent Tor (SX 471804)
B16 Greystone Quarry (SX 364807)
B17 Pitts Cleave Quarry (SX 501761)
B18 Trusham Quarry (SX 846807)
B19 Ryecroft Quarry (SX 843847)

LITHOLOGICAL AND CHEMICAL VARIATION

All the Devonian and early Carboniferous volcanics in Cornwall and Devon appear to be submarine in origin, with water depth varying from relatively shallow (reef/platform environment) to deep (basin and basin/slope margin environment). Although dominated by basaltic material, the volcanics represent a bimodal basic–acid suite comprising lavas, high-level intrusives and abundant volcaniclastics, often in close stratigraphical association.

The following summarizes some of the different lithological and eruptive associations that are readily identified, together with an outline of their primary chemical variation (see also Floyd, 1983).

Basaltic pillow lavas and pillow breccias

The original petrography of these submarine lavas is commonly obscured by the secondary products of low-grade regional metamorphism (or contact metamorphism when located within the granite aureoles). Primary mafic phases are generally lacking or rare, and low-temperature albite invariably replaces original plagioclase. However, relict textures and secondary assemblages indicate that the pillow lavas were predominantly plagioclase-phyric, variably vesicular, quenched basalts subsequently metamorphosed in the prehnite–pumpellyite or lower greenschist facies. They are typical 'spilites' in terms of their mineral assemblages and were referred to as such in the early literature (for example, Dewey, 1914), whereas today we recognize them as metabasalts, without reference to special spilitic magmas or crystallizing conditions. Intimately associated with some pillow-lava sequences are pillow breccias (derived via brecciation along cooling fractures in the pillows) and rarer hyaloclastites which may show the effects of current reworking.

The primary chemical variation displayed by the Devonian and Carboniferous basaltic lavas reflects not only different chemical suites and magma types, but can be related to broad tectonic units. This can be illustrated with reference to late Devonian pillow lavas developed within basinal sequences throughout south-west England. As seen in Figure 4.2, the basalts define three main geographical areas that broadly relate to different tectonic domains. Those with high Zr/Nb ratios are tholeiites (akin to mid-ocean ridge basalts (MORB)), from the south Cornish *mélange*, together with Start greenschists and also the Lizard dykes, and are representative of thrust slices containing ocean-floor fragments. The next group, from south and west Cornwall (Penwith Peninsula and Camborne areas), are tholeiites within the parautochthonous Mylor Slate Formation; they have chemical features intermediate between enriched-type MORB and basalts found in the interior of plates. The third, a

Lithological and chemical variation

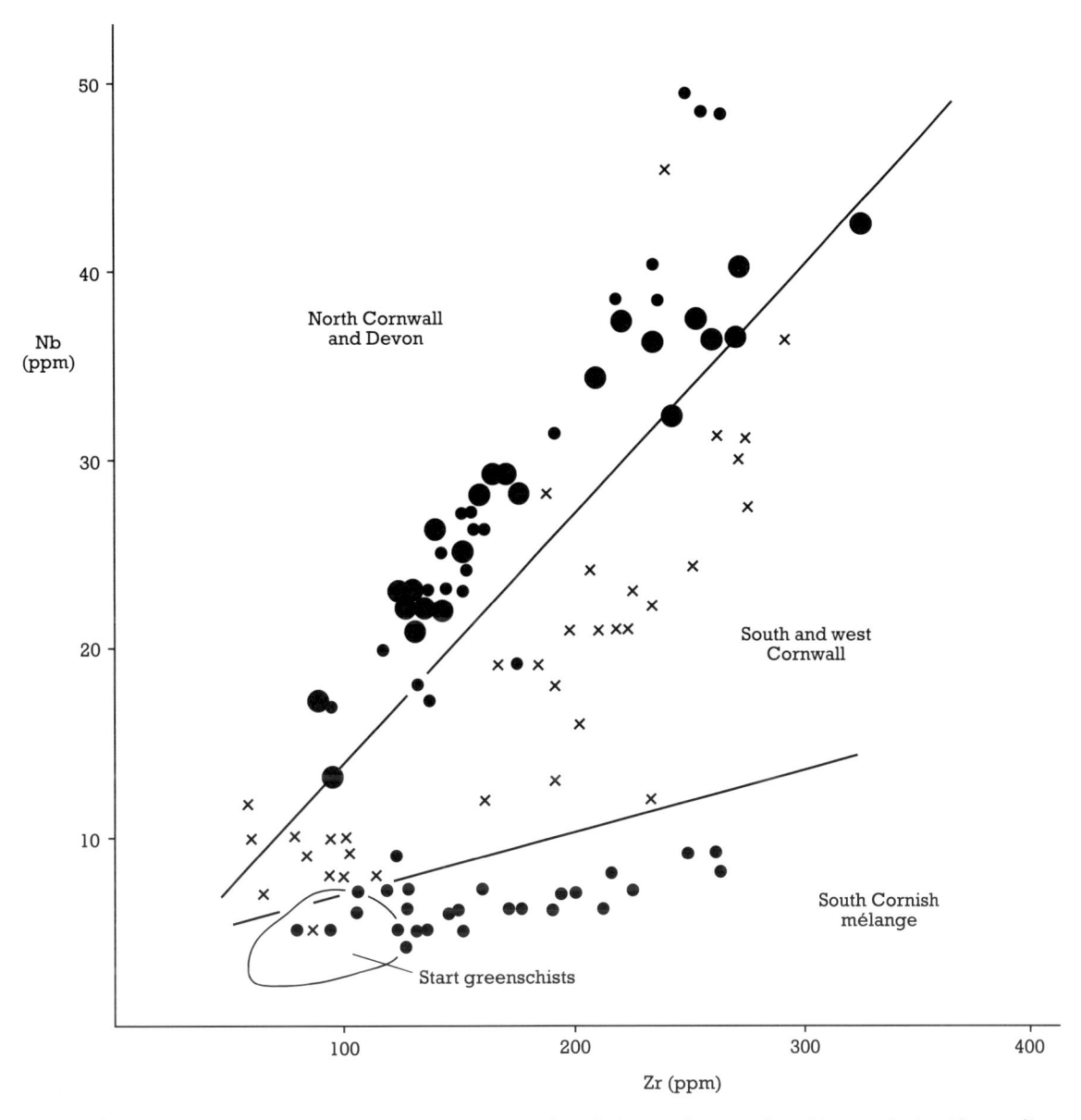

Figure 4.2 Variation of Zr and Nb in Upper Devonian (small dots and crosses) and Lower Carboniferous (large dots) basaltic lavas relative to different geographical regions. Data largely from Floyd et al. (1983) and unpublished.

low Zr/Nb group, are all alkali basalts and common to the rest of Cornwall and all of Devon. They are divided from the other tectonic units by the Perranporth–Pentewan–Start Line. Alkali basalts present in different nappe units show little distinctive chemical variation in this major area. Furthermore, early Carboniferous lavas from the same geographical location are also alkaline and they exhibit similar Zr/Nb ratios (Rice-Birchall and Floyd, 1988) and normalized incompatible-element patterns to the late Devonian lavas. The assumed MORB-like compositions of some early Carboniferous basalts within the Greystone Nappe reported by Chandler and Isaac (1982) are not substantiated by this data.

As seen in a chemical-based, tectonic environment, discrimination diagram (Figure 4.3) there is a progressive change from MORB-like tholeiitic compositions to within-plate alkaline compositions across south-west England that relate to major units which are now geographically juxtaposed by thrust–nappe tectonics.

Pre-orogenic volcanics (Group B sites)

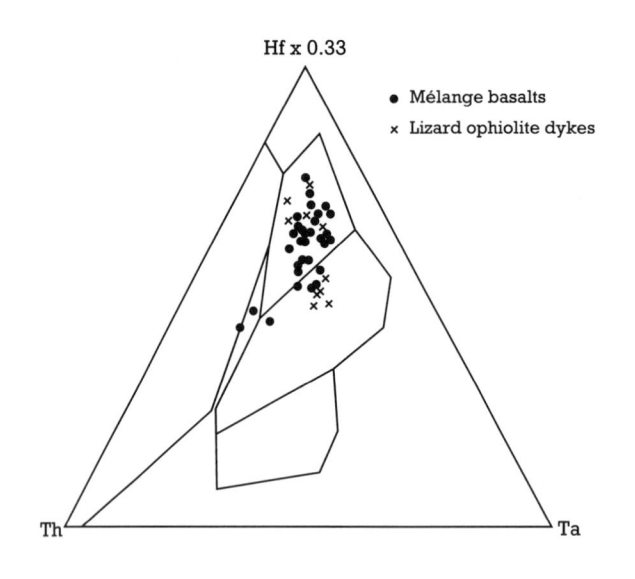

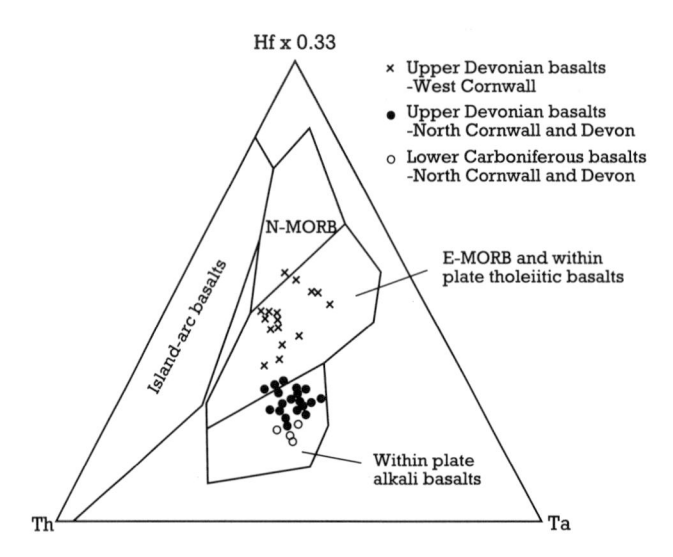

Figure 4.3 Th–Hf–Ta variation in Devonian and Carboniferous basaltic rocks from different tectonic units and regions in south-west England. Tectonic discrimination fields from Wood (1980).

Rhyolitic lavas

Owing to their variably altered and sheared nature these acidic lavas were often referred to as (quartz-)keratophyres or 'felsites' in the earlier literature (for example, Ussher, 1904). However, they retain relict textures and an observed (or inferred) primary mineralogy that clearly suggest that they were quenched rhyolitic or rhyodacitic flows.

Early acidic flows (early Devonian) in south Devon were nodular, quartz–alkali-feldspar-phyric rhyolites with a spherulitic quenched matrix, and they crystallized close to the ternary eutectic of the wet granite system (Durrance, 1985a). They are highly siliceous, with high K, Rb and Th, but chemically distinct from the Cornubian granites, granite porphyries and Permian rhyolites. Determination of their possible tectonic setting using chemical data suggests an active margin environment (Durrance, 1985a).

Lithological and chemical variation

High-level dolerite intrusives

Often associated with the basaltic pillow lavas are massive intrusive dolerite sheets and sills emplaced at various depths in the variable, consolidated sedimentary sequence. Some very thick sills can have pillowed and vesicular tops or irregular contacts with evidence of soft-sediment deformation in the adjacent adinolized sediments; others, intruded at greater depths, exhibit sharp contacts and narrow spotted thermal zones in undisturbed sediments. The larger sills may show considerable grain-size variation with gabbroic centres and pegmatitic zones, as well as differentiation to ferrodolerite and granophyre.

In a similar manner to the lavas, the intrusives have been affected by low-grade alteration but to a lesser extent, and they retain primary assemblages and textures which are only partly masked by secondary minerals such as albite, chlorite, epidote, amphibole, sphene and pumpellyite. In general, two main primary dolerite suites can be recognized:

1. a common anhydrous suite with calcic augite–plagioclase–ilmenite and sometimes olivine;
2. a rarer hydrous suite with olivine–titanaugite–plagioclase–brown amphibole–biotite–ilmenite.

Dolerites of the latter suite were often referred to as 'proterobases' in the early literature, to distinguish them from the dominant 'greenstones' or 'diabases' of normal dolerite composition as represented by the first suite. The hydrous-suite amphibole and biotite are both magmatic in origin, showing replacement relationships with the pyroxene, and are characterized by being highly titaniferous (Floyd and Rowbotham, 1982).

To some extent the intrusives show a similar chemical polarization between south and west Cornwall relative to north Cornwall and Devon, as is seen in the lavas. However, the lavas and intrusives in the two regions also may not be directly related either spatially (the intrusives are not necessarily the deep-seated equivalents of the lavas) or genetically (via a magmatic process such as crystal fractionation) (Floyd, 1983). The dolerites/gabbros of south Cornwall are generally isolated, massive, differentiated tholeiitic bodies with olivine–pyroxene cumulates (Floyd and Al-Samman, 1980). They are relatively primitive chemically, with low Zr/Y and high Zr/Nb ratios, but dissimilar to the Lizard gabbro in having progressively enriched REE patterns. In the rest of south-west England, the dolerites belong to an alkali-basalt lineage and they include both anhydrous and hydrous suites associated together in the same region (for example, around Padstow). The Carboniferous dolerites are generally more evolved than those intruding the Devonian, with high incompatible-element contents and characterized mineralogically by an abundance of apatite (Floyd, 1976). Massive sill-like bodies are found in the Tavistock area and the Teign Valley, to the west and east of the Dartmoor Granite respectively, and may often show internal differentiation (Morton and Smith, 1971).

Ultramafic bodies

Spatially associated with a small number of the massive dolerites and pillow lavas are thick ultramafic sheets referred to as augite picrites (for example, Reid *et al.*, 1911). Two major occurrences associated with Devonian-aged rocks are seen at Polyphant (north-east of Bodmin Moor) and Clicker Tor (Liskeard area). These bodies are invariably altered and tectonized with the development of variable secondary serpentine–amphibole–chlorite–talc–carbonate assemblages within elongate sheared lenses and at the thrust-terminated boundaries of the body. However, their generally massive nature has enabled primary minerals to be preserved, and these suggest that many of the ultramafics are hydrated spinel peridotites with cumulate textures and an alkaline chemistry. Olivine (sometimes fresh)–titanaugite–spinel is a common assemblage, although brown amphibole and biotite may also be present, suggesting a petrological link with the hydrous dolerite suite. No modern mineral or chemical analytical techniques have been applied to these rocks, so the actual compositions of the phases are unknown relative to the dolerites. Equally, little is known about their bulk-rock chemistry, although (as seen in Figure 4.4), Ni and Cr contents for both the Polyphant and Clicker Tor localities fall in the stratiform cumulate field. They are chemically distinct from the spinel peridotite of the Lizard and ophiolitic cumulates in general: the latter point suggesting that the Polyphant body is unlikely to form part of an ophiolite as postulated by Chandler and Isaac (1982).

Pre-orogenic volcanics (Group B sites)

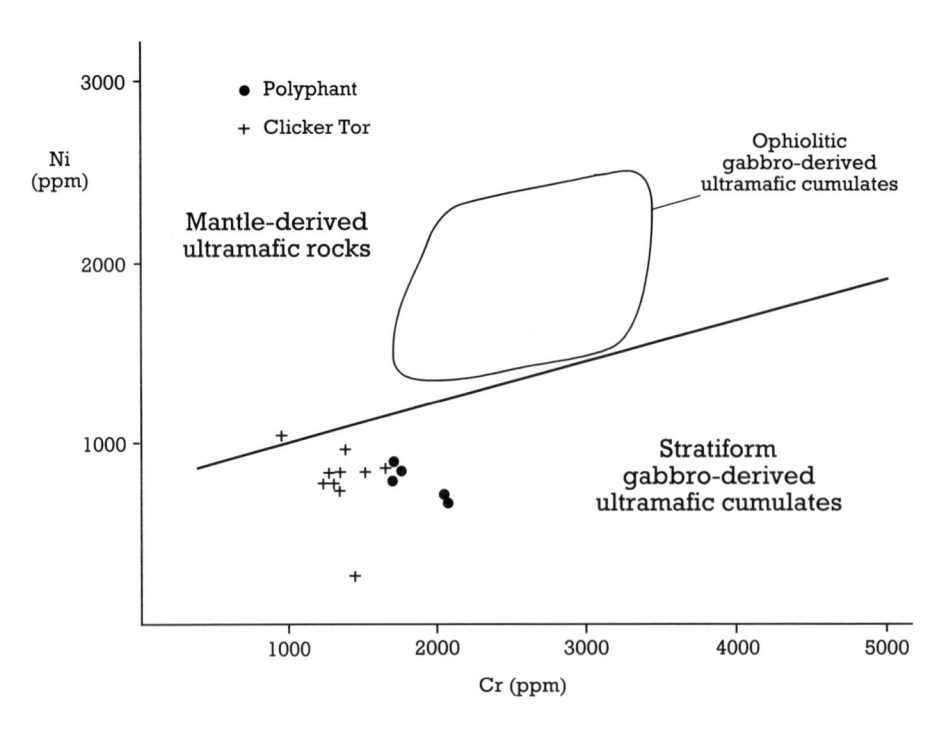

Figure 4.4 Distribution of Ni and Cr in Variscan ultramafic bodies associated with dolerites and pillow lavas relative to ophiolitic and stratiform cumulates (boundaries from Figure 3.3).

Volcaniclastic rocks

Largely due to poor outcrop and the high degree of alteration exhibited, little modern volcanological work has been attempted on the tuffaceous volcaniclastic horizons. In general both acid and basic tuffs ('keratophyric' and 'spilitic', respectively, in previous accounts) are common, with an admixture of lithic clasts, altered glass, broken crystals and pumiceous lava fragments. Coarser-grained volcaniclastics could be represented by part of the early Carboniferous Tintagel Volcanic Formation (Freshney et al., 1972), which contains large tectonized fragments that might once have been a basaltic agglomerate or lava breccia. Many of the commoner laminated tuffs within the mid and late Devonian appear to have settled through the water column, although there is also evidence for current reworking and sorting (Holwill, 1966). It is generally assumed that the volcaniclastics are the explosive products of submarine volcanism which were sometimes redistributed away from the vents by slumping down volcanic slopes. There is no reason to suppose that some are not air-fall tuffs subsequently deposited through water, although most volcaniclastic accumulations are associated with submarine lavas. However, the presence of shallow-water fauna associated with middle Devonian tuffs in south Devon could imply near-surface volcanic edifices. Some tuff horizons have been interpreted as subaqueous pyroclastic flows (Burton, pers. comm., 1986) with basal zones of local, soft-sediment rip-up clasts and flow alignment of large volcanic fragments.

B1 PORTHLEVEN
(SW 628254–SW 634250)

Highlights

The time relationships between sedimentation, dolerite sill intrusions, and the earliest Variscan deformation, are well displayed at this type locality of the Mylor Slate Formation.

Introduction

The section of low cliffs and beach outcrops to the south-east of Porthleven Harbour (from Little Trigg Rocks to about Eastern Tye) is often considered the type area for the laminated argillites and sandstones of the Upper Devonian Mylor 'Series' or Mylor Slate Formation (Lever-

Porthleven

Figure 4.5 Apparently discordant relationship between a basic intrusive body (on the right) and adjacent foliated sediments of Lower Devonian age (on the left). Porthleven, Cornwall. (Photo: P.A. Floyd.)

idge and Holder, 1985; Holder and Leveridge, 1986). The intrusive bodies which here affect the Mylor beds are typical of the smaller greenstones in west Cornwall. Little detailed work has been done on them, although they are mentioned in early descriptions of the general area (Phillips, 1876; Flett, 1903, 1946).

Description

Within the sediments at Porthleven are small, relatively thin (typically <1–2 m in thickness), high-level, intrusive sill-like bodies of basic composition. They are typical Cornish 'greenstones' or 'diabases' – regionally altered or metamorphosed doleritic intrusives featuring secondary albite, chlorite, actinolite, iron oxides and minor epidote, quartz, mica and carbonate. No chemical work is available, although they are probably tholeiitic in composition, in common with other bodies intruding Devonian strata in south and west Cornwall.

Intrusive relationships with the adjacent sediments are variable, although generally concordant contacts with the sedimentary lamination indicate that most of the bodies are sill-like in attitude. However, on the small scale, contacts may be slightly transgressive or highly irregular, with lobate junctions and cross-cutting intrusive tongues (Figure 4.5). Some contacts suggest that intrusion occurred at a high level into wet material with the development of attendant, soft-sediment deformation features, although thorough mixing between magma and sediment (peperite) was not achieved. Adjacent sediments may be locally baked and bleached at the contacts, although adinolization is rare; a c. 1 m pale-grey bleached zone is seen at the (upper?) contact of the largest body near the end of the beach section.

Relative to the deformation, there is evidence that some intrusives were pre-cleavage, and that they have been conformably folded with the sediments, probably during the early F_1 fold phase of the late Devonian (Taylor and Wilson, 1975). Local shearing and minor faulting affect some masses, together with later extensive net veining by massive quartz.

In thin section, the intrusive greenstones are

Pre-orogenic volcanics (Group B sites)

fine-grained, non-vesicular, subophitic metadolerites, typically replaced by various green, hydrous secondary minerals. The only primary phases remaining are (rare) clinopyroxene and apatite, the former showing partial replacement by a uralitic fringe of actinolite or pseudomorphed by chlorite. The more altered varieties are dominated by albite–chlorite (up to 60–70 vol.%), with variable carbonate, quartz, muscovite, epidote, iron oxides, sphene and rare biotite. Contact zones may show chilling in the thicker sills, but are often foliated and composed of green chlorite.

Interpretation

Accompanying soft-sediment deformation implies that some of these small bodies were intruded at a very high level into a wet sediment pile. It is interesting to speculate that the small greenstones did not actually form pillows because they were intruded into a sequence of rapidly deposited distal turbidite-generated muds and sands, and the magma did not actually reach the sediment–water interface. Penecontemporaneous intrusion and sedimentation is one of the more interesting aspects of the site, together with the sills being involved in the first phase of folding in south Cornwall. They are also typical of the smaller greenstone bodies of this region: illustrating the effects of low-grade regional metamorphism and the development of a variety of secondary assemblages characterized by albite–chlorite. The grade of regional metamorphism is generally assumed to be lower greenschist (chlorite zone), as pumpellyite has not been recorded and the chemical composition of the chlorite is undetermined.

Conclusions

That the dolerite sills are only slightly younger than the rocks in which they are found, is shown by the evidence that these sediments were still soft when intrusion took place. Further, the earliest structural element in the sediments, a cleavage, can also be found in some of the sills, indicating that these at least were emplaced before the onset of the Variscan Orogeny.

B2 CUDDEN POINT–PRUSSIA COVE (SW 548275–SW 555278)

Highlights

The massive, zoned metadolerite/gabbro at this site is the best exposed of the south Cornish sills; its chemical composition is unique within that group. Low-grade regional metamorphism is typical, but an unusual, axinite-bearing, vein assemblage probably reflects the influence of the nearby Godolphin Granite.

Introduction

The site covers the rocky rib of Cudden Point headland and the adjacent coves and gullies from Arch Zawn in the west to Bessy's Cove in the east, as well as small exposures inland. The intrusive greenstone seen at Cudden Point is not only representative, but one of the best exposed of the massive sill-like basic bodies seen along the south Cornish coast. It is situated within the Porthleven Breccia Member of the Upper Devonian Mylor Slate Formation (Holder and Leveridge, 1986). Although once interpreted as one limb of a broad fold on the older Geological Survey maps, this is no longer considered to be the case, as the intrusion is terminated by an approximate east–west tectonized contact with the adjacent sediments behind the headland (Figure 4.6). Previous work has largely involved petrographic and geochemical studies (Floyd and Lees, 1972; Floyd and Fuge, 1973; Floyd, 1976; Floyd and Al-Samman, 1980), which revealed the primitive tholeiitic nature of the sill relative to the alkaline composition of many south-west England greenstones.

Description

This intrusive body is typical of massive greenstones, with the preservation of good relict textures and primary minerals that indicate that it was originally an internally differentiated dolerite. Dark-green olivine–clinopyroxene cumulates are present, together with gabbroic pegmatites within the central portion of the sill, and minor granophyric pods.

The lower tectonized contact is exposed in Arch Zawn and Piskies Cove adjacent to deformed and foliated Mylor phyllites and semi-

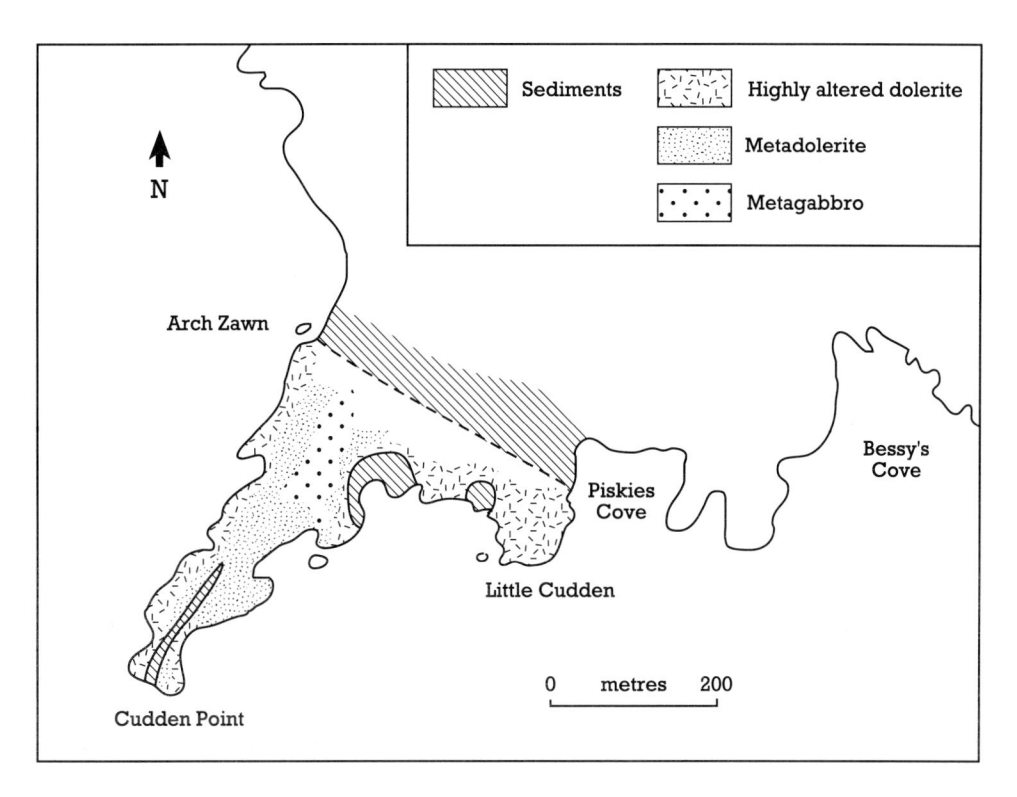

Figure 4.6 Simplified map of the Cudden Point greenstone body.

pelites. At Arch Zawn, the southward-dipping sheared contact is subparallel to the dominant axial-planar cleavage within the metasediments; a number of small shear zones within the dolerite body have a similar trend. In the east near Little Cudden, the greenstone is highly sheared, with about 30–40 m of a pale-green, albite–chlorite–actinolite schist showing an irregular, possibly folded, contact with the adjacent metasediments. At the extremity of Cudden Point proper is a pale-grey semipelitic metasediment raft with near-vertical tectonized margins, that was probably incorporated during intrusion, as the dolerite is chilled against its margin. Other interesting contact features include various mineral veins developed parallel, as well as oblique, to the main metasediment foliation, with the assemblage axinite–calcite–tremolite–chalcopyrite, and bi-mineralic veins of tremolite with either quartz, plagioclase or actinolite. The Ca required for these veins was probably mobilized from the dolerite during shearing. Axinite, however, is a Ca- and B-bearing mineral typical of contact metamorphism adjacent to potassic granites, and it is rare outside the visible contact aureoles of south-west England. Thermally developed chlorite spots may also be seen in the more pelitic metasediments near the contacts, although this may not have been induced by the dolerite, as spotting is often developed on a regional scale in west Cornwall, reflecting the influence of hidden granite ridges.

The foliated margins of the sill are now pale-green albite–chlorite–actinolite schists, whereas the major part of the body is a metadolerite that grades into a metagabbro. Central portions of the sill exhibit large relict plates of primary augite that are invariably fringed by fibrous actinolite (Figure 4.7). Sometimes, very thin exsolution lamellae of pigeonite may be observed within the larger clinopyroxenes. Towards the sill margins, clinopyroxene is invariably replaced, either partially or totally, by actinolite, in common with a general increase in the degree of alteration. Chlorite ovoids within the augite and matrix probably represent pseudomorphs after early olivine. Apart from the augite relicts and skeletal ilmenite, the only other primary phase is a rare, brown magmatic amphibole that is always replaced along its margins by zoned blue-green to colourless secondary actinolite. The presence of a primary hydrous phase is unique in these greenstones, and proves that the original magma contained some water and was not a typically anhydrous tholeiitic melt. Alteration domains of secondary minerals are common within the outer

Pre-orogenic volcanics (Group B sites)

Figure 4.7 Photomicrograph of the coarser facies of the Cudden Point greenstone showing primary augite partly replaced by a fringe of actinolite (cross polars). (Photo: P.A. Floyd.)

portion of the sill and are dominated by chlorite–epidote–white mica with all original plagioclase now replaced by albite. Pumpellyite has not yet been recorded.

Interpretation

This greenstone is unique among the south Cornish intrusives in having a chemically primitive tholeiitic composition with low incompatible-element and high Mg, Ni, Cr and Sc abundances. These features suggest that the magma was probably derived by relatively high degrees of melting (c. 25%). Incompatible-element ratios, such as Zr/Y and Zr/Nb, can be matched with the extrusive suites which make up the south Cornwall magmatic province (Floyd, 1984; see Figure 4.2 above). The wide range of Cr (c. 700–50 ppm) and Ni (c. 600–50 ppm) values indicate that chemical variation was governed by mafic fractionation and dominated by the accumulation of olivine and clinopyroxene (Floyd and Al-Samman, 1980). One further feature of chemical interest is the strong enrichment of the marginal foliated facies in K, Rb, Cs, Li, F and Cl (Floyd and Lees, 1972; Floyd and Fuge, 1973), which, together with the presence of spotting and veins exhibiting typical aureole minerals (for example, axinite), suggest the effects of contact metasomatism. It is likely that granite-derived fluids, possibly from a hidden, shallow extension of the nearby Godolphin Granite, penetrated the sill via the channelways afforded by the sheared margins.

Conclusions

The Cudden Point Sill is typical of the massive greenstone intrusives found in south Cornwall, but it has some unique chemical and mineralogical features. It is a chemically primitive, olivine-bearing tholeiite, internally differentiated, with a textural variation governed by the differential cooling of a massive body. That the magma became hydrated on crystallization, is indicated by the presence of rare brown amphibole – until recently only recorded in the alkali dolerites of the north Cornish coast near Padstow. The degree of alteration was governed by distance from the locally sheared contact, although alteration domains are patchily developed through the

Penlee Point

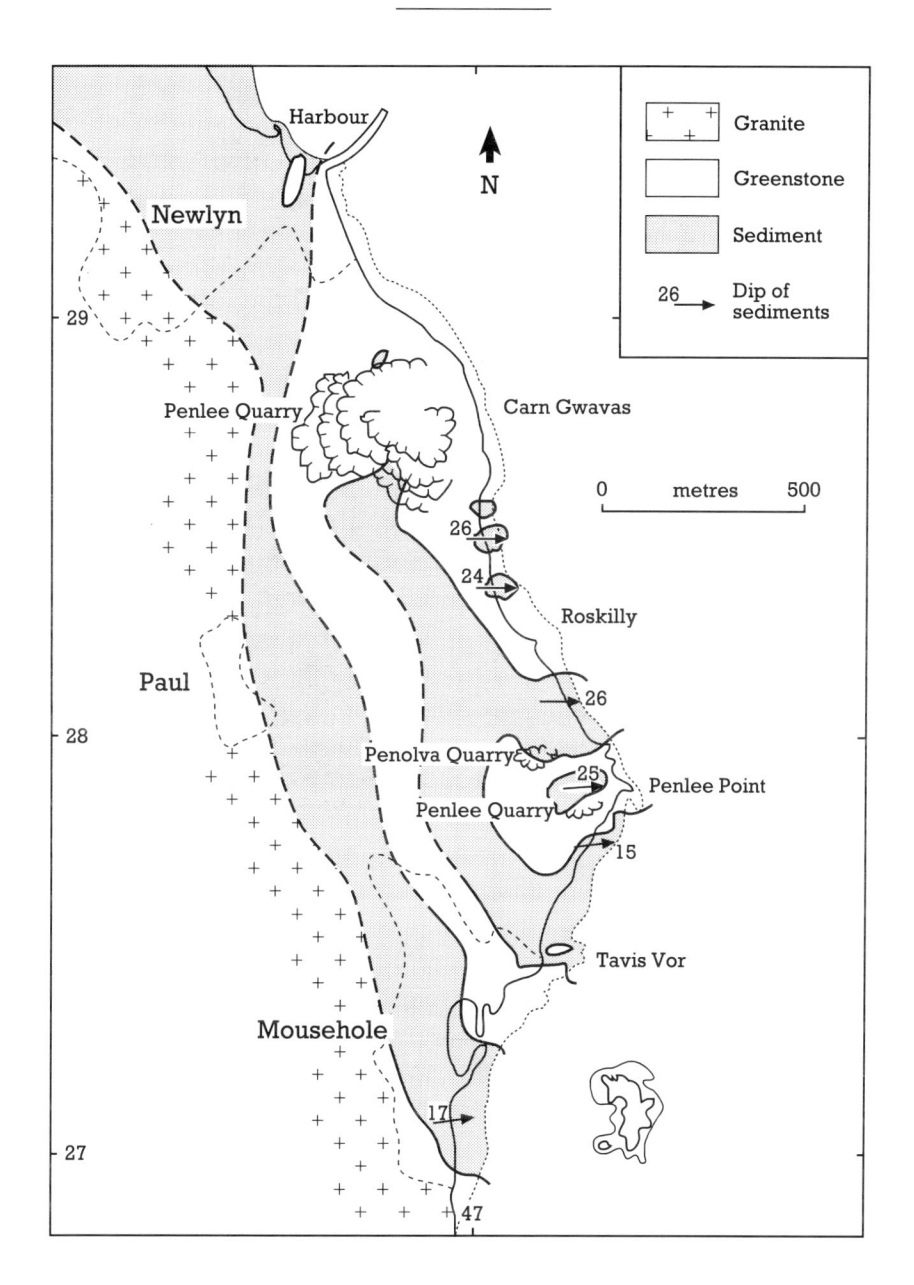

Figure 4.8 Geological map of the Mousehole–Newlyn section of the Land's End Granite aureole, showing the distribution of the dolerite sills around Penlee Point (after Floyd, 1966a).

body. The marginal facies provides evidence for alkali-element and F–B metasomatism from a nearby granite source.

B3 PENLEE POINT (SW 474269)

Highlights

At Penlee Point, spectacular contact-metamorphic and metasomatic mineral assemblages are superimposed on a dolerite sill and its sediment rafts.

Introduction

The site is located within the Land's End Granite aureole. It is topographically 600–700 m from the inferred contact, and it exhibits the effects of low-grade contact alteration. Apart from brief descriptions within the Survey Memoir (Reid and Flett, 1907), Floyd (1966a) recognized two sills in this section of the aureole, the uppermost of which (Penlee Sill) constitutes this site; the lower sill (Gwavas sill) is exposed along the foreshore to the north (Figure 4.8). It is generally assumed

Pre-orogenic volcanics (Group B sites)

that both these massive greenstone sills represent offshoots from a parent body in the huge, working Penlee (or Gwavas) Quarry near Newlyn, the rocks of which might represent a feeder pipe to all local sills (Floyd, 1966a), although compositionally it is more differentiated than the strictly basic sills and no direct contact is seen with the Penlee Sill of this site.

Description

The site includes about 300 m of rocky foreshore around Penlee Point and the disused Penolva and old Penlee quarries just inland (Figure 4.8). The central mass of the Penlee Sill is seen on the foreshore, while the lower contact is within Penolva Quarry and the upper is exposed in Penlee Quarry. A relatively uniform sill-like body (about 30 m thick) is indicated, with narrow flinty adinoles at the contacts with the normal aureole pelites whose dominant bedding foliation dips eastward at about 25°.

On the foreshore opposite Penlee Point, a small biotitized and heavily spotted pelitic raft is apparently enclosed within the sill. The basic rock is foliated at the contact, with the development of a cummingtonite–cordierite assemblage (Lacy, 1958) that replaces the usual actinolite–plagioclase association of other altered parts of the sill margins.

The sill is within the granite aureole; it shows the superimposed effects of albite–epidote hornfels facies of contact metamorphism on a variably textured and differentiated basic intrusive. The marginal zones are fine-grained, actinolite–albite–ilmenite-dominated hornfelses derived from a doleritic precursor, whereas the central portion is a coarser metagabbro with relict pyroxene fringed by actinolite, and calcic plagioclase partially replaced by albite. Other secondary minerals include chlorite, minor epidote, quartz, white mica and sphene. Owing to the low-grade nature of the alteration, these effects are similar to those exhibited by the regionally metamorphosed, massive greenstones seen in south Cornwall. The direct effects of contact alteration, however, are seen in the patchy development of metamorphic biotite replacing fibrous actinolite or nucleating on ilmenite grains, and this gives the normally greenish rock a brown tinge. Later, granite-induced hydrothermal effects are the variable

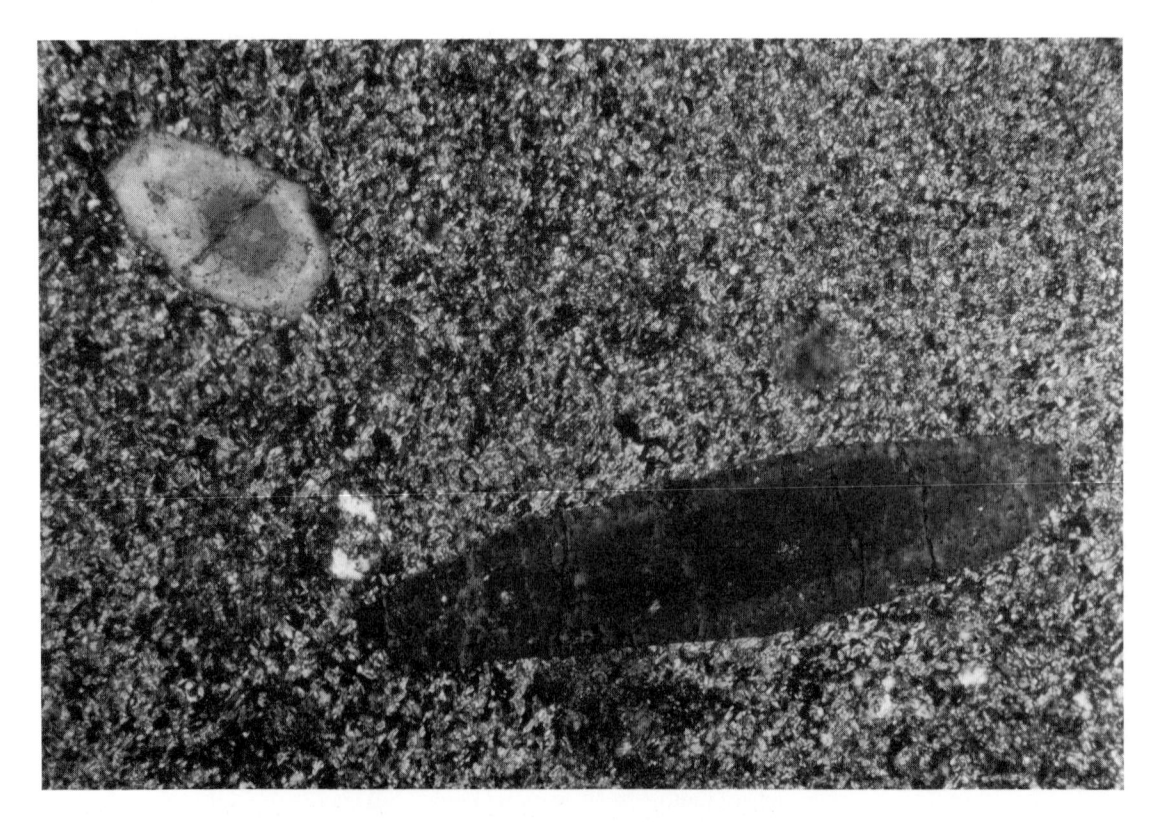

Figure 4.9 Photomicrograph showing late zoned tourmaline replacing chloritic matrix of contact metamorphosed Penlee dolerite (cross polars). (Photo: P.A. Floyd.)

chloritization of mafic constituents and the development of large, zoned, blue tourmaline (schorlite) crystals replacing altered matrix (Figure 4.9). Asbestiform actinolite veinlets are also common, together with small irregular pockets of late amphibole within the metagabbroic portion of the sill. In the small cove next to the lifeboat house, a large quartz vein with smaller apophyses intrudes a variably biotitized greenstone which has been almost totally chloritized adjacent to the vein.

Interpretation

The site provides a compact example of a small greenstone sill that has undergone low-grade contact metamorphism by the Land's End Granite. Contact-metamorphic biotite, together with relatively late-stage hydrothermal effects can be demonstrated. Assemblages are typified by albite–actinolite, whereas nearer the granite (Penlee Quarry at Newlyn) albite–hornblende is characteristic for this part of the aureole (Floyd, 1966a). It has a mildly alkaline composition, not dissimilar to the evolved basic metadolerites on the coast near Penzance and, in this respect, is different from other more massive, chemically primitive, dolerite/gabbro intrusives typical of south Cornwall generally (for example, Cudden Point, described below). Its relationship to the pipe-like (feeder?) intrusion at Penlee Quarry is not clear, but the latter has a different composition, even allowing for hydrothermal alteration effects.

An unusual feature, at Penlee Point proper, is the development of a cummingtonite–cordierite assemblage in the marginal zone of the intrusive adjacent to the enclosed pelitic raft. This probably developed from contact metamorphism of an altered and foliated Ca-poor marginal facies of the dolerite generated during initial regional deformation.

Conclusions

The alkali-dolerite sill at Penlee Point displays a coarse, gabbroic inner zone and raft of pelitic sediment. When the later Land's End Granite was emplaced nearby, spectacular low-grade metamorphic (albite–actinolite and cummingtonite–cordierite) and metasomatic (tourmaline-bearing) mineral assemblages were developed across the sill, its rafts and the enclosing country rock.

B4 CARRICK DU–CLODGY POINT (SW 507414–SW 512410)

Highlights

Pillow lavas and lava–sediment relationships within the Mylor Slate Formation are very well displayed here, as well as low-grade contact metamorphism and late-stage hydrothermal alteration of basic extrusives by Land's End Granite.

Introduction

Prior to the microfaunal discoveries of Turner *et al.* (1979), which confirmed a late Devonian age, the Mylor 'Series' of the Land's End Memoir was considered to be of early Devonian age on structural grounds (Dearman *et al.*, 1969). On the basis of the abundance of pillow lavas in this part of the Land's End aureole, however, Lacy (1958) had equated them with the well-documented Pentire Point pillow lavas of late Devonian age. Although it is now recognized that the two pillow-lava groups have different chemistries (Floyd, 1983, 1984) and tectonic settings, the lavas within the Mylor Slate Formation are, nevertheless, considered to belong to the parautochthonous Upper Devonian, and they are associated with the deep-water argillaceous facies within the Gramscatho Basin (Holder and Leveridge, 1986)

Description

This site is located along a stretch of about 1200 m of scenic coastline just to the north-west of St Ives and Porthmeor beach, in the outermost reaches of the Land's End Granite aureole. It comprises two low-lying headlands of small cliffs and gullies separated by a pebble beach, as well as a disused quarry in the pasture behind Carrick Du.

The site displays both vertical sections (Carrick Du) and northward-dipping platforms (Clodgy Point) of pillow lavas (Figure 4.10) together with some pelitic sediment and tongues of more massive lava. Many of the pillows are small, with cross-sections of only 0.2–0.6 m, although some horizons have clearly undergone a degree of tectonic flattening and shearing into phacoidal forms. Interpillow relationships show that they

Pre-orogenic volcanics (Group B sites)

Figure 4.10 View of the pillow-lava sequence at Clodgy Point, Penwith, Cornwall. (Photo: P.A. Floyd.)

are the correct 'way-up', generally concordant with the sedimentary laminations, and that they dip about 20° to the north. The elongate tube-like structure of typical pillow-lava piles is not well developed, as many appear to be either of short length or individually bun-shaped, possibly developed towards the base of a submarine slope. There is little interpillow sediment within the main pillow horizons, although some discontinuous sediment horizons contain small, irregular, lava globules representing squirts of basalt magma into wet, partially consolidated, sediment (Figure 4.11). Although the pillow matrix is fine-grained, chilled margins are rarely preserved and vesicles uncommon. Some pillows, however, exhibit small pits or 'spots' which in thin section appear to represent the weathering out of cordierite porphyroblasts, rather than true vesicles. Polygonal and transverse cooling cracks may be seen on the exposed top surfaces of some small bun-like pillows (Figure 4.12).

Apart from the pillow lavas and their sheared analogues, thin (1–3 m), high-level, sheet-like intrusives are also present which invariably show a tectonized contact with the metasediments. Hard, white massive and mylonitized adinoles, together with thin shear zones containing an admixture of brecciated and rolled adinole, pelite and greenstone fragments may be observed at sediment–greenstone contacts (Figure 4.13).

All the rocks now display a typical, low-grade, albite–epidote hornfels facies of contact-metamorphism mineralogy. The pelitic/semipelitic sediments are mainly laminated quartz–mica–chlorite and cordierite–biotite hornfelses, whereas the pillow lavas are fine-grained albite–actinolite

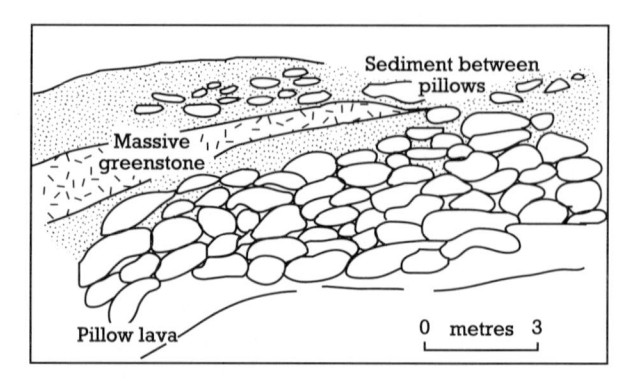

Figure 4.11 Relationship between Upper Devonian pillow lavas and interlayered pelitic sediment, Clodgy Point, Penwith Peninsula.

Carrick Du–Clodgy Point

Figure 4.12 Polygonal cooling cracks on pillow-lava sequence at Clodgy Point, Penwith, Cornwall. (Photo: P.A. Floyd.)

Figure 4.13 Sketch of the tectonized contact between adinolized sediments and greenstone, Clodgy Point, Penwith Peninsula.

hornfelses with no original magmatic phases left. Contact-metamorphic biotite may replace the actinolite and impart a purplish colour to the rock. The most distinctive feature of the basic rocks is the effect of granite-derived, late-stage hydrothermal fluids. This takes the form of pale, bleached patches and zones often closely associated with networks of pale-green amphibole veinlets. Mineralogically the 'bleached' matrix shows chloritization of amphibole and biotite, leucoxenization of ilmenite, and replacement of plagioclase by kaolinite. Amphibole veinlets may also exhibit radiate groups of blue-green to dark-blue zoned tourmaline, both of which may be partially replaced by late epidote, calcite and rare axinite (Figure 4.14). These mineralogical replacements suggest that the initial Na–Mg–Fe-rich fluids subsequently became more Ca-rich.

Chemically, the pillow lavas of Clodgy Point are tholeiitic basalts with subhorizontal chondrite-normalized REE patterns and, in this sense, are not

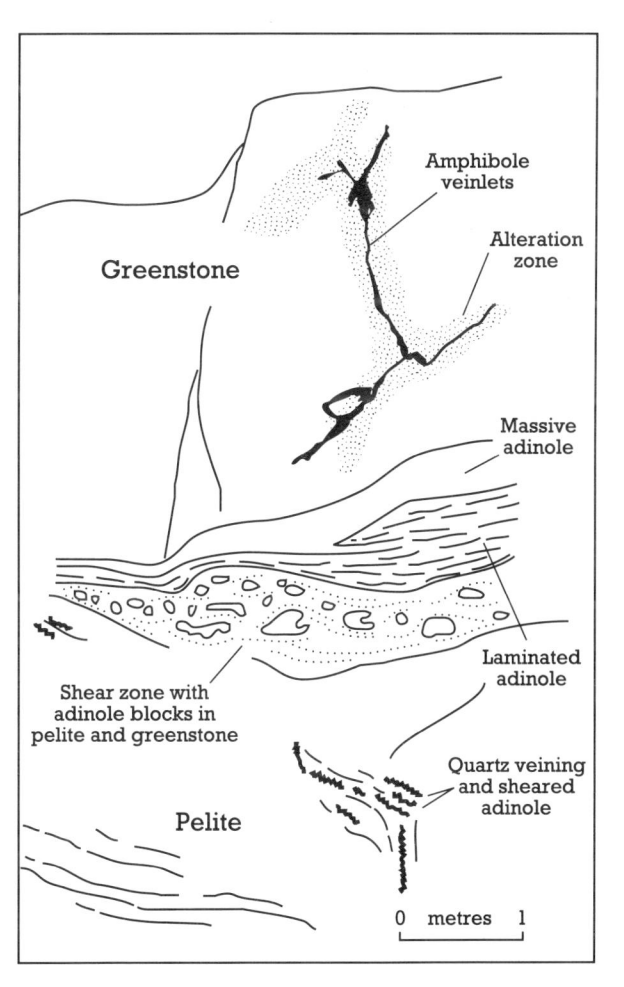

Pre-orogenic volcanics (Group B sites)

Figure 4.14 Photomicrograph of mineral relationships in late amphibole-rich hydrothermal veins, near Clodgy Point, Penwith Peninsula (cross polars). (Photo: P.A. Floyd.)

directly comparable with normal-type MORB (Floyd, 1984).

Interpretation

The importance of this site rests on the excellent examples of pillow lavas exhibited that typify the extrusives within the parautochthonous Upper Devonian of west Cornwall. They are evidence for submarine volcanism associated with the development of the deep-water facies of the Gramscatho Basin. Their limited stratigraphic occurrence and non-MORB chemistry indicates that true oceanic crust was not developed in this particular area of the Gramscatho Basin. Although they belong to the same magmatic province as other volcanics in south Cornwall, they are chemically distinct from the metabasalts of the *mélange* zone (Chapter 2) (Floyd, 1984). Moreover, their distinctive morphology and chemistry, in terms of specific incompatible-element ratios (e.g. Zr/Y), relative to other pillow-lava horizons at about the same stratigraphical level elsewhere within the aureole (e.g. Kenidjack Castle), suggest that a number of separate volcanic centres were active at this time.

The other major feature of the site is the superimposed contact-metamorphic and late-hydrothermal effects, consequent upon the emplacment of the Land's End Granite, that have completely replaced the primary mineralogy and texture.

Conclusions

The sedimentary rocks here were originally deposited as clays and silts on a sea-floor during the late Devonian Period, around 370 million years ago. Contemporaneous with sedimentation are piles of submarine lavas that formed superimposed masses of bulbous tubes (or 'pillows') as they escaped from the vent or fissure on the seabed. The original basalt lavas have been subsequently altered chemically and mineralogically and now bear the imprint of contact metamorphism by the Land's End Granite. However, chemical data indicate that their original eruptive environment was probably in a basin underlain by continental crust rather than oceanic crust like the Lizard ophiolite.

B5 GURNARD'S HEAD (SW 432387)

Highlights

This site is one of the few that shows a continuous gradation within a dolerite body from a massive, lower part to a pillowed top. Metasomatic effects, associated with the nearby Land's End Granite, are also well developed.

Introduction

The rocky peninsula of Gurnard's Head shows rocks in the metamorphic aureole of the Land's End Granite. Although much of the volcanic development within the Upper Devonian is recorded by classic exposures of pillow lavas; they can often be spatially associated with more-massive, intrusive, sheet-like doleritic bodies (Taylor and Wilson, 1975). Only in some cases, such as at Gurnard's Head, can these be seen to be directly related via vertical gradation from massive sheets to pillowed top.

Description

Gurnard's Head is topographically divided into two by a wide grassy hollow underlain by a metasedimentary horizon that separates the two greenstone masses to the north and south (Figure 4.15). At the site, the contrast between the type and level of emplacement of the two bodies may be seen. The southern greenstone is massive throughout, and although its contacts with the adjacent sediments are heavily sheared, it cuts the sedimentary lamination and was emplaced as a relatively late intrusive sheet. The northern greenstone rests conformably on the metasediments, below which are partially adinolized pelites and semipelites. The contact can be traced across the neck of the headland dipping at about 25° NNW; it is typically sheared and impregnated with radiating groups of hydrothermal, green amphibole. The base of the body is composed of a massive sheet-like intrusive which changes upwards to a crudely pillowed and vesiculated top. This implies emplacement at a high level, with the topmost magma batch in contact with water. The lenticularity of some of the pillows is partly a consequence of later shearing (Figure 4.16).

Interpretation

Both of the Gurnard's Head greenstones (which originally ranged from dolerite to basalt) lie within the Land's End Granite aureole, and they are now fine-grained ilmenite–plagioclase–actinolite hornfelses. The later replacement of contact-metamorphic amphibole by biotite implies the migration of K-bearing hydrothermal fluids across the width of the aureole from the granite source about 1000 m away. This example of externally derived fluids metasomatizing the greenstone is in contrast to the localized movement (within restricted shear zones) of Ca-rich fluids initially derived from the greenstone. This latter effect is represented by horizons within the pillowed greenstone, which contain white bands of diopside that replace the actinolite and probably represent the mobilization of Ca from original calcite-filled vesicles.

The few chemical analyses of the northern greenstone body indicate that it was originally a moderately incompatible-element-enriched tholeiite, typical of the southern Cornwall magmatic group (Figure 4.2). It is chemically similar to other metatholeiite lavas within the Penwith Peninsula with a composition indicative of an intraplate eruptive setting. However, differences in some incompatible-element ratios (for example, Zr/Y) imply that the Gurnard's Head greenstones and the nearby pillow lavas at Clodgy Point and Botallack Head are not comagmatic, but represent separate volcanic centres (Floyd, 1984).

Conclusions

The main interest in the site concerns the emplacement of the two greenstone masses that were intruded at different depths relative to the contemporaneous sediment–water interface. The change from sheeted to pillowed morphology indicates that the northern greenstone was emplaced near the interface, whereas the southern greenstone was intruded at a greater depth and within the sediment pile. Although the volcanics of this site are similar chemically to other metatholeiites in the magmatic province of south Cornwall, they probably represent a separate

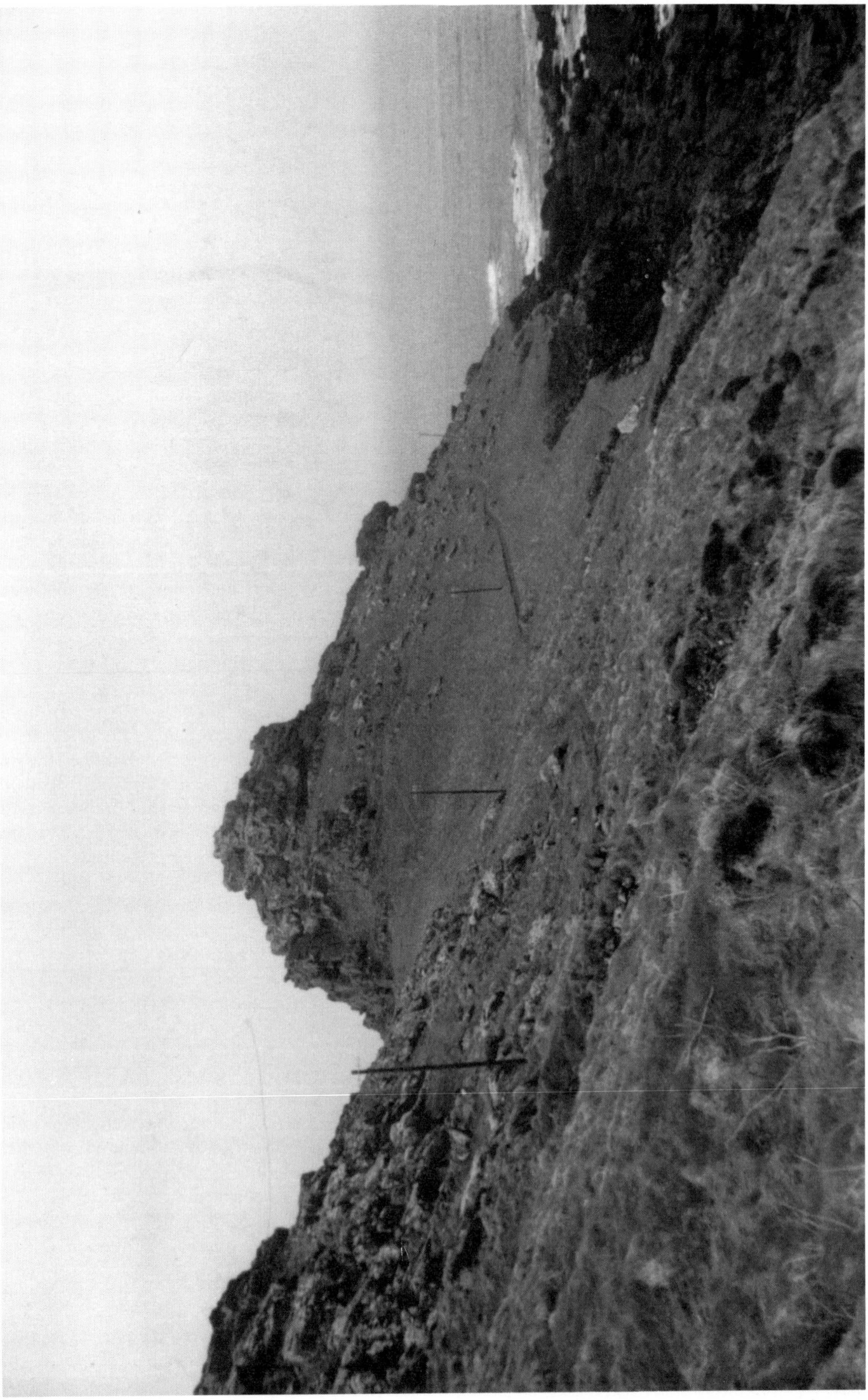

Botallack Head–Porth Ledden

Figure 4.16 Sheared and flattened Upper Devonian pillow lavas associated with the massive greenstone body at Gurnard's Head, Cornwall. (Photo: P.A. Floyd.)

volcanic centre. This is a common situation, with many outcrops representing isolated, small volcanic edifices at about the same stratigraphical level within a deep-basinal sequence.

The other feature of interest concerns the metasomatic replacement of basic contact-metamorphic assemblages by the minor development of biotite and diopside. These minerals are representative of two different metasomatic processes seen elsewhere within the Land's End aureole: the K (for biotite) being derived 'externally' from the granite, whereas Ca (for diopside) was derived 'internally' via mobilization of greenstone constituents.

Figure 4.15 The two greenstone masses of Gurnard's Head. The intervening hollow is underlain by metasediments. Gurnard's Head, Cornwall. (Photo: P.A. Floyd.)

B6 BOTALLACK HEAD–PORTH LEDDEN (SW 362339–SW 355322)

Highlights

Within the Land's End aureole, this classic site provides the best section of basic hornfels formed by the metamorphism and metasomatism of massive dolerite sills and basaltic pillow lavas. Uniquely, these pre-Variscan basalts contain granitoid xenoliths, possibly indicating an underlying continental crust at the time of their extrusion. This site also contains industrial relicts of some of the most famous and productive tin and copper mines in the mining history of South-west England.

Introduction

Scenically and geologically, this site is one of the best coastal sections within the aureole of the Land's End Granite. Steep cliffs and deep, inset, narrow gullies (locally called 'zawns'), together with a few exposures on the grassy cliff-top platform, provide excellent examples of both metamorphic and metasomatic hornfelses derived from basaltic pillow lavas and massive greenstones.

This area represents a classic example of contact metamorphism of basic volcanic rocks by the Land's End Granite, although gross morphology, relict textures and mineralogy often enable their original composition to be deduced. In general, contact metamorphism produced both typical hornblende-bearing hornfelses, as well as more unusual hornfelsic assemblages whose origin has been often debated and for which this area is famous.

The 'normal' basic hornfelses mainly display ilmenite–plagioclase–hornblende ± biotite assemblages and have often been intensively sheared into bands and lenses. It is not always possible to determine whether they were originally extrusive or intrusive, although the degree of heterogeneity within the sequences and the sheared outline of pillow lavas indicates that the majority of the normal, as well as the exotic, hornfelses were probably extrusive. A petrographic relict of the intensive deformation prior to contact metamorphism, is seen in the granulation of primary ilmenite into parallel trains across which new contact minerals have grown. All aureole hornfelses (including the metasediments) have been metasomatized by the granite and exhibit enhanced contents of Sn, Zn, Be, B, F, Cl, U and the light-REE (Floyd, 1966b, 1967; Wilson and Floyd, 1974; Alderton and Jackson, 1978; Mitropoulos, 1982, 1984; van Marcke de Lummen, 1985).

This part of the aureole initially became known for the presence of two groups of hornfelses with unusual mineral associations:

1. Mg-rich assemblages with anthophyllite–cordierite and cummingtonite–plagioclase, and
2. Ca-rich assemblages with sphene–diopside–hornblende, diopside–garnet and garnet–epidote–calcite.

Early work suggested that the two unusual hornfelsic groups represented the metasomatically derived end products of the normal hornblende-bearing hornfelses, under the influence of hydrothermal solutions emanating from the granite (Tilley and Flett, 1930; Tilley, 1935). However, rather than an origin invoking purely Mg and Ca metasomatism, it is now considered that they were developed by the isochemical contact metamorphism of previously altered basaltic volcanics (Vallance, 1967; Chinner and Fox, 1974). Alteration would have taken place during earlier, low-grade regional metamorphism (late Devonian), with the patchy development of variably degraded assemblages ranging from chlorite-rich (isochemically producing the Mg-rich assemblages) to epidote–carbonate-rich (eventually producing the Ca-rich assemblages). During contact metamorphism, Ca-rich solutions, however, were mobilized from Ca-bearing degraded areas and together with granite-derived mineralizing fluids (containing Sn, B, etc.) also locally metasomatized the adjacent normal hornfelses (Floyd, 1965; Jackson and Alderton, 1974; Floyd, 1975; Alderton and Jackson, 1978; van Marcke de Lummen, 1985). During the redistribution of elements within the aureole volcanics on contact metamorphism, it is clear that Sn-rich fluids were also circulating so that ore deposition within the aureole is often marked by Ca-rich skarn-type alteration (Jackson, 1974). However, oxygen and hydrogen isotope studies indicate that fluids in equilibrium with the skarn minerals were not purely magmatic, but, as temperatures fell, were mixed with an increasing meteoric component (van Marcke de Lummen, 1985).

Description

The site comprises the coastal strip between Botallack Head in the north to Porth Ledden, near Cape Cornwall, in the south. An outline geological map of the area is shown in Figure 4.17. It contains a prehistoric cliff castle at the Kenidjack Castle headland and is particularly famous for its mining history that covered an active period of nearly 200 years. Botallack Mine, probably one of the oldest in the St Just mining area, commenced mining in 1721 and became one of the richest tin mines by the early nineteenth century. Between 1815 and 1905, about 20 000 tons of copper ore and 14 000 tons of tin ore were won, with workings extending 800 m beyond the cliffs under the sea at a maximum depth of about 500 m (240 fathom level, Figure 4.18) (Barton, 1965; Embrey and Symes, 1987). Two derelict engine-houses still stand perched on the cliff edge at The Crowns, just south of Botallack Head (Figure 4.19), from which the famous, diagonal Boscawen shaft descended into the mine. Inland from Kenidjack Cliff was the extensive Wheal Owles Mine that, apart from tin, was one of the

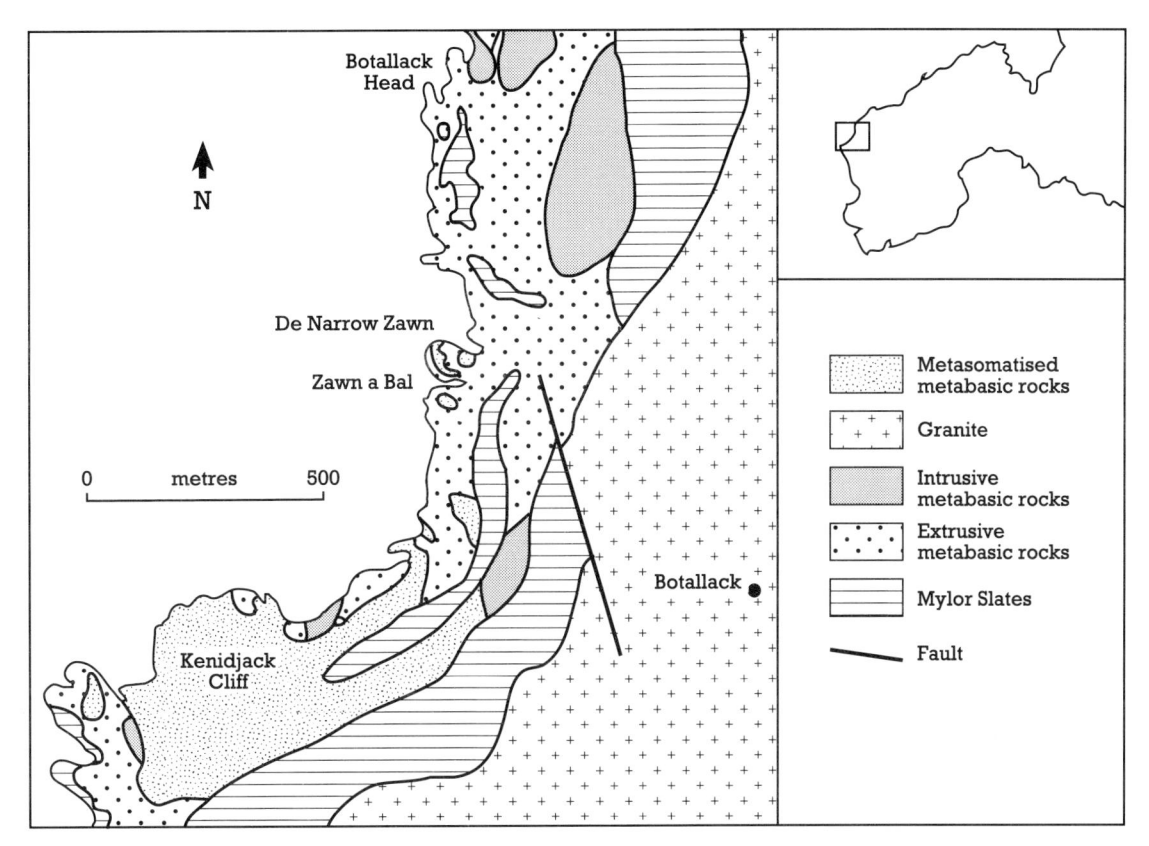

Figure 4.17 Geological map of the Botallack–Cape Cornwall section of the Land's End aureole, Penwith Peninsula (after Goode and Merriman, 1987).

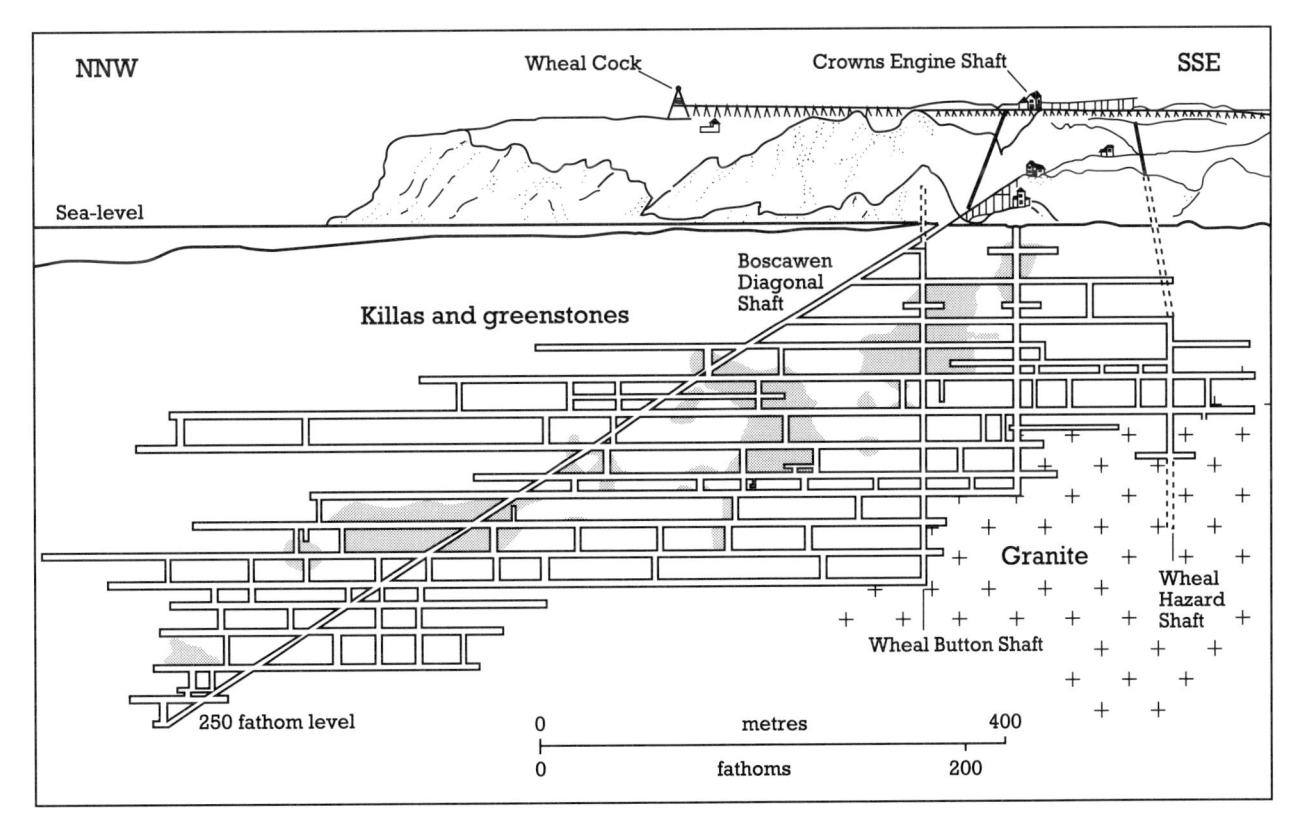

Figure 4.18 Section through the Botallack Mine, showing the sub-sea-floor workings and famous diagonal shaft, near St Just, Penwith Peninsula (after Embrey and Symes, 1987).

Pre-orogenic volcanics (Group B sites)

Figure 4.19 Line drawing of the cliff-edge engine-houses of the Botallack Mine and the beginning of the diagonal shaft at The Crowns, near St Just, Penwith Peninsula (reproduced from Barton, 1965).

few mines to produce uranium (pitchblende, zeunerite and various secondary uranium minerals) on the Penwith Peninsula. Nearby, to the north of Botallack Head, is the Wheal Cock Mine, where a number of rare beryllium minerals within a hydrothermal sulphide-bearing skarn have been recorded, as well as botallackite (a Cu chloride) that derives its name from the local area (Kingsbury, 1961, 1964).

The intimate association of both sediments and volcanics, as well as normal and unusual basic hornfelses, are well displayed in the cliffs adjacent to Botallack Head and The Crowns (SW 362336, Figure 4.17). The sedimentary wedge (mainly a retrogressed chlorite-bearing biotite–cordierite pelitic hornfels) has been locally metasomatized to a pale-grey adinole at the junction with the metavolcanics. In the vicinity of The Crowns the basic metavolcanics are variably biotitized plagioclase–hornblende hornfelses, although below the upper engine-house, they are replaced by hard, dark cordierite–anthophyllite hornfelses with cordierite porphyroblasts aligned along the major foliation. Thin-section examination shows stellate groups of anthophyllite needles, developed parallel to the ilmenite granule foliation, traversing large cordierite porphyroblasts which often enclose relicts of replaced plagioclase. Figure 4.20 is a composite drawing of the mineralogical relationships often seen in cordierite–anthophyllite assemblages: it illustrates two growth periods for both anthophyllite (stellate and prismatic) and biotite. Cummingtonite-bearing Mg-rich hornfelses are intimately associated with biotite–cordierite–anthophyllite-bearing bands on the promontory (SW 362333) between De Narrow Zawn and Zawn a Bal. Grey cordierite–anthophyllite hornfelses, with some bands of brown biotite-bearing variants, are well exposed at Kenidjack Cliff headland (SW 355326), and, in the adjacent quarries, grade into normal hornblende-bearing hornfelses. Both anthophyllite- and cummingtonite-bearing hornfelses may sometimes exhibit silica-deficient assemblages containing green spinel (pleonaste) and diaspore (Figure 4.20).

Various skarn-type Ca-rich hornfels are well exposed below and around the lower engine-house at The Crowns; they form pale-coloured grey, pink and green masses. The banded foliation

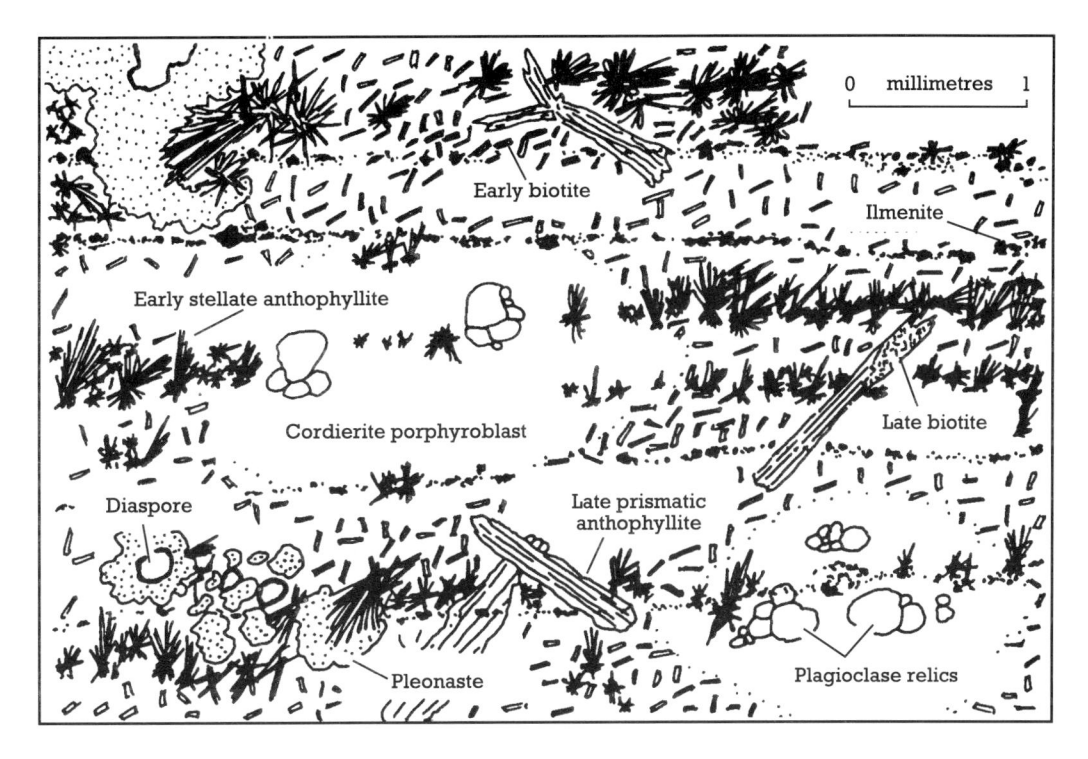

Figure 4.20 Composite drawing of mineral relationships in the biotite–cordierite–anthophyllite assemblage, based on exotic hornfelses from the Zawn a Bal to Kenidjack area, Land's End aureole, Penwith Peninsula.

of the adjacent hornblende hornfelses may be lost where the skarn replaces it, although the latter often develops almost monomineralic horizons subparallel to the main hornfels banding. The main calc-silicate minerals are green diopside, red grossularite garnet (anisotropic and zoned) and amphibole, with minor idocrase, epidote, axinite, tourmaline, calcite, chlorite, spinel, sphene and sulphides (largely chalcopyrite). A number of stages for the migration of Ca-rich fluids are indicated by the cross-cutting of the skarn masses by a 0.3–0.35 m thick tourmaline–diopside-bearing garnet vein.

Massive volcanics and pillow lavas composed of the more normal hornfelsic assemblages are present on the rocky platforms of Kenidjack Castle. The pillow lavas are generally thin (<1 m) with interpillow spaces filled with silica. Secondary amphibole veinlets are common throughout the volcanics. Chemical data for the pillow lavas indicate that they are tholeiitic and form part of the same group as the other lavas within the south Cornish magmatic province of the Penwith–Camborne area (Floyd, 1982a, 1984). Although chemically similar to the coeval Clodgy Point and Gurnard's Head basalts, differences in their Zr/Y ratios and degree of light-REE-enrichment implies that the lava occurrences are not comagmatic, but represent three separate volcanic centres generated by variable partial melting of a common mantle source.

Within the basic hornfelses near Zawn a Bal are rare, small (0.1–0.4 m diameter), pink, weathered, feldspar-rich crystalline xenoliths (Goode and Merriman, 1987). These unusual xenoliths, some of which exhibit a pre-Hercynian tectonic fabric, range from intermediate to acid in composition and were derived from granitic precursors. The significance of these granitic xenoliths, carried upwards by late Devonian basic magmas, lies in the possibility that they represent remnants of continental crust lying below the Penwith Peninsula, and as such might have provided a source for the Cornubian granites (Goode and Merriman, 1987).

Interpretation

The Botallack area has long been famous for its history of tin and copper mining, with the engine-houses and dumps along this coastal strip now providing silent testimony to past endeavours. Apart from its past economic significance, the site provides excellent examples of the different hornfelsic types produced during the contact

Pre-orogenic volcanics (Group B sites)

metamorphism of variably degraded basaltic volcanics. Initially, studies by Tilley (1935) identified the possible derivation of exotic anthophyllite–cordierite- and cummingtonite-bearing assemblages by variable Fe + Mg metasomatism of the adjacent normal hornblende hornfelses. This was one of the first detailed petrographic descriptions of these rocks within a granite aureole in Britain, and Tilley (1935) compared them with occurrences within the Precambrian crystalline terranes of Scandinavia. For example, early studies by Eskola (1914) indicated that the anthophyllite–cordierite rocks of the Orijarvi region had been derived by the wholesale Mg metasomatism of siliceous 'leptites' (acidic lavas and tuffs). Tilley (1935) considered the Botallack exotic hornfelses to have been formed by the same metasomatic process, but with the important difference that the former rocks were derived from basic parents via the internal redistribution of Mg (together with the loss of Ca, etc.), and not by the addition of Mg from an external granitic source as at Orijarvi. The origin of these unusual rocks is important, not only in terms of metamorphic paragenesis and parental composition, but as economic guides because they are often associated with massive sulphide deposits in many parts of the world. Although the theory of a metasomatic origin held sway for some time, work on the regional degradation of basaltic volcanics has indicated that they could also be generated by the isochemical metamorphism of the low-grade secondary products of such rocks (Vallance, 1967; Chinner and Fox, 1974). However, as indicated by Floyd (1975), removal of Ca, etc. from the degraded assemblages is still required to produce a suitably Mg-rich precursor. In this context, Floyd (1975) linked the derivation of the skarn hornfelses to the Mg + Fe-rich hornfelses, with the former representing the repository of the released Ca. Thus, both groups of exotic hornfelses can be related to the variable degradation of basaltic volcanics and the differential migration of Ca-rich fluids. The normal hornblende hornfelses were developed isochemically from metabasalts containing relatively small proportions of secondary Ca-bearing phases. It is interesting to record, however, that Reynolds (1947), noting the intimate association of the exotic hornfelses, suggested that the Mg + Fe-rich group were derived from calcareous sediments (remnants becoming the Ca-rich group) which suffered Mg metasomatism from an advancing basic front produced by local granitization.

The other major significant feature of this site is the discovery of possible continental basement as xenoliths within the basaltic lavas. Apart from some crustal fragments within Permian lavas, no deep-seated xenoliths, either crustal or mantle, have been found in the Variscan basaltic lavas of south-west England. The granitic xenoliths at Botallack may represent the only example we have of continental crust underlying this area, although Goode and Merriman (1987) speculate that some of the foliated granites of Haig Fras might be comparable with and not related to the Cornubian batholith as previously thought.

Conclusions

This site is world famous for its copper and uranium minerals and the mining activity which they attracted in the last century. These minerals occur within rocks, which were originally marine sediments and lavas formed in late Devonian times, around 370 million years ago. These rocks were subsequently mineralized by the action of hydrothermal solutions emanating from the Land's End Granite that caused the mobilization and redistribution of economically important elements. Prior to granite emplacement, the basaltic lavas had undergone low-grade alteration with the development of two different chemical groups – one Fe + Mg-rich, the other Ca-rich. On contact metamorphism by the granite, these two groups developed a unique set of exotic mineral assemblages for which the area is famous. Within the Devonian lavas are found inclusions of granitic rocks that are considered to represent fragments of continental crust through which the basalt magmas passed. Their presence is the only direct evidence for the existence of continental crust below the submarine basins of south-west England.

B7 TATER-DU (SW 440230)

Highlights

At this excellent contact with the Land's End Granite, the progressive change from basaltic

Figure 4.21 Massive cliff section composed of various banded, amphibole-bearing, basic hornfelses of volcanic origin. In the foreground is a small irregular raft of metasediment caught up during the emplacement of the basalts. Tater-du, Cornwall. (Photo: P.A. Floyd.)

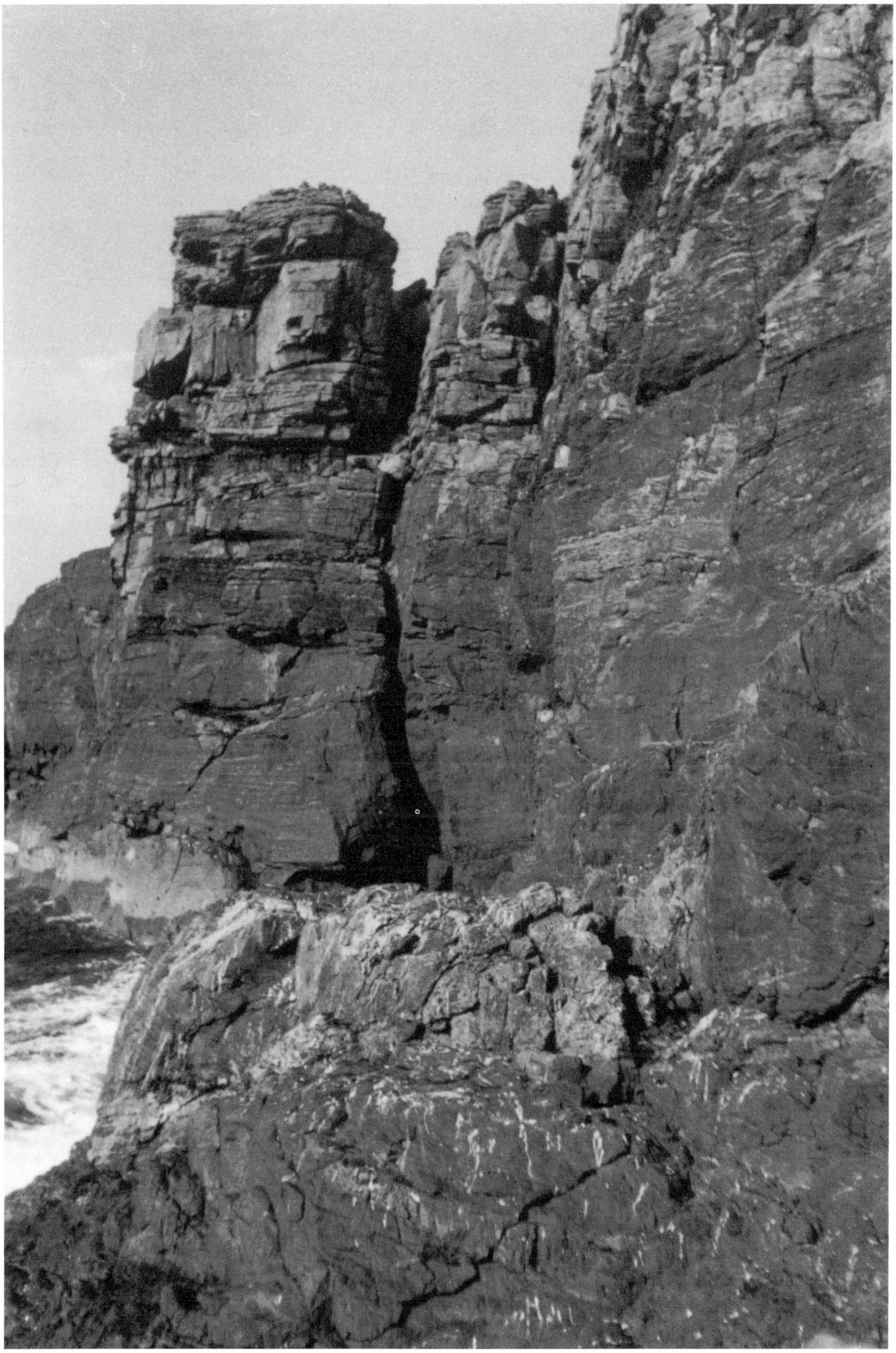

Introduction

The dark, steep cliffs and sloping rock platforms of Tater-du at the southern tip of the Penwith Peninsula (SW 440230) are a small erosional relict of aureole rocks adjacent to the Land's End Granite (Figure 4.21). The area from Tol Toft in the east to Zawn Gamper in the west, including the crags and quarry just inland, are perched on the southerly dipping contact zone of the surrounding megacrystic granite. The eastern end is represented by a faulted gully, while to the west an excellent contact between the aureole rocks and the marginal facies of the granite is exposed in the wall of Zawn Gamper. Apart from a petrographic description of the hornfelses (with some chemical data), little research has been done on this aureole segment (Floyd, 1965, 1975), although it represents a microcosm of the contact alteration of basic volcanics in the hornblende hornfels facies.

Description

Directly above the granite contact at Zawn Gamper, is a thin, variably retrogressed cordierite–biotite pelite that can be traced intermittently along the lower rock ledges below the main cliffs of Tater-du. At the granite contact it is tourmalinized and shows the development of late, randomly orientated, muscovite flakes. The rest of the site is mainly composed of massive, well-banded, dark hornblende-bearing hornfelses, with minor horizons of unusual Mg-rich and Ca-rich hornfelses. These are best developed below the main cliffs near one of the metasedimentary lenses; this has a thin, partly developed, cummingtonite-bearing adinolized contact zone with the adjacent metavolcanics.

The basic volcanics here were intensely sheared prior to contact metamorphism, so that there is little direct evidence as to their original nature. The presence of adinolized sediment implies that some of the volcanics were probably intrusive sheets, as adinole development is only seen adjacent to dolerite sills in other less-tectonized parts of the aureole. However, the majority of the Tater-du basic hornfelses exhibit small diopside-rich lenses and spots, now drawn out parallel to the foliation, which may have originally been infilled vesicles in a sequence of lavas. The mimetic growth of contact-metamorphic minerals accentuates the early shear foliation and locally this produces a banded rock composed of various, nearly monomineralic, layers – typically amphibole, biotite or diopside (Figure 4.22). Some of the more lenticular structured hornfelses show weather-resistant amphibole phacoids surrounded by deeply eroded, pale-purple, biotite rims. The hard phacoids might have originally represented the more crystalline cores of metamorphosed pillow lavas; the altered glassy rims (altered to chlorite) were subsequently replaced by biotite during contact metasomatism.

The normal basic hornfelses are mainly composed of the ilmenite–plagioclase–hornblende assemblage with variable replacement of the amphibole by biotite, representing the introduction of K from the granite. Some bands feature calciferous zones with the development of diopside and labradorite in lenses and the replacement of primary ilmenite by sphene in the matrix. This feature can be seen throughout the site, but is well displayed in the small quarry behind the lighthouse, where the calciferous lenses may represent metamorphosed vesicle infillings. The normal assemblages were largely developed by the isochemical metamorphism of basic volcanics under hornblende hornfels facies conditions.

In addition to these dominant assemblages, unusual hornfelses are well developed on the rock platforms under the main cliffs and towards Tater-du Point. Two groups may be observed: Mg-rich hornfelses and Ca-rich hornfelses. The lower portion of the cliffs consists of lenticular, variably biotitized hornblende hornfelses with large white lenses of diopside. Below and interbedded with this horizon is a conspicuous highly foliated biotite–cummingtonite hornfels with black lenses of cummingtonite-rich rock weathering less than the biotitic matrix. Traced towards Tater-du Point, these rocks become more massive, lustrous black in colour, and are represented by cordierite–anthophyllite hornfelses. This small section demonstrates the lateral gradation from normal hornfelses through to cummingtonite-bearing and finally anthophyllite-bearing types. On the same rock platforms, but below the pelitic wedges, is a small outcrop of the Ca-rich hornfels or skarn. This highly weathered mass replaces both the hornblende hornfelses below and the

Tater-du

Figure 4.22 Typical, banded, basic hornfels of volcanic origin, composed of dark layers of hornblende and biotite, with light-coloured, segregation lenses of diopside. Tater-du, Cornwall. (Photo: P.A. Floyd.)

adjacent pelites, such that fragments of adinole are found within the skarn deposit. The skarn is dominated by grossularite and diopside, although sphene, zoisite, clinozoisite, axinite and calcite are also present. In the contact zone with the basic hornfelses, diopside and garnet replace hornblende and plagioclase, and sphene nucleates on ilmenite.

Interpretation

The significance of this locality concerns the paragenesis and range of hornfelses produced by the contact metamorphism and metasomatism of originally basaltic volcanics. Not only have normal hornblende-bearing contact hornfelses been produced, but also two groups of unusual hornfelses that are genetically complementary and chemically linked. In particular, they demonstrate both the importance of original composition and the effects of local metasomatism on the final metamorphic assemblage. Their development is the same as similar exotic hornfelses from Botallack (discussed above), with contact-metamorphic effects acting on variably degraded basaltic volcanics, together with localized element migration under hydrothermal conditions. The lateral gradation of hornblende hornfelses into both the Mg-rich and Ca-rich hornfelses is important field evidence for the origin of the unusual hornfelses from the initial basic assemblages. However, mineralogical relationships and replacements suggest two different, but complementary, processes. The normal hornfelses represent basic volcanics that have been only slightly altered by previous regional metamorphism, whereas the Mg-rich hornfelses were developed isochemically from highly degraded, carbonate-poor but chloritic metabasics, as direct mineral replacements are rare. On the other hand, the skarn deposits show ample evidence for replacement of the normal hornfels assemblage, as well as progressive internal replacement of previous calc-silicates. After the initial assemblage of sphene–diopside–grossularite was produced under hornblende–hornfels–facies conditions, a retrogressive phase started with the replacement of garnet by clinozoisite and zoned axinite, and finally all minerals by calcite (Floyd, 1965). As at Botallack, the involvement of granite-derived fluids is indicated here by the presence of B-bearing axinite during the later phases of skarn development.

Conclusions

Tater-du shows rocks of late Devonian age (around 370 million years old) that have been so severely affected by later geological events that there is little direct evidence of their original composition, although they are believed originally to have been basalt lava flows. They have been altered to different hornfelses, the product of mineralogical and chemical alteration largely induced by contact with the emplacement of the adjacent Land's End Granite. The different types of mineral assemblages within the hornfelses indicate different phases of replacement both prior to and during granite intrusion. This is a key site for studying the complexities of chemical and mineralogical change brought about by element mobility and the reactions that occur in rocks of differing composition and origin.

B8 PENTIRE POINT–RUMPS POINT (SW 923805–SW935812)

Highlights

This classic British site for the study of pillow lavas, is one of the best exposed in Cornwall. The alkali-basalt lavas are only weakly metamorphosed and a massive metadolerite (greenstone) is also well exposed.

Introduction

This scenic section of the north Cornish coast includes the steep 80–100 m high cliffs around Pentire Point and also the rocky crags of the The Rumps Peninsula. The latter has a prehistoric cliff castle and is joined to Pentire Point proper by a neck of land showing the now subdued remains of an ancient earthwork.

During the late Devonian, the main expression of magmatism associated with basin development was the production of localized pillow-lava sequences. The north Cornish coast from Pentire Point westwards to Port Isaac shows many isolated examples of pillow lavas, although the best-exposed sequences are found along the Pentire cliff section. These rocks were some of the first lavas of submarine origin to be recognized in Britain (Whitley, 1849). The Pentire Point pillow lavas were described in detail by Reid and Dewey (1908) and Dewey and Flett (1911) and provided evidence for early ideas as to their mode of formation (Reid et al., 1910; Dewey, 1914). To early workers like Dewey (1914), pillow lavas were individual spheroids, each representing a 'thick-walled bubble of lava' and, together with their highly vesicular nature, he suggested that each lava droplet welled up independently to form a buoyant, gas-filled, floating pillow. Another interesting feature of the initial work on the Pentire pillows was the notion that 'vapours' or 'juices' trapped within the lava reacted with the hot, partially crystallized rock to produce the alteration assemblages we recognize today as the effects of post-consolidation metamorphism (Vallance, 1965).

Description

The steep east–west orientated cliffs of Pentire Point and the rock platforms below, show the underside of the pillow lava sequence in longitudinal section. Here the lavas can be seen to be elongate tubes drapped over each other, with the pillow interspaces filled with chert or sometimes calcareous argillite.

On the other hand, cross-sections (seen on joint faces at right angles to the cliffs) are characteristically ovoid in shape with each pillow moulded over the ones below, providing evidence for 'way-up' and a southwards-younging direction. The total cliff section is composed of three main pillow-lava domes, separated by argillite that laps on to the sides of the domes as well as eventually enveloping them completely (Figure 4.23).

Stratigraphically, the pillow lavas are Frasnian in age and they form the basal part of a sequence of Upper Devonian slates that are slightly overturned and which young southwards, away from the coast. They comprise the Pentire Pillow Lava Group (Gauss and House, 1972), which is about 450 m thick and includes various pillow-lava horizons and subordinate tuffaceous sediments and agglomerates in the local area. The Upper Devonian here is allochthonous, the pillow lavas forming part of the Port Isaac Nappe (Selwood and Thomas, 1986a). They are chemically distinc-

Figure 4.23 View of Pentire Point cliffs showing Upper Devonian pillow-lava mounds. (Photo: P.A. Floyd.)

Figure 4.24 Pillow-lava breccia formed by fragmentation on cooling soon after submarine extrusion. Pentire Point, Cornwall. (Photo: P.A. Floyd.)

Pre-orogenic volcanics (Group B sites)

tive, relative to the west Cornish tholeiitic lavas of similar age in the Penwith Peninsula, in being alkali basalts with intraplate chemical features (Floyd, 1982a, 1983).

The pillows are generally 0.3–0.6 m in diameter, rarely over 1 m, and are highly vesicular. Some show a series of vesicular zones separated by massive lava and a central vacuole. The vesicles are now filled with chlorite at the margins and carbonate and/or silica in the interior which appears to have replaced earlier chlorite or smectite. Originally glassy margins are no longer seen, either having spalled off during extrusion or been completely replaced by secondary chlorite. Small, black, wispy fragments of laminated argillite may be seen within some pillows, and these represent partly consolidated sediment entrapped during extrusion. Although relatively uncommon, pillow breccias (Figure 4.24) are present, situated near the margins of the lava domes. These demonstrate the partial fragmentation of the lavas along both radial and concentric cooling joints within the pillows (Figure 4.25). The breccias have suffered minimum downslope movement, as individual fragments can often be fitted together jigsaw fashion.

Although now metabasalts, the pillow lavas still display features that can be used to infer their primary characteristics. The lavas are plagioclase-phyric, with lath-shaped microphenocrysts and skeletal microliths, now converted to secondary albite. The microliths may be curved and sometimes show tuning-fork terminations indicative of rapid quenching. The matrix is largely composed of chlorite, carbonate and oxidized materials which probably replaced original glass. No primary mafic minerals remain. Stable-element chemical data demonstrate that the Pentire lavas form a single differentiated suite of basalts which were not comagmatic with the intrusive dolerite at The Rumps. One interesting chemical feature concerns the location of stable elements in the altered pillow basalts. A study by Williams and Floyd (1981) showed that elements like Ti, Zr and Nb are relocated on alteration into stable secondary phases like rutile and zirconolite. On the other hand, U migrated throughout the matrix, eventually concentrating around vesicle margins and sometimes, within the infilling calcite, giving it a cloudy appearance.

The pillow-lava sequence is underlain by the Middle Devonian Trevose Slate Formation which

Figure 4.25 *In situ* autobrecciation of a lava pillow. Pentire Point, Cornwall. (Photo: P.A. Floyd.)

has been intruded by a massive greenstone sill at The Rumps. Separated from Pentire by a small fault, this body is typical of differentiated dolerite-gabbro intrusives often associated with lavas, although in this case there is no evidence for high-level emplacement or chemical association with the lava sequences. The body exhibits a sharp concordant junction with the sediments which are bleached and indurated, but not adinolized. One feature of interest shown by the greenstone is the effect of intense shearing that has granulated and foliated the body into highly oxidized chloritic schist zones.

Interpretation

The interest of this site is that it not only represents one of the classic pillow-lava locations in Britain, but is typical of the major episode of extrusive late Devonian activity within the Variscan fold belt, such as that exhibited in the Harz Mountains and Lahn-Dill in Germany. It was here that some of the early ideas concerning pillow-lava development were formulated and enabled the correlation with similar, but more extensive, volcanism of similar age within the Rhenohercynian Zone of Germany to be made. In tectonic terms, the volcanic activity represents the onset of basinal development, with magma penetrating marginal fractures that mark the site of the change from platform to slope. They are also typical of Variscan spilites, characterized almost entirely by secondary assemblages, which were at one time considered to have been produced by late primary magmatic (deuteric) processes. We now recognize these lavas as low-grade metamorphosed basalts.

This site is significant in the magmatic history of Variscan south-west England, as the lavas are plagioclase-phyric alkali basalts with enriched incompatible-element patterns, forming a province quite distinct from temporally analogous pillow lavas in south Cornwall. They exhibit chemical features typical of intraplate volcanics that are also characteristic of many lavas within the Variscan fold belt of Northern Europe.

Conclusions

This is a key site for the study of Upper Devonian submarine lavas and has been studied since the middle of the nineteenth century, when it was realized that the peculiar pillow-shaped masses of lava, normally up to a metre across, were the product of eruption of basalt lava into sea-water. Rapid chilling by the sea-water produced glassy margins which often spalled off or completely fragmented the pillow into small angular blocks. Similar pillow lavas may be seen forming today where submarine volcanism occurs, producing rounded or sausage-shaped extrusions on the sea-bed, venting bubbles of gas. At Pentire Head, the lavas and subordinate fragmented volcanic rocks make up a 450-m-thick pile locally. The lavas were originally basalts, although they have been altered subsequently under conditions of low-grade metamorphism during the Variscan Orogeny. Pillow lavas are characteristic of a major phase of volcanic activity in the Devonian and Carboniferous seas of Variscan Europe; this is one of the classic sites where they were first studied and attempts made to explain their form, chemistry and origins.

B9 CHIPLEY QUARRIES (SX 807712)

Highlights

This is a classic site for the study of pillow lavas. Unusually, these vesicular, shallow-water, pillowed alkali basalts can be dated precisely here from their association with fossiliferous sediments.

Introduction

This site includes the two small disused quarries on the wooded hillside adjacent to the Chipley–Bickington road. The site is situated in a classic late Devonian region which is now recognized as allochthonous and as forming part of the southerly derived Chudleigh Nappe (Selwood and Thomas, 1986a).

Some of the earliest descriptions of the volcanic rocks of this region of south Devon were made by Champernowne (1889). The detailed petrography of the Chipley pillow lavas was initially described by Flett (*in* Ussher, 1913), and he included a single major-element analysis. The lava sequence here is situated within the grey Gurrington Slate Formation which has yielded ostracods (*hemis-*

Pre-orogenic volcanics (Group B sites)

phaerica–dichotoma biozone) indicative of a Famennian age (Middleton, 1960). Although part of the major late Devonian extrusive activity in south-west England this contrasts with the Frasnian age of similar pillow lavas at Pentire Point (discussed above). Geochemical work indicates that the Chipley lavas are also intraplate alkali basalts, although they can be distinguished from the Pentire sequence by different incompatible-element ratios and light-REE-enrichment patterns (Floyd, 1982a, 1983). Both occurrences, however, are characteristic of the 'spilite suite' which has counterparts within the Rhenohercynian zone of northern and central Europe, and whose origin and composition have long been debated (for instance, Juteau and Rocci, 1974; Wedepohl *et al.*, 1983).

Description

The Chipley pillow lavas (Figure 4.26) are similar in many respects to those at Pentire Point and are excellent examples of extrusive submarine activity. Both of the quarries in the site exhibit a *c.* 30 m sequence of tectonically flattened, but well-formed, closely fitting lava tubes with negligible interpillow sediment. Good ovoid cross-sections are visible, displaying the highly vesicular (up to 50% volume) nature of the lavas (Figure 4.27). Vesicles are smaller at the margins than towards the core, although central vacuoles are uncommon. The high degree of vesicularity of the pillows suggests extrusion into relatively shallow water (<500 m depth). However, the basalts were originally alkaline and, as such magmas often contain a relatively high volatile content, the degree of vesiculation may not always be a reliable guide to the depth at which extrusion occurred.

The pillow lavas were originally plagioclase-phyric alkali basalts, but are now completely degraded to low-grade metamorphic assemblages with no primary minerals remaining. Both phenocrystic and matrix plagioclase are now converted to secondary albite, set in a matrix composed of variable chlorite, quartz, carbonate, epidote, prehnite, sphene, magnetite and leucoxene – a typical spilite assemblage. Plagioclase microlites may be skeletal or form long, curved crystals, sometimes with tuning-fork terminations; features which are indicative of rapid quenching in a submarine environment. Much of the fine-grained chloritic matrix probably represented original glass. Vesicles are infilled with chlorite or more rarely prehnite, both of which may be replaced by quartz or carbonate.

Interpretation

Owing to the low-grade, altered nature of the lavas, it is only possible to determine their original magma type by using incompatible elements that are stable during secondary alteration. These indicate that the lavas are alkali

Figure 4.26 Upper Devonian pillow lavas of alkali-basalt composition. Chipley Quarries, Devon. (Photo: P.A. Floyd.)

Chipley Quarries

basalts with an overall composition similar to other extrusives and intrusives within the north Cornwall and Devon magmatic province (Floyd, 1982a; Rice-Birchall and Floyd, 1988).

Late Devonian basaltic activity was often associated with the development, especially the deepening of basins, although there were environmental differences between north Cornwall (with the Pentire lavas) and south Devon (with the Chipley lavas) during this period. For example, the latter area saw the extensive development of carbonate-dominated facies, with thin, condensed sequences of Upper Devonian, pelagic, nodular carbonates and mudstones developed on submarine rises (House, 1963, 1975). These grade into adjacent, deeper-water, basinal argillites which contain the local pillow lavas encompassed by this site. Their development near a submarine rise is different from broadly similar lavas at Pentire Point in Cornwall, which developed on the northern margin of a deepening fault-controlled trough (Selwood and Thomas, 1986b). Chemically, the pillow lavas are also characteristic of the north Cornwall and Devon magmatic province of alkali volcanics (Chapter 2).

Conclusions

This locality has been studied since the late nineteenth century when it was realized that the peculiar pillow-shaped masses were the result of the eruption of basalt lava into sea-water. At Chipley the pillow lavas are shot through with voids (vesicles) which were originally the site of gas bubbles. As a result of the sudden release of pressure and rapid chilling on the extrusion of individual pillow flows, gas was forced out of solution in the volatile-rich lava, leaving the solidified rock full of voids. The Chipley basaltic lavas have been altered and their original mineralogy replaced by secondary minerals (cf. Pentire Point), although textures still preserve evidence of rapid chilling. Laterally the lavas may be traced out into the sedimentary rocks which were deposited as mud on the deep sea-bed. Volcanic activity occurred here at the very end of the Devonian period around 360 million years ago. Chipley differs from the pillow-lava site at Pentire Head (of about the same age) in terms of the chemical composition of the basalts and its eruptive setting.

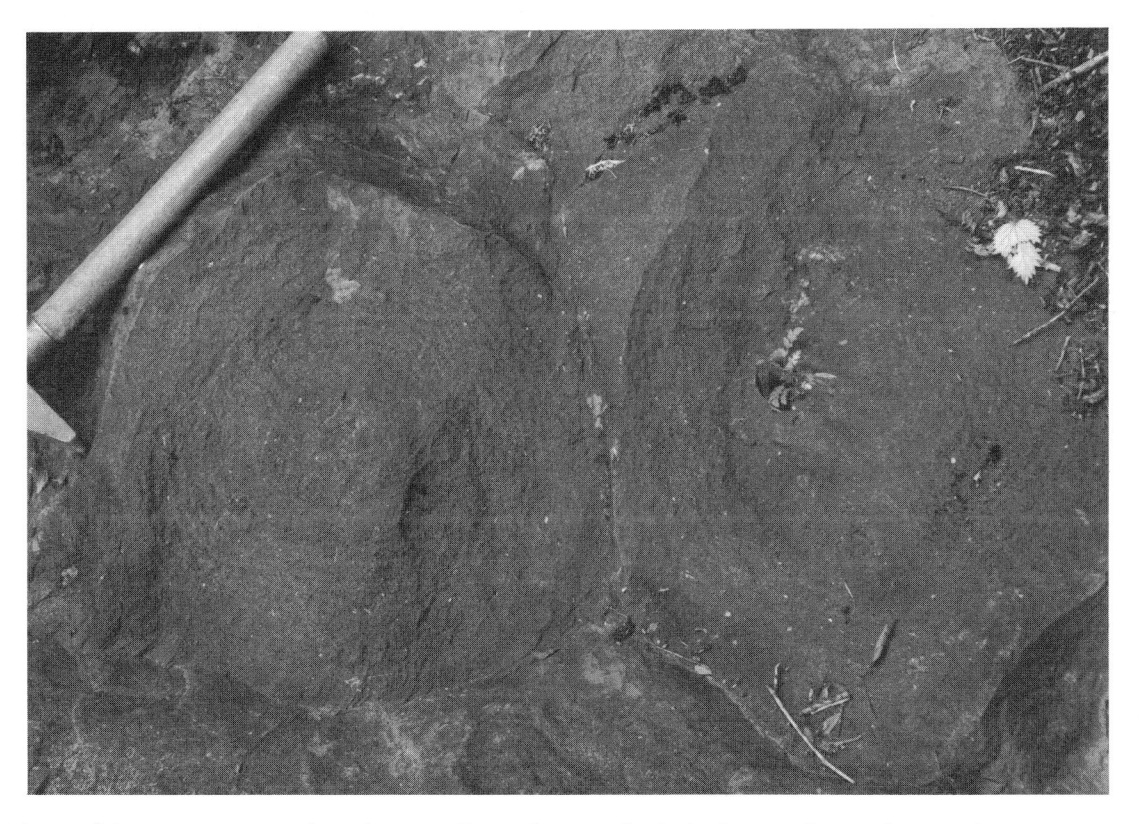

Figure 4.27 Cross-section through two pillows showing the high degree of vesicularity and its concentric disposition. Chipley Quarries, Devon. (Photo: P.A. Floyd.)

Pre-orogenic volcanics (Group B sites)

B10 DINAS HEAD–TREVOSE HEAD (SW 847761–SW 850766)

Highlights

This site is the best in south-west England and a classic in Britain for the study of the progressive development of adinole at the contact between a dolerite intrusion and its enclosing sediments.

Introduction

This scenic coastal site includes the rocky, terraced cliffs of Dinas Head from Mackerel Cove to Stinking Cove, and the rock platforms around Trevose Head lighthouse (Figure 4.28). The old quarry at the back of Stinking Cove is also included.

Apart from the massive intrusive greenstone which was described by Reid *et al.* (1910) and later by Agrell (1939), interest in this site has centred on the contact effects of the body on the adjacent sediments. One of the characteristic features of some large, sill-like intrusions, is the development of a narrow contact aureole of intensely Na + Si-metasomatized sediments called spilosites and adinoles. Fox (1895) initially noted the development of these adinoles as metasomatized sediments, whereas their particular pseudospherulitic texture was commented on by McMahon and Hutchings (1895). The Dinas Head adinoles were also mentioned by Dewey (1915) who demonstrated the importance of sediment composition in the production of adinoles, with Fe^{3+}-poor, grey and black argillites being readily altered to adinoles. A more extensive survey of adinolization in the area (Agrell, 1939, 1941) showed that the abnormally large metasomatic zone at Dinas Head was a function of the intimate penetration of the sediments by many sill-like offshoots from the top of the intrusive mass.

Description

The sedimentary rocks here are laminated Upper Devonian black and blue-grey argillites; they form part of the Port Isaac Nappe (Selwood and Thomas, 1986a). The sediments occupy the rib of Dinas Head, although as sea-level is approached, they become progressively metasomatized and penetrated by tongues of greenstone (metadolerite). The sedimentary rocks/greenstone junction is often irregular with random ramifications and entrapment of sedimentary wedges. The first effects seen of contact metamorphism are the development of chloritic and incipient cordierite spotting, as well as the growth of andalusite prisms (Agrell, 1939). Partially metasomatized argillites (spilosites) retain the original slaty cleavage, but show the development of quartz–chlorite and porphyroblastic albite 'spots', together with tourmaline and leucoxene. Completely metasomatized argillites (adinoles) consist largely of quartz–albite assemblages with variable carbonate content. Original spots and andalusite are now pseudomorphed by quartz–chlorite or carbonate–chlorite. The progressive metasomatic changes are well displayed in a section from the top of Dinas Head down to sea-level towards the intrusive. However, spilosites and adinoles may often be interbedded, reflecting changes in the original composition of the sediments rather than distance from the intrusive contact. A number of different textural types have been recognized in the metasomatized sediments by Agrell (1939).

The majority of the greenstone is a fine- to medium-grained, subophitic alkali metadolerite with occasional relicts of primary clinopyroxene set in an altered low-grade matrix. The pyroxene may be partially replaced by a fringe of uralitic actinolite or completely pseudomorphed by actinolite and chlorite. Plagioclase is now albite, but may also be replaced by sericite, epidote, prehnite and carbonate. Chlorite, albite and carbonate are common secondary minerals throughout the finer-grained marginal facies. Skeletal ilmenite may be leucoxenized or rimmed by replacive sphene granules. Apatite needles are relatively rare.

The general shape of the greenstone mass appears sheet-like, but its junction with the sediments is very irregular on the medium scale (several metres) and a number of separate large tongues penetrate the sediments to the north and south of Dinas Head. In the quarry, which is topographically above the Dinas Head part of the greenstone, the very high-level nature of the upper surface can be demonstrated. Much of the greenstone here is very fine grained and contains

Figure 4.28 (Opposite) Wedge of argillite (pale-coloured cliffs) resting on dark intrusive dolerite near sea-level. Trevose Head, Cornwall. (Photo: P.A. Floyd.)

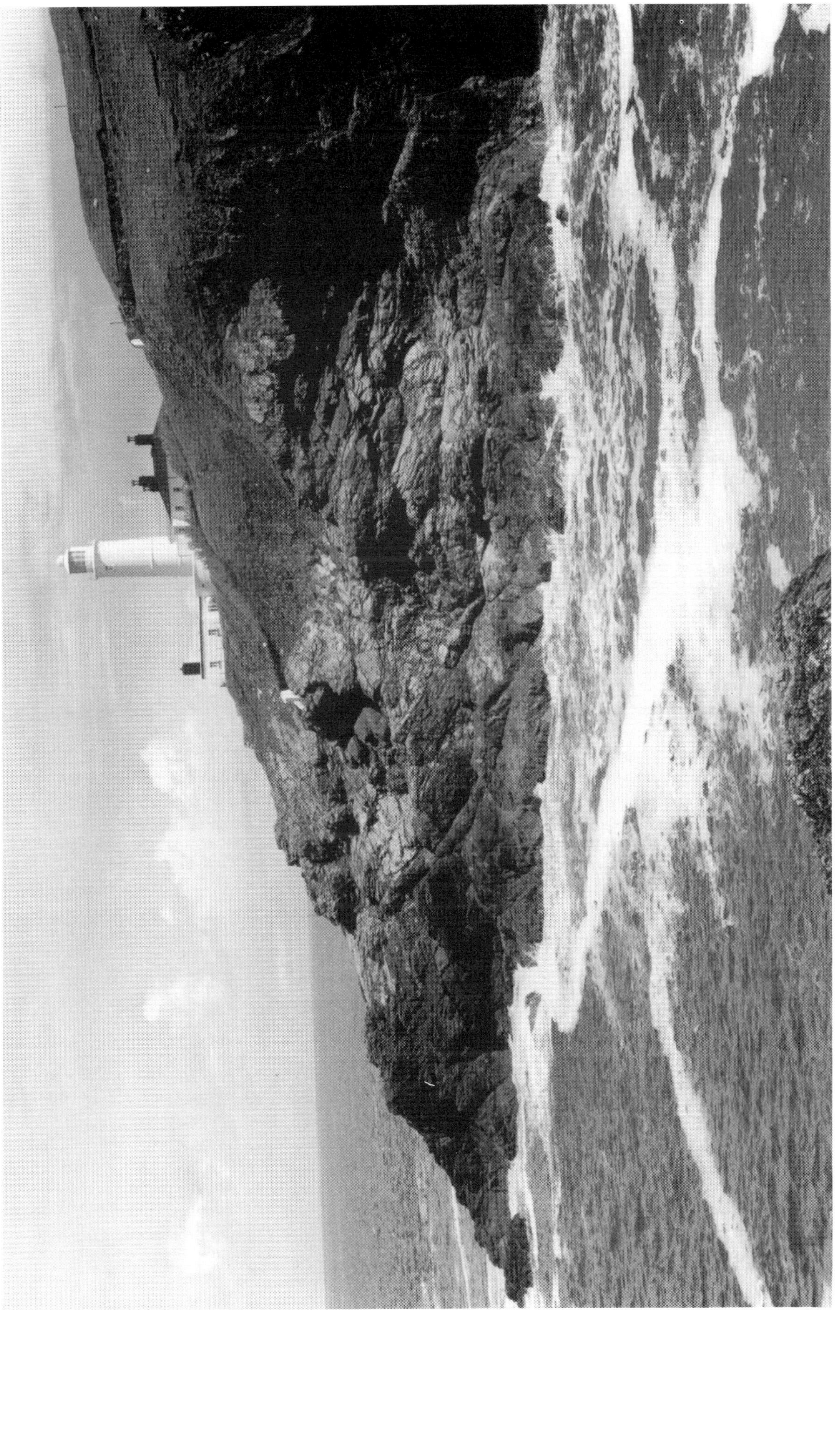

flattened and aligned chlorite-filled ovoid vesicles. Small wedge-shaped rafts of laminated and adinolized sediments are also present within the greenstone, which shows irregular cuspate margins indicative of intrusion into water-bearing, partly consolidated, sediment. Fallen blocks lying on the quarry floor exhibit pillow structure with small, surface-aligned vesicles in the rim zone and larger, circular vesicles in the pillow core. The top of the quarry also exhibits a number of thrusts that separate the fine-grained marginal facies of the greenstone from black, unaltered (non-adinolized) laminated sediments. The thrusts trace an irregular surface over the resistant greenstone, below which may be heavily quartz veined and sometimes foliated at the junction.

Interpretation

Agrell (1939) considered the Na + Si-rich metasomatizing fluids were derived from the basic igneous rock, driving all other constituents out of the adjacent sediments and precipitating albite and quartz. However, one feature not mentioned is that there is evidence for soft-sediment deformation at the margins of the body, suggesting that intrusion was into only partly consolidated sediments. These would still have retained sea-water in sediment pore spaces, which could have become an active metasomatizing fluid on heating by the intrusion. In this model, the metasomatizing agents are obtained from the sediments and not the basic intrusion. In general terms, therefore, extensive adinolization appears only to be developed adjacent to high-level massive bodies where there is evidence for intrusion into partly consolidated wet sediments. In other cases, where the intrusion took place at a deeper level and the sediments are relatively 'dry', only isochemical thermal effects are observed adjacent to massive bodies.

One of the main features illustrated by this greenstone is its irregular nature and evidence for high-level intrusion into partly consolidated sediments. The large size of the body and the degree of lateral penetration into wet sediments are probably the main reasons for the extensively developed contact zone of adjacent metasomatized sediments. The development of the adinoles and spilosites is of considerable significance as they represent one of the few classic British occurrences of such metasomatized rocks and certainly the best example in the Variscan of south-west England. However, it has been generally assumed that the metasomatic fluids were derived from the igneous body itself, whereas a more likely mechanism is that they were mobilized by the heat of the intrusion from within the wet, partly consolidated, sediments. In this context the nature of the contacts with the sediments and the general shape of the body is considered important.

Conclusions

Here is seen a large irregular, high-level intrusive 'greenstone' body (originally a dolerite) that was emplaced into and altered the still wet and unconsolidated enclosing sediments. The hot magma not only deformed the sediments but caused localized chemical changes in their bulk composition (metasomatism). Some of the less-altered or metasomatized sedimentary rocks still retain their original sedimentary laminations (termed spilosites), whereas others have nothing remaining of their original texture or minerals (termed adinoles) and are totally replaced by new phases. This is an important site at which to study the effects of thermal and chemical change caused by basic intrusions on adjacent fine-grained sediments.

B11 TREVONE BAY (SW 890762)

Highlights

This locality provides the best exposure of one of a suite of unusual, hydrous, potassic metadolerites found in north Cornwall. The pumpellyite facies of regional metamorphism is also well seen.

Introduction

The intrusive body constituting the interest of this site is located in the small cliffs and rock platforms along the east side of Porthmissan beach towards Roundhole Point.

The majority of the massive intrusive greenstones of the Variscan were originally subophitic dolerites with a primary anhydrous mineralogy. Both tholeiitic and alkaline types are present, with the latter restricted to north Cornwall and Devon. However, apparently confined to the Upper Devonian are another set of greenstones that are mineralogically and chemically distinct

Trevone Bay

and referred to as 'proterobases' or 'minverites' (after the type locality at St Minver, north of Wadebridge) in the early literature (Reid *et al.*, 1911; Dewey, 1914). This suite is characterized by having a primary hydrous assemblage and an overall alkaline composition and features (in fresh samples) high levels of large-ion-lithophile elements, such as K, Rb, Ba and the light-REE (Dewey, 1914; Floyd and Rowbotham, 1982). Relative to the commoner greenstone type of anhydrous sodic metadolerites, these rocks can be classified as hydrous potassic metadolerites.

The site illustrates an example of this greenstone type; it is situated within the grey, Upper Devonian slates of the Port Isaac Nappe. The presence of secondary prehnite and aluminous pumpellyite, along with other alteration minerals, in the greenstone indicate that the area reached the pumpellyite facies of regional metamorphism during the Variscan Orogeny (Floyd and Rowbotham, 1982).

Description

The sill-like basic body here is very variable in texture and grain size, grading from fine-grained, granular dolerite to subophitic gabbro containing large, lustrous black clinopyroxenes, that often weather out on joint surfaces. The central portion of the body may have irregular pegmatitic zones, these were initially composed of large clinopyroxene prisms (5–10 mm long), but are now invariably replaced by green chlorite. Although altered to varying degrees, petrographic relationships between the main primary phases are easily discernible, and these illustrate a crystallization history under falling temperature and increasingly hydrous conditions. Minor original olivine (now occurring as chloritized ovoids) is totally enclosed in large plates of titaniferous clinopyroxene (salite) that have been selectively replaced, in a magmatic reaction relationship, by brown kaersutite (Figure 4.29). The amphibole may also be fringed with titaniferous, dark-brown biotite of magmatic origin. Microprobe analyses of the

Figure 4.29 Photomicrograph of a hydrous dolerite showing large irregular crystal of dark, primary, kaersutitic amphibole replacing colourless clinopyroxene (bottom right); long needle-like apatite crystal traverses the amphibole unaltered (top). Trevone Bay, Cornwall. (Photo: P.A. Floyd.)

Pre-orogenic volcanics (Group B sites)

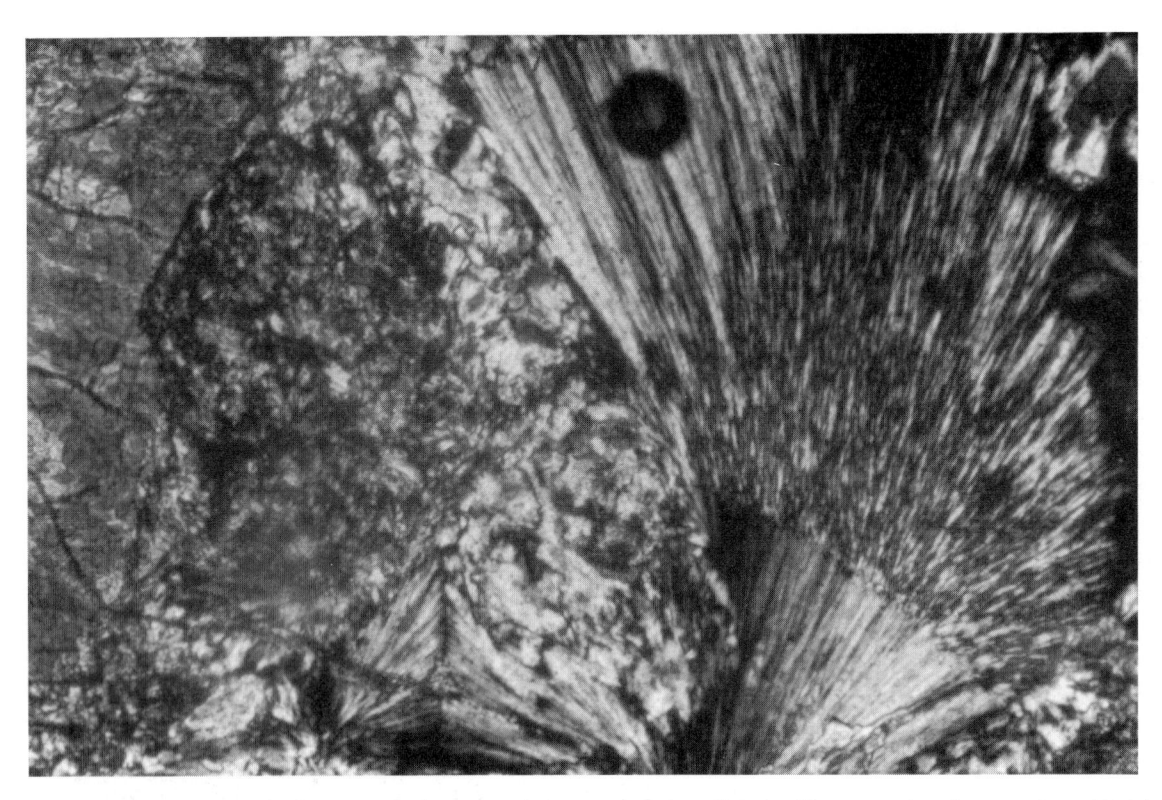

Figure 4.30 Photomicrograph of a hydrous dolerite showing the fan-like growth of secondary Al-rich pumpellyite that replaced the original plagioclase. Trevone Bay, Cornwall. (Photo: P.A. Floyd.)

primary amphibole by Floyd and Rowbotham (1982) showed that it was a kaersutite, rather than 'barkevikite' or brown hornblende as reported in the earlier literature (for example, Dewey, 1914). Mineralogically, the rock is characterized by the titaniferous nature of the major mafic phases, as well as the abundance of large, cored apatite crystals. Apart from the primary mineralogy, the metadolerite also exhibits a secondary assemblage typical of the pumpellyite facies of regional metamorphism (Floyd and Rowbotham, 1982). Together with the characteristic prehnite and colourless pumpellyite (Figure 4.30), albite, chlorite, actinolite, sphene, muscovite and epidote are also present.

Although the greenstone has a sill-like intrusive form, the upper fine-grained contact at Roundhole Point appears to be pillowed, with minor adinolization of the adjacent sediment. However, the contact zone is complex and both greenstone and sediments have been sheared and brecciated. The greenstone is often tectonized, with the development of a schistose fabric within which are resistant phacoids of coarser metadolerite, whereas the sediments show rolled adinole and rusty, pyritous sandstone phacoids set in a grey fine-grained matrix. However, where an actual contact is visible it is very irregular and often cuspate, suggesting intrusion into wet sediment.

Interpretation

This site illustrates one of the more unusual intrusive greenstone types found within the Variscan of south-west England. It is mineralogically and chemically distinct from the more normal alkaline sodic dolerite of north Cornwall and Devon, being characterized by a primary hydrous assemblage (amphibole and biotite) and enrichment in the large-ion-lithophile elements (K, Rb, Ba, light-REE). In particular, the major mafic minerals are highly titaniferous, the ore phase is ilmenite and abundant cored apatites are features typical of this potassic dolerite suite. These intrusives appear to be emplaced only in Upper Devonian strata and are particularly common in the north Cornish area of the Port Isaac Nappe. It is speculated here that some of the water available for the primary crystallization may have been sea-water that penetrated the original magma chamber which fed these high-level intrusions.

The other main feature of this site is the presence of a secondary assemblage which indicates the area underwent relatively low-grade pumpellyite-facies metamorphism during the Variscan. The secondary alteration of basic rocks can be used to determine metamorphic facies and in this respect shows that the grade here was lower than the greenschist facies of the Tintagel area just to the north-east (Primmer, 1982), although similar to south Cornwall.

Conclusions

At Trevone Bay, a massive greenstone (altered dolerite) intrusion can be seen, which still preserves evidence of the original mineral assemblage that crystallized from the magma. In addition to the usual olivine–pyroxene–plagioclase, this assemblage is distinct from other intrusive dolerites in containing primary hydrous phases (amphibole and biotite) and abundant apatite. These features indicate that the melt contained abundant water possibly derived from sea-water that leaked into the magma chamber situated at a high level in the crust. The other primary characteristic is the enrichment in many trace elements (K, Rb, Sr, Nb, Zr and P) that indicates that these rocks were derived from a distinct mantle source relative to other greenstones.

B12 CLICKER TOR QUARRY (SX 285614)

Highlights

The ultramafic rocks of this site are unique in the Variscan of south-west England in that they contain fresh olivine. They differ greatly from the ultramafic rocks of the Lizard Complex.

Introduction

The site consists of the walls of the disused and flooded elongate quarry at Lower Clicker, just south of Menheniot. The distinctive, blue-green rock has been extensively quarried for road metal since the early nineteenth century.

The Clicker Tor ultramafic body (Figure 4.31) occurs within the Middle Devonian (largely Givetian) Milepost Slate Formation (Burton and Tanner, 1986) and is closely associated with basic intrusives, as well as pillowed lavas and volcaniclastics (Ussher, 1907). Little recent work has been done on this body, which is referred to in the early Geological Survey literature as an augite picrite (e.g. Reid *et al.*, 1911), although texturally it resembles an ultramafic cumulate. Chemical analyses have been given by Parker (1970).

Description

The mapped form of ultramafic body here and its close association with gabbros (north quarry wall) suggests that it was originally intrusive or part of an intrusion, although the actual contacts with the adjacent sediments are no longer visible. The general lack of internal and marginal shearing could imply that the ultramafic rock unit was not thrust into place as a cold slab, but formed by crystal accumulation from a larger intrusive basic body after emplacement. There has been some minor internal movement, however, as seen by the development of slickensiding and the polishing of joint surfaces coated with asbestiform amphibole.

The brittle, dark-blue rock was originally composed of subrounded cumulate olivine, interstitial clinopyroxene and numerous magnetite granules (Figure 4.32). Plagioclase appears to be absent, although small, highly altered, lath-shaped areas subophitically enclosed by the pyroxene suggest its former presence in very minor amounts. A rare feature for a Variscan ultramafic rocks is the preservation of a little fresh olivine. Much of the rock has been altered or serpentinized, with olivine replaced and veined by serpentine minerals, a colourless amphibole (tremolite?) and chlorite, as well as being peppered with magnetite. A strongly pleochroic fibrous stilpnomelane is associated with or replaces the pyroxene. The overall primary composition and texture of the Clicker Tor mass suggest that it was a cumulate phase produced by the mafic fractionation of a large basic body prior to partial serpentinization.

Interpretation

The importance of this site concerns the presence of a relatively rare ultramafic unit probably

Pre-orogenic volcanics (Group B sites)

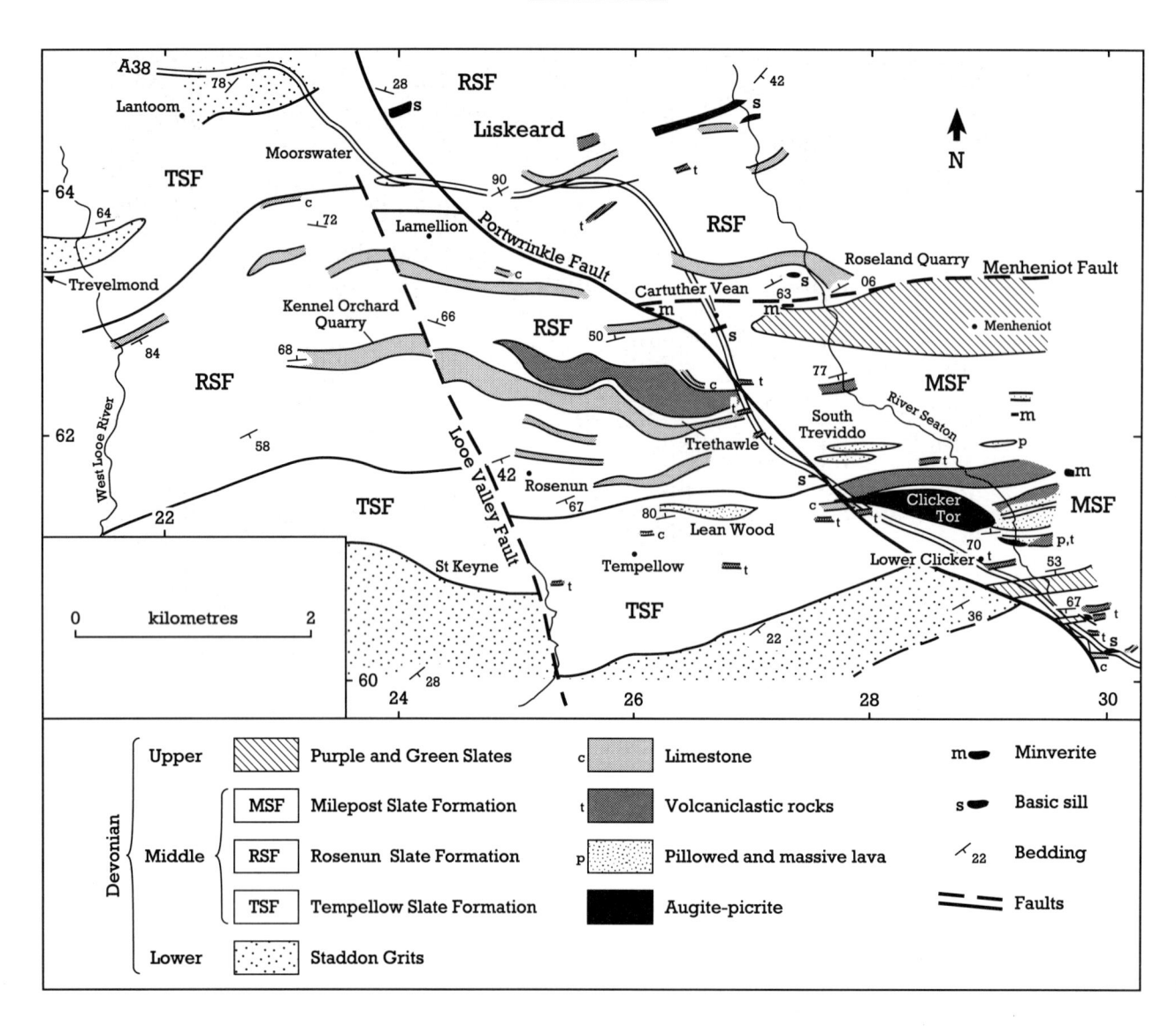

Figure 4.31 Geological map of the area to the south of Liskeard showing the location of the Clicker Tor ultramafic body (after Burton and Tanner, 1986).

derived from associated basic intrusives by mafic fractionation. Texturally, it is an olivine-rich cumulate and it has a similar chemical composition to the ultramafic cumulates of major, fractionated, stratiform basic massifs. It is probably unique for Variscan ultramafic rocks in retaining fresh olivine after partial serpentinization. It is chemically and mineralogically distinct from the ophiolitic Lizard peridotite and in no way resembles a fragmented or tectonized portion of ultramafic rocks commonly associated with ophiolites which are often high Cr–Ni harzburgites. It is also mineralogically different to the Polyphant Complex ultramafics in that it does not contain any primary hydrous phases as sometimes seen at Polyphant (as discussed below). This feature indicates that the cumulates were derived from a fractionating, anhydrous basic body different from those that crystallized primary hydrous phases.

The presence of primary olivine, a single pyroxene and possibly plagioclase indicate that the rock is probably a picritic cumulate rather than a peridotite or pyroxenite. Chemical analyses by Parker (1970), who was principally interested in the chemical and mineralogical effects due to weathering, showed that the fresh rock had high, but variable, MgO (23–28 wt. % range) and TiO_2 contents, but low Ni values (<1000 ppm) that are generally typical of stratiform-related ultramafic cumulates.

Figure 4.32 Photomicrograph of partly altered olivine crystals (with veins) and intercumulus pyroxene in the ultramafic body at Clicker Tor, Cornwall. (Photo: P.A. Floyd.)

Conclusions

Here a body of ultramafic rock associated with dolerite intrusives is emplaced into Middle Devonian (380-million-year-old) sediments and lavas. The rock is thought to have been formed within a large magma chamber as an accumulation of crystals (cumulate) which settled out during the early phases of cooling of the magma that subsequently crystallized as the adjacent dolerites. It is unique among Variscan ultramafics in showing some fresh olivine, which in similar rocks (Lizard peridotite) is invariably replaced by serpentine minerals. However, much of the body had been altered by the process of serpentinization whereby primary Mg–Fe silicates are variably replaced by serpentine together with amphibole and chlorite. The chemistry of the Clicker Tor body is distinct from that of the ultramafics at Polyphant in this area and in the Lizard Complex.

B13 POLYPHANT (SX 262822)

Highlights

This is the type locality for the assemblage of mineralogically distinctive ultramafic and mafic rocks known as the Polyphant Complex. This complex continues to play a central role in interpretations of Variscan plate tectonics in this region.

Introduction

This site, just to the north-west of Polyphant village, includes the old Polyphant Quarry and the adjacent hillside outcrops to the south of the River Inny. The quarry was the type locality for the famous 'Polyphant stone', a highly altered talcose rock that has been used for ornamental carvings since the eleventh century.

The locality is representative of part of the Polyphant Complex which is composed of an association of ultramafics, gabbros and dolerites intruded into Upper Devonian slates of the Tredorn Nappe (Stewart, 1981).

Geological mapping of the Polyphant Complex shows it is a fault-bounded, NW–SE-extending, ultramafic body with a maximum exposure width of 0.5 km (Stewart, 1981). The actual form of the body is difficult to determine geophysically because it is within the steep gravity gradient of the Bodmin Moor Granite. However, interpretations by Chandler *et al.* (1984) indicate that it is, in fact, a thin slice, about 32 m thick, and not the exposed part of a major, deep-rooted intrusion.

The mineralogy and alteration of the Polyphant ultramafic body was described by Dewey (*in* Reid *et al.*, 1911) who noted the presence of brown amphibole, and also the close association of 'proterobases' (potassic alkaline metadolerites) containing a similar hydrated primary mineralogy. In the early literature these ultramafic rocks were called picrites, although in recent times they have been referred to as peridotites and, according to Chandler and Isaac (1982), they are dominated by lherzolitic compositions.

Unlike some of the other minor ultramafic–mafic associations (see Clicker Tor) the Polyphant Complex as a whole has assumed importance in the study of the tectonic development of central south-west England. According to a pre-thrusting reconstruction for this region and the tectonic model of Isaac *et al.* (1982), the complex was immediately below the Greystone Nappe assemblage which contains Lower Carboniferous lavas and intrusive dolerites with a MORB-type chemistry (Chandler and Isaac, 1982). The apparently close geological and temporal association of Upper Devonian ultramafics and Lower Carboniferous oceanic basalts leads to the suggestion that they were part of a dismembered ophiolite which originally represented a small, short-lived ocean or marginal basin (Chandler and Isaac, 1982; Isaac, 1985). However, according to Selwood and Thomas (1986b) the facies reconstruction does not imply the presence of an ocean basin, and the interpretation of the basic rocks as MORB by Chandler and Isaac (1982), based on chemical features such as Ti–Y–Zr distributions, is open to question (Floyd, unpublished data).

Description

Although most of the ultramafic rocks in the site are serpentinized to some degree, two main lithologies can be conveniently recognized and depend on the relative degree of the secondary alteration. On fresh faces in the old quarry, a variably foliated and highly altered, blue-grey metaperidotite composed of a serpentine–chlorite–talc–carbonate assemblage can be seen. Zones rich in granular magnetite have been oxidized to brown limonite (Figure 4.33). This extreme alteration has apparently developed only at the margins of the ultramafic body due to extensive shearing as it was thrust into place. Less-altered rocks, on the slopes just south of the River Inny, show a primary assemblage of abundant olivine (invariably replaced by serpentine and tremolite) enclosed within purplish clinopyoxene or brown amphibole; the latter of which often exhibits a reaction relationship with the pyroxene (Figure 4.34). Other primary minerals include deep-red biotite, apatite and magnetite; no plagioclase appears to be present. Both the primary amphibole and biotite may be replaced by secondary tremolite and chlorite. Some of these rocks were layered cumulates, composed of mainly olivine and clinopyroxene with intercumulus hydrous fluids that subsequently crystallized amphibole and biotite. A fine banding can also be seen on some weathered surfaces. The Polyphant ultramafic rocks are mainly cumulates associated with basic intrusives that are unlikely to have formed part of a dismembered ophiolitic complex.

Interpretation

Apart from representing an example of the alteration of Variscan ultramafic rocks, the most interesting mineralogical feature at this site is the presence of primary hydrous phases. In this respect they resemble the so-called 'proterobases', or hydrous potassic suite of greenstones, in containing magmatic amphibole and biotite. The significance of the presence of similar hydrous phases could imply that the hydrous ultramafic Polyphant rocks represent the olivine + pyroxene-rich cumulates of a differentiated hydrous greenstone body. Although there are little supportive chemical data available, the Ni and Cr contents (unpublished data, Floyd, 1988) are comparable with cumulates genetically associated with layered or stratiform mafic–ultramafic bodies (Figure 4.4). The mineralogical and chemical evidence mentioned above is important in the light of some studies which interpret these rocks as part of a dismembered ophiolite complex (Chandler

Figure 4.33 Weathering of the Polyphant ultramafic body (hydrous picrite) showing a core boulder of serpentinite within a highly oxidized, degraded matrix. Polyphant, Cornwall. (Photo: P.A. Floyd.)

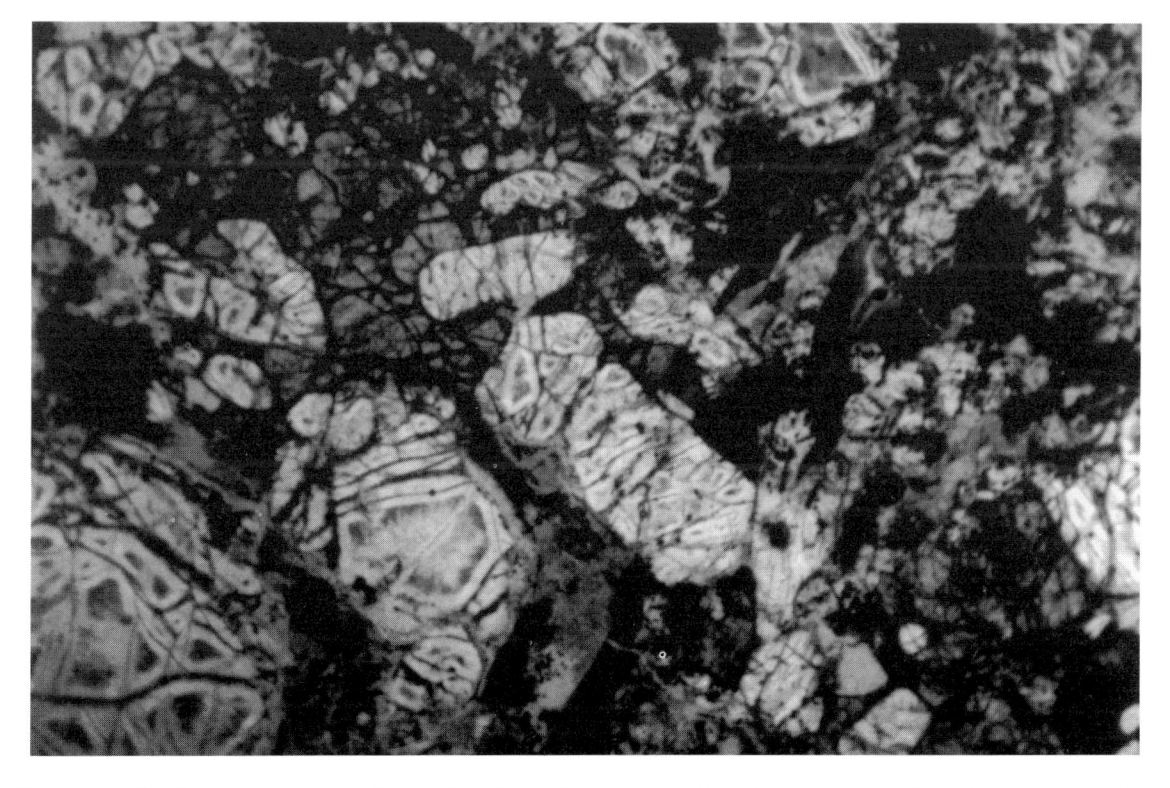

Figure 4.34 Photomicrograph of the Polyphant hydrous picrite, showing serpentinized olivine crystals, pyroxene and dark kaersutitic amphibole (top left). Polyphant, Cornwall. (Photo: P.A. Floyd.)

and Isaac, 1982). This seems unlikely on the available data and derives also from the misinterpretation of associated Lower Carboniferous lavas and intrusives as MORB – rather than intraplate alkali basalts.

Conclusions

This locality exposes an 0.5-km-wide igneous sheet which, subsequent to its emplacement, has been fragmented and sliced by low-angled thrusts. The mass has been transported on one of these thrusts into its present position and thus detached from its original roots. The rock is ultramafic in composition, a variably serpentinized peridotite, containing primary olivine, clinopyroxene, amphibole, biotite, apatite and magnetite. The presence of water-bearing primary minerals (amphibole and biotite) distinguishes it from other south-west England peridotites such as in the Lizard Complex. Associated with the peridotite are gabbros and dolerites intruded into late Devonian slates. Below the Polyphant sheet is another thrust slice containing early Carboniferous volcanic rocks and intrusions, with a chemistry possibly akin to oceanic basalts. It was suggested that the Polyphant ultramafic rocks together with the Carboniferous basalt and sediments, represented a segment of ancient oceanic crust (ophiolite), now dismembered by thrusting. However, further work on the Carboniferous volcanics and sediments has led to their interpretation as oceanic crust being questioned.

B14 TINTAGEL HEAD–BOSSINEY HAVEN (SX 047892–SX 066895)

Highlights

This is the type locality for the Tintagel Volcanic Formation, a late sequence of alkali basalt, basinal volcanics of early Carboniferous age. It is also important because the metamorphic grade is higher than elsewhere in this region.

Introduction

This scenic coastal section of north Cornwall extends from The Island at Tintagel Head via the headlands of Barras Nose and Willapark to the deep, narrow inlet of Bossiney Haven. It is historically famous for the folklore of King Arthur that is attached to ancient Tintagel Castle and its prehistoric earthworks.

The site includes examples of the early Carboniferous Tintagel Volcanic Formation that occur within the Tredorn Nappe (Figure 4.35; Selwood and Thomas, 1986a) and whose volcanic equivalents can be traced southwards to Trebarwith Strand and inland to Lewannick, north-east of Bodmin Moor (Selwood, 1961, 1971; Stewart, 1981). Exposures typical of the Tintagel Volcanic Formation just inland and along the north Cornish coast show a series of variably foliated metabasic lavas and volcaniclastics (Dearman et al., 1970; Freshney et al., 1972) whose original nature is often difficult to determine (Figure 4.36). The local Tredorn Nappe stratigraphy (Stewart, 1981) comprises a Dinantian sequence with the Tintagel Volcanic Formation sandwiched between a lower Barras Nose Formation and an upper Trambley Cove Formation, all of which are overlain by Upper Devonian slates transported into place on the Willapark Thrust. The main part of the Tintagel Volcanic Formation forms a north–south band running inland from Tintagel to Treknow and cropping out on the coast at Trebarwith Strand (Figure 4.37). This forms a single structural unit, whereas due to further thrusting and normal faulting, it is repeated on the coast around Tintagel Head to form a separate overriding slice (Freshney et al., 1972).

Recent work on the Tintagel Volcanic Formation and associated metasediments has been concerned with the grade of metamorphism and their geochemical composition and petrogenesis. Early work described the occurrence and preponderance of low-grade phyllosilicates in both pelites and volcanics, as well as chloritoid-bearing phyllites and Mn-garnet-bearing slates (Tilley, 1925; Phillips, 1928). Freshney et al. (1972) recognized the low-grade and polyphase nature of the metamorphism and, in particular, considered late biotite porphyroblasts within the Tintagel Volcanic Formation to be the effect of superimposed contact metamorphism. Mn-rich garnets have been found in the low-grade Delabole Slates, in graphitic slates associated with the Tintagel

Figure 4.35 (Opposite) Map and section of north Cornwall, showing the distribution and relationship of the major nappes (after Selwood and Thomas, 1986a). The Tintagel Volcanic Formation occurs in the Tredorn Nappe.

Legend

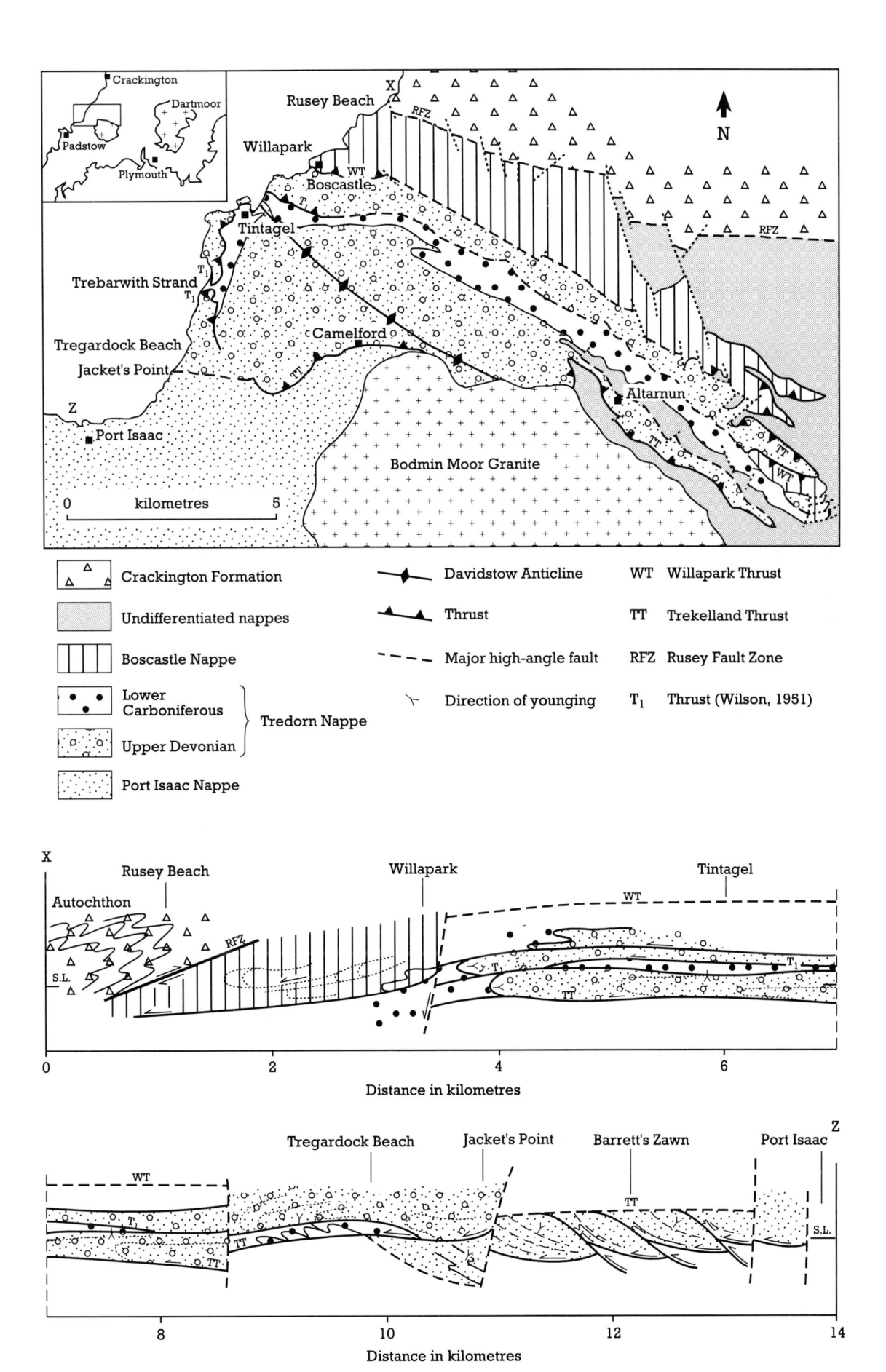

- Crackington Formation
- Undifferentiated nappes
- Boscastle Nappe
- Lower Carboniferous ⎤
- Upper Devonian ⎦ Tredorn Nappe
- Port Isaac Nappe

- Davidstow Anticline
- Thrust
- Major high-angle fault
- Direction of younging

- WT Willapark Thrust
- TT Trekelland Thrust
- RFZ Rusey Fault Zone
- T_1 Thrust (Wilson, 1951)

Pre-orogenic volcanics (Group B sites)

Figure 4.36 Contorted greenschists belonging to the Tintagel Volcanic Formation at Gullastem, north of Tintagel, Cornwall. (Photo: P.A. Floyd.)

Volcanic Formation and in zoned phosphatic nodules within the Transition Group of the local area (Phillips, 1928; Primmer, 1982; Andrews and Power, 1984). Illite crystallinity studies and pelite mineral paragenesis indicate that the Trebarwith Strand to Boscastle area is characterized by the greenschist facies relative to the lower-grade regional metamorphism ('anchizone') to the north and south (Brazier *et al.*, 1979; Primmer, 1983a, 1983b).

Description

Within the coastal site are located two tectonic slices containing various metavolcanic rocks of the Tintagel Volcanic Formation – the area around The Island, Tintagel Head and Barras Nose form a thin, upper, thrust-bounded slice capped with late Devonian slates, whereas Smith's Cliff and Gullastem to Bossiney Haven are part of the main, but lower, thrust segment of the metavolcanics exposed inland and at Trebarwith Strand (Figure 4.37).

The Island at Tintagel Head shows a thin horizon of the Tintagel Volcanic Formation sandwiched between dark, late Devonian slates, with the metavolcanics dipping gently to the north-west and forming the base of the headland to seaward (Figure 4.38). The low-lying neck of land joining The Island to the mainland is a low-angle fault zone which continues north-eastwards across Barras Nose headland. The metavolcanics within this area are heavily sheared vesicular lavas and various cleaved volcaniclastics (tuffs and possibly agglomerates). On the west side of Barras Nose the volcanics are boudinaged with resistant vesicular lava phacoids set in a foliated, calcareous greenschist matrix; they have a gradational contact with the underlying Barras Nose Formation slates. A NNW lineation is common here and lava fragments are also aligned in the same direction (Freshney and McKeown, *in* Dearman *et al.*, 1970). Magnetite octahedra may be found concentrated into lenses and thin bands within highly carbonated pale grey-green metavolcanics, showing the extreme degradation of the original volcanics.

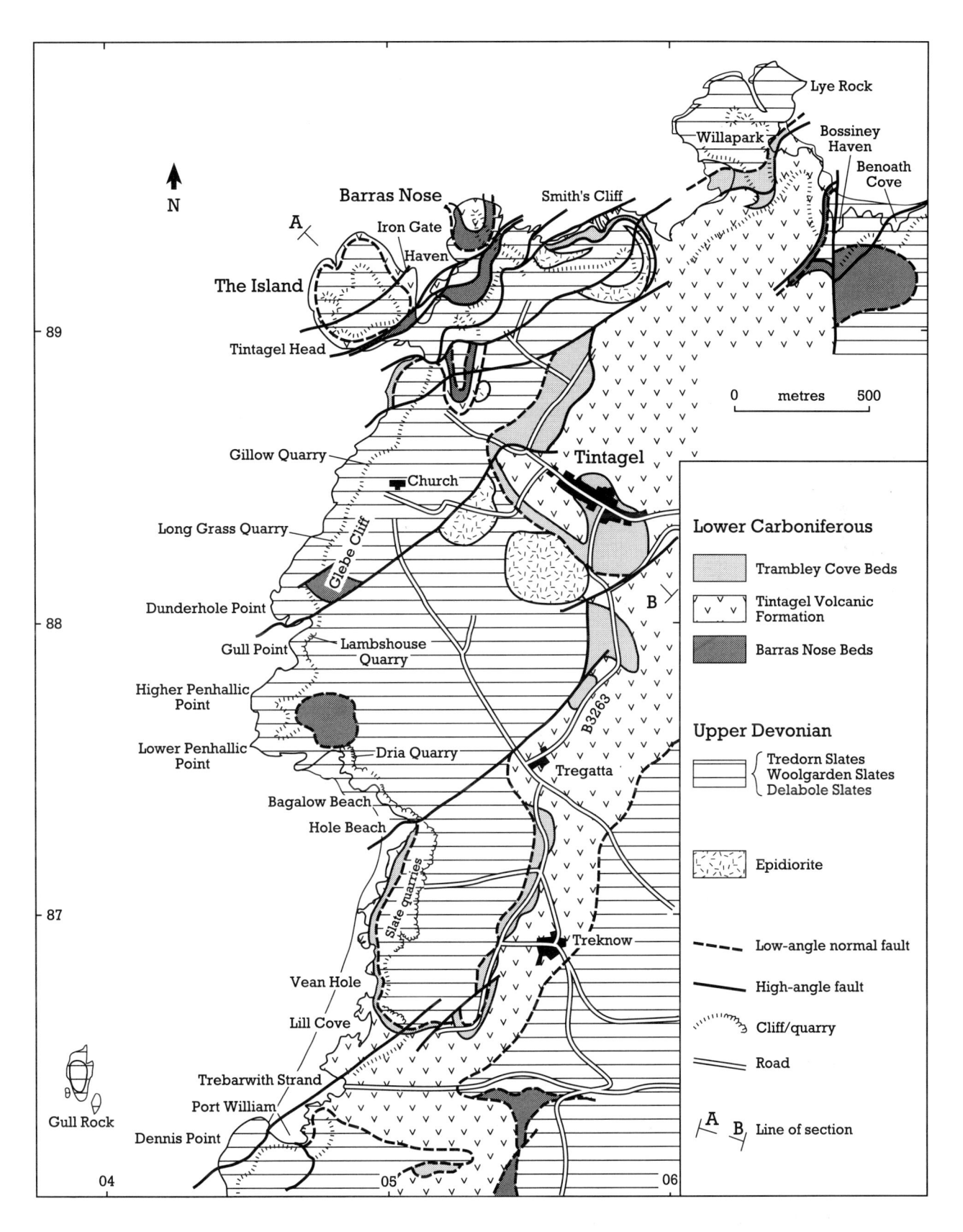

Figure 4.37 Distribution of the Tintagel Volcanic Formation between Bossiney Bay and Trebarwith Strand, north Cornwall (after Freshney and McKeown, *in* Dearman *et al.*, 1970).

Pre-orogenic volcanics (Group B sites)

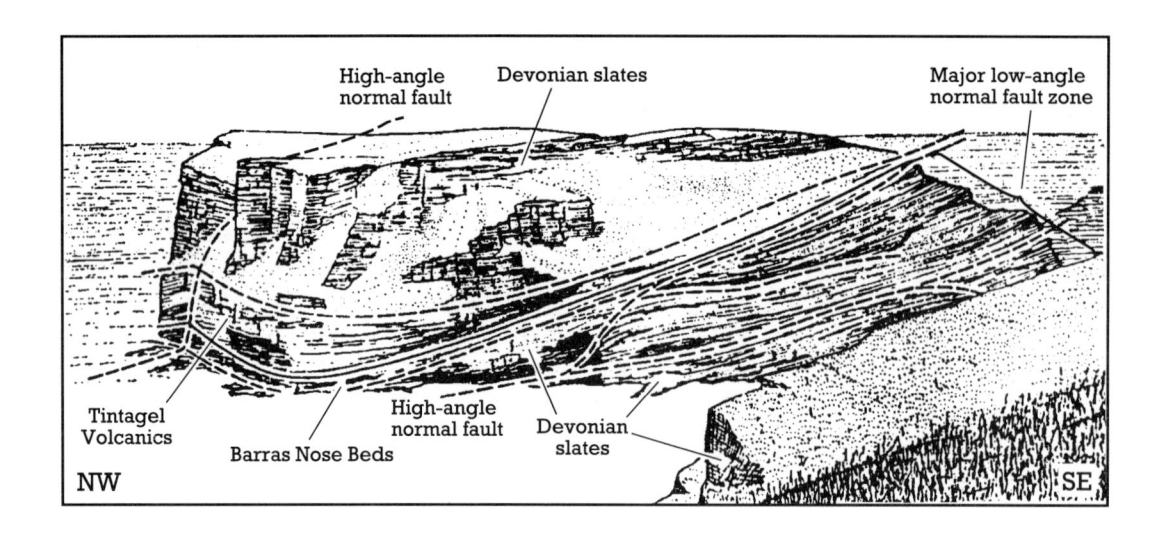

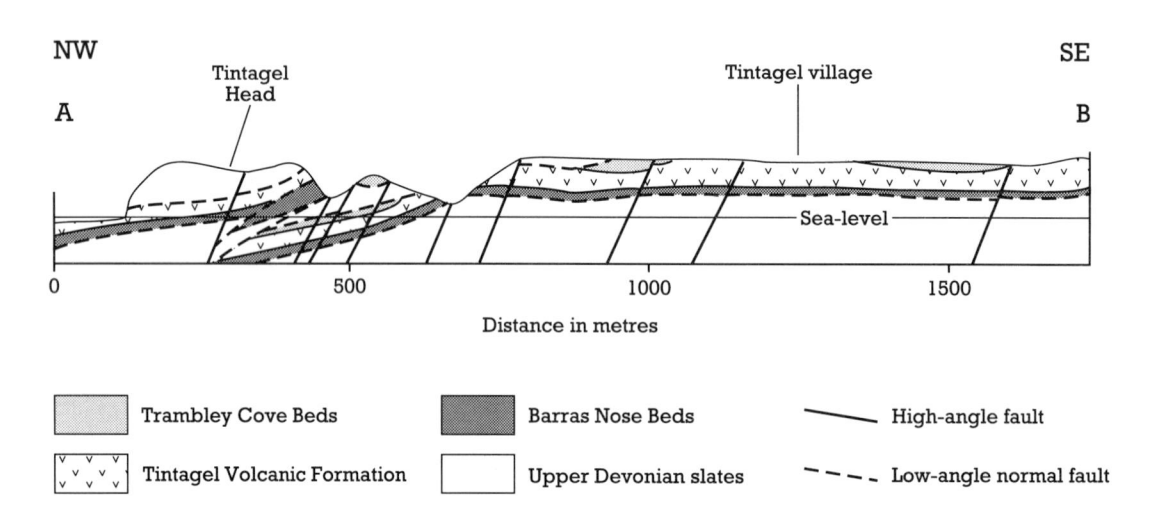

Figure 4.38 Sketch and section of Tintagel headland, north Cornwall, showing the thin upper slice of the Tintagel Volcanic Formation truncated by thrusts and cut by later normal faults (after McKeown, in Dearman *et al.*, 1970).

Farther along the coast at the cliffs of Gullastem, about 100 m of typical metavolcanic greenschists are seen, 'interbedded' with harder, sometimes boudinaged and contorted, crystal tuffs and lavas or possibly thin intrusive horizons. The typical assemblage consists of albite, chlorite, epidote, carbonate and sometimes actinolite. The greenschists may enclose lenses of vesicular lava which could represent sheared pillow lavas with the cores now forming the resistant lenses and the rims, the foliated matrix. On the rocky beach below the cliffs, are many blocks of massive and variably sheared gabbro traversed by greenschist zones. Although these blocks have been derived from an unknown locality, a sheet-like intrusive mass forms the crags near the cliff-top path at Smith's Cliff. This is often referred to as an epidiorite, but it resembles other Cornish metadolerites and metagabbros with low-grade secondary minerals replacing ophitically related primary pyroxene and plagioclase. To the south of Willapark (through Rocky Valley) to Bossiney Haven, the Tintagel Volcanic Formation is largely composed of volcaniclastics with various crystal–lithic tuffs and basalt-bearing agglomerates.

The highly foliated greenschist assemblages of the Tintagel Volcanic Formation also indicate greenschist facies (biotite zone) developed under high PO_2 equilibrium conditions (Robinson and Read, 1981; Primmer, 1982). The Tintagel Volcanic

Formation exhibits two periods of mineral growth – syntectonic with the development of the dominant albite–epidote–chlorite assemblage which forms the main foliation, and post-tectonic with the static, random overgrowth of biotite and albite porphyroblasts. In this context, Robinson and Read (1981) considered the late biotite overgrowths to be regional in origin, rather than due to contact metamorphism as suggested by Freshney *et al.* (1972). Although the greenschists and less-foliated members of the Tintagel Volcanic Formation have long been recognized as a metavolcanic suite, geochemical data indicate that they were originally a single, differentiated, comagmatic association of alkali-basalt composition with intraplate chemical characteristics, not dissimilar to their Devonian analogues in the same general area (Robinson and Sexton, 1987; Rice-Birchall and Floyd, 1988).

Interpretation

Although situated within a highly tectonized zone of north Cornwall, the Tintagel Volcanic Formation is representative of the continuing volcanic pulses associated with the deepening of basins, this time during the early Carboniferous. Chemically, these rocks are not dissimilar (in terms of incompatible-element ratios and abundances) to their late Devonian analogues (Rice-Birchall and Floyd, 1988) and form part of the same alkaline north Cornwall–Devon magmatic province (this chapter, 'Lithology and chemical variation' section).

The volcanic products appear to be more varied at this level than in the late Devonian, with the apparent extensive development of volcaniclastics. This could indicate a more explosive mode of volcanism, near shallow submarine vents, relative to the 'quiet' effusion of deep submarine lava flows. However, many of the metavolcanics have been extensively foliated (to greenschists) and it is not always possible in the field to determine if the enclosed resistant lava phacoids were originally part of a pillow-lava sequence or the larger fragments within a series of volcaniclastics of variable grain size. There is good evidence to show that the enclosed lenses are mainly quenched and holocrystalline plagioclase-phyric basalt and that the bulk chemical composition of the greenschists is that of an alkali basalt (Rice-Birchall and Floyd, 1988). Relatively homogeneous agglomerates are mentioned by Freshney *et al.* (1972) within the Tintagel Volcanic Formation, although these have also suffered shearing, and could in some cases represent low-strain zones originally composed of pillow lavas or pillow breccias.

The other feature of this site is the nature and grade of metamorphism exhibited by the metavolcanic greenschists. The textures and mineral assemblages support evidence derived from work done on the associated sediments of the region and demonstrate that the Tintagel–Boscastle area is representative of biotite-zone greenschist facies of regional metamorphism. This is significant in a regional context, as the areas to the north and south are lower grade 'anchizone' which typifies much of north Cornwall. This is approximately equivalent to the pumpellyite facies exhibited by metadolerite intrusives near Padstow (Floyd and Rowbotham, 1982). Two phases of metamorphic mineral growth are recognized in the greenschists, although the late, static growth of biotite is now considered to be regional in origin (it is in equilibrium with earlier phases), rather than due to the effects of superimposed contact metamorphism by the Variscan granites.

Conclusions

The early Carboniferous rocks (350 million years before the present) seen here originally consisted of basalt lavas and fragmentary volcanic material deposited under water. The volcanic episode was essentially a continuation of the submarine eruptions which typified the late Devonian, and the products of the two volcanic events share similarities of composition – both being composed of alkaline basalts enriched in specific trace elements. They have all been altered by later deformation and metamorphism during the Variscan mountain-building event and foliated into green schists which exhibit two phases of secondary mineral growth.

B15 BRENT TOR (SX 471804)

Highlights

This locality shows a unique example of an early Carboniferous basaltic pillow lava and hyaloclastite seamount, or mound, with a reworked volcaniclastic apron.

Pre-orogenic volcanics (Group B sites)

Figure 4.39 The conical knoll of Brent Tor is composed of Lower Carboniferous basaltic pillow lavas and hyaloclastites which formed a near-emergent seamount with a reworked volcaniclastic apron. Brent Tor, Devon. (Photo: P.A. Floyd.)

Introduction

This site covers the conical knoll of Brent Tor capped by its historic chapel (Figure 4.39). It is composed of early Carboniferous volcanics within the flysch-dominated Blackdown tectonic unit (Isaac, 1981) or Heathfield Nappe (Isaac *et al.*, 1982) and forms an isolated klippen occupying the high ground above the Greystone Nappe below. In the scheme of Selwood and Thomas (1986a) it is part of the Blackdown Nappe. The Brent Tor volcanics are generally recognized as being early Carboniferous as they rest on cherts and slates assigned to this age. However, Selwood (1974) suggested that the volcanics and associated radiolarian-bearing black slates might be late Devonian, because the sediments are lithologically similar to strata of this age north of Tavistock. The problem is compounded by the structural complexity; much of the area is composed of thin thrust slices of both Upper Devonian and Lower Carboniferous strata.

The detailed petrography of the volcanic rocks of the region to the west of Dartmoor has been described by Reid *et al.* (1911), as well as the mineralogical effects of contact metamorphism by the granite. The rocks of Brent Tor, which are outside the granite aureole, have long been recognized as the products of a volcanic eruption and, together with other magmatic rocks in the immediate vicinity, were annotated and examined in thin-section by Rutley (1878). This work was one of the first to figure hand-painted thin-section drawings of magmatic rocks from south-west England by the Geological Survey. Little modern work has been done on this interesting volcanic edifice, although a few of our unpublished chemical analyses indicate that the lavas are alkali basalts, in common with many of the extrusives and intrusives found to the west of Dartmoor.

Description

The Brent Tor volcanics are mainly composed of

coarsely bedded volcaniclastics that have a southerly dip. Near the base of the section are variably foliated, platy, light and dark-grey fine tuffs upon which rest a series of basaltic hyaloclastites and pillow-lava breccias that comprise the main outcrop. The grain size of the hyaloclastites and the distribution of pillow fragments varies considerably. The crags directly below the chapel near the main path are composed of small, elongate, dark basaltic fragments set in a greyish-green, speckled tuffaceous matrix. Occasionally, larger (0.1–0.2 m), often highly vesicular, broken fragments of pillow lava with curved surfaces and chilled margins may be present. Near the chapel and towards the top of the volcanic sequence are numerous small, red (highly oxidized), scoriaceous lava fragments set in a hyaloclastite matrix. There is also some suggestion of autobrecciation of a reddened lava flow.

On the slopes to the south of the chapel are hyaloclastites containing closely packed, large fragments (up to 0.25 m long) of dark, non-vesicular basalt interbedded with foliated tuffs containing broken, vesicular pillows. The smaller, dark, non-vesicular fragments that make up the majority of the hyaloclastite matrix are irregular in shape and were probably glassy. Further downslope from the chapel, but high in the volcanic sequence are found graded hyaloclastites and pillow breccias which represent the reworking and slumping of debris down the sides of the volcanic mound. The reworked volcanic debris probably travelled some distance away as, 4 km to the south in a small quarry near Kilworthy, is a lithic–crystal tuff with rounded, oxidized fragments of Brent Tor-type lava.

All the lava fragments are highly altered and oxidized basaltic material. In thin section some contain replaced microlitic and quenched skeletal plagioclase in an oxidized, magnetite-rich glassy matrix. Small, disrupted, often filamentous, originally glassy scoria may be bounded by internal vesicle walls and exhibit curving lines reminiscent of perlitic cracking. The finer hyaloclastite matrix appears to have been generally glassy, although much is now replaced by secondary hematite, prehnite, carbonate, sericite and sphene.

Interpretation

The importance of this early Carboniferous volcanic site is that it represents an excellent example of a hyaloclastite–pillow breccia mound and, relative to other volcanic localities (dominated by pillow lavas and minor volcaniclastic sheets), it is unique in this respect. The general shape and limited extent of the hyaloclastite deposit suggests a localized submarine eruption which built a high-level mound of largely unsorted, basaltic, glassy fragments and pillow breccias. The highly reddened or oxidized character of the lavas is unusual for Cornish volcanics and might indicate that the mound was built to a high level prior to penetration by oxygenated sea-water. The upper portion of the mound was reworked by current action and unconsolidated volcanic debris slumped downslope to form an apron. Thus, this volcanic structure is different from both late Devonian and early Carboniferous volcanic forms, which invariably exhibit pillow lavas, shallow sills or thin volcaniclastic sheets. The structure exposed probably represents the top part of a small seamount-type edifice on the floor of the basin, whereas pillow lavas in the area are typically small, domal bodies developed in a much deeper-water environment. Although its form is different from other volcanic products, it is chemically compatible in having an enriched alkali-basalt composition similar to both late Devonian and early Carboniferous volcanics in the same area.

Conclusions

This locality shows unique evidence for the presence of an early Carboniferous (around 350-million-year-old) volcanic seamount. Basaltic lavas and fragmentary volcanic material were erupted at a localized centre and built a small submarine volcano up to quite shallow depths. The rocks provide evidence consistent with eruption in a submarine setting: pillow lavas (see Chipley and Pentire Head above), glassy fragments formed by the rapid chilling of the lava by sea-water and irregular clinker-like clasts (scoria) full of voids produced by the evacuation of gases on cooling, together with cemented ashes (tuffs). The fact that they are now seen to be reddened is the result of exposure of the top of the seamount to the oxygenated upper levels of the sea and indicate its growth well above the contemporaneous sea-floor.

Pre-orogenic volcanics (Group B sites)

B16 GREYSTONE QUARRY (SX 364807)

Highlights

Early Carboniferous, high-level dolerites were here intruded into unconsolidated, fossiliferous sediments penecontemporaneously with early deformation. The presence of a series of thrust units, including exposure of the tectonically important Greystone thrust, is also regionally important.

Introduction

The site includes the working quarry cut into the valley side and situated on the west bank of the River Tamar.

Recent structural interpretations of central south-west England, between Bodmin Moor and Dartmoor, have demonstrated the importance of 'thin-skinned' thrust–nappe tectonics involving late Devonian and early Carboniferous rocks (Isaac *et al.*, 1982). Within the quarry site a number of thrust surfaces have been identified (Turner, 1982), of which the tectonically highest and youngest is the regionally significant Greystone Thrust. This thrust carries late Devonian green slates over an early Carboniferous allochthonous sequence of basinal cherts and argillites within the Greystone Nappe (Isaac *et al.*, 1982). The early Carboniferous succession is also characterized by thick doleritic intrusions and spatially associated basic lavas. Faunal dating of the enclosing sediments (Stewart, 1981) indicates a period of volcanism that lasted from latest Tournaisian to mid-Viséan times (Chandler and Isaac, 1982).

Description

The quarry site is largely composed of a number of doleritic intrusions and early Carboniferous basinal argillites and cherts. At the highest level (varying from about 50 m to 100 m OD) is the undulating Greystone Thrust surface, on top of which the late Devonian green slates are exposed around the western side of the quarry. A geological map and cross-sections of the quarry (Turner, 1982) are shown in Figure 4.40, and these illustrate a number of sub-Greystone thrusts (T2–T5) with highly undulating surfaces within dolerites and along chert horizons. The generally competent nature of the dolerites produced thrust ramping, both parallel and perpendicular to the direction of movement (Turner, 1982).

Although some of the massive dolerites appear to have an intrusive sill-like relationship with the sediments, the development of pillowy tops implies intrusion at a high-level into wet sediments. Also, detailed examination of contact relationships and enclosed sedimentary xenoliths shows that the intrusives cut deformed, early compaction and burial fabrics in the sediments. This implies that although the sediments were wet and only partly consolidated, the volcanics were intruded contemporaneously with the start of gravity-induced deformation (Isaac *et al.*, 1982; Chandler and Isaac, 1982).

The dominant intrusives are mildly altered, subophitic-textured metadolerites with variable secondary assemblages of chlorite, white mica, albite and carbonate. The larger sills, however, have suffered most alteration in the marginal zone in contact with the sediments and can be extensively carbonated. Material derived from cores in the volcanics show greater variability with vesicular (infilled with chlorite and calcite) and plagioclase-phyric (now albitized or replaced by carbonate or rarely epidote) flows and minor intrusives with flow banding. The ubiquitous and typically low-grade assemblages and apparent lack of actinolite is possibly indicative of the pumpellyite facies of regional metamorphism (pumpellyite was tentatively identified by Chandler and Isaac (1982) from the Launceston area to the north); this is supported by the clay mineralogy (illite–chlorite) of the local sediments (Grainger and Witte, 1981).

Interpretation

The environmental importance of the intrusions (as seen in the site) lies in their association with the lavas, both of which are said to have stable-element compositions akin to ocean-floor basalts (Chandler and Isaac, 1982). The particular chemical signature for the volcanics and the facies

Figure 4.40 Map and cross-sections of Greystone Quarry, showing the development of undulating thrust surfaces cutting dolerite and the transportation of Upper Devonian sediments over Lower Carboniferous volcanics by the major Greystone Thrust (after Turner, 1982).

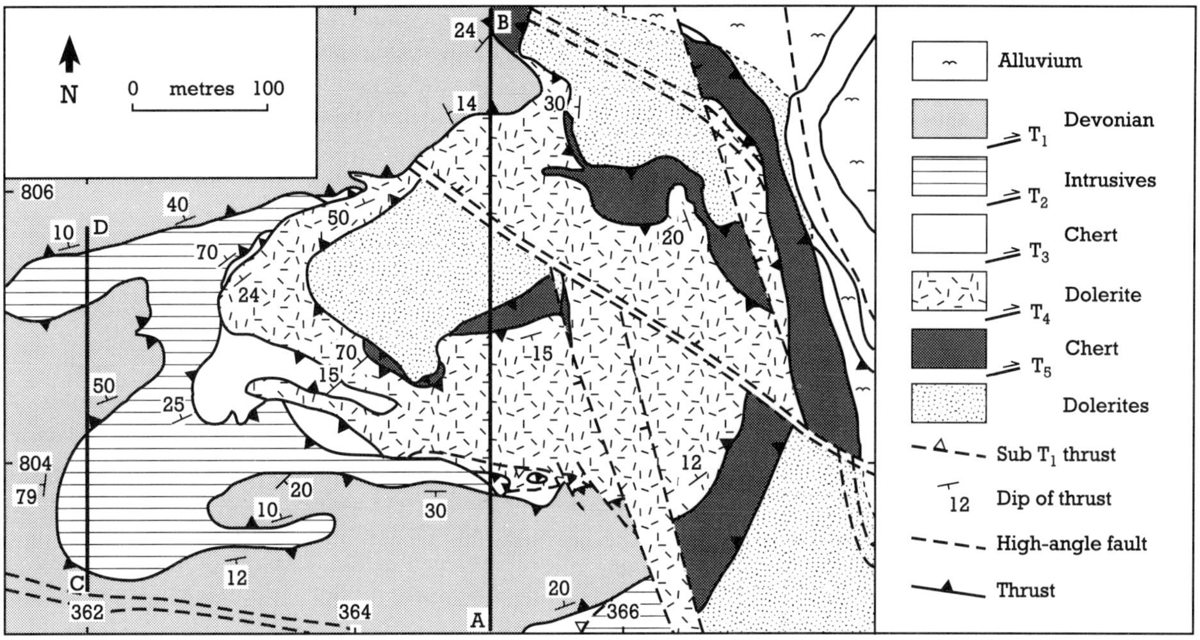

N
0 metres 100
Alluvium
Devonian
T_1
Intrusives
T_2
Chert
T_3
Dolerite
T_4
Chert
T_5
Dolerites
Sub T_1 thrust
12 Dip of thrust
High-angle fault
Thrust
A
B
C
D
806
804
362
364
366

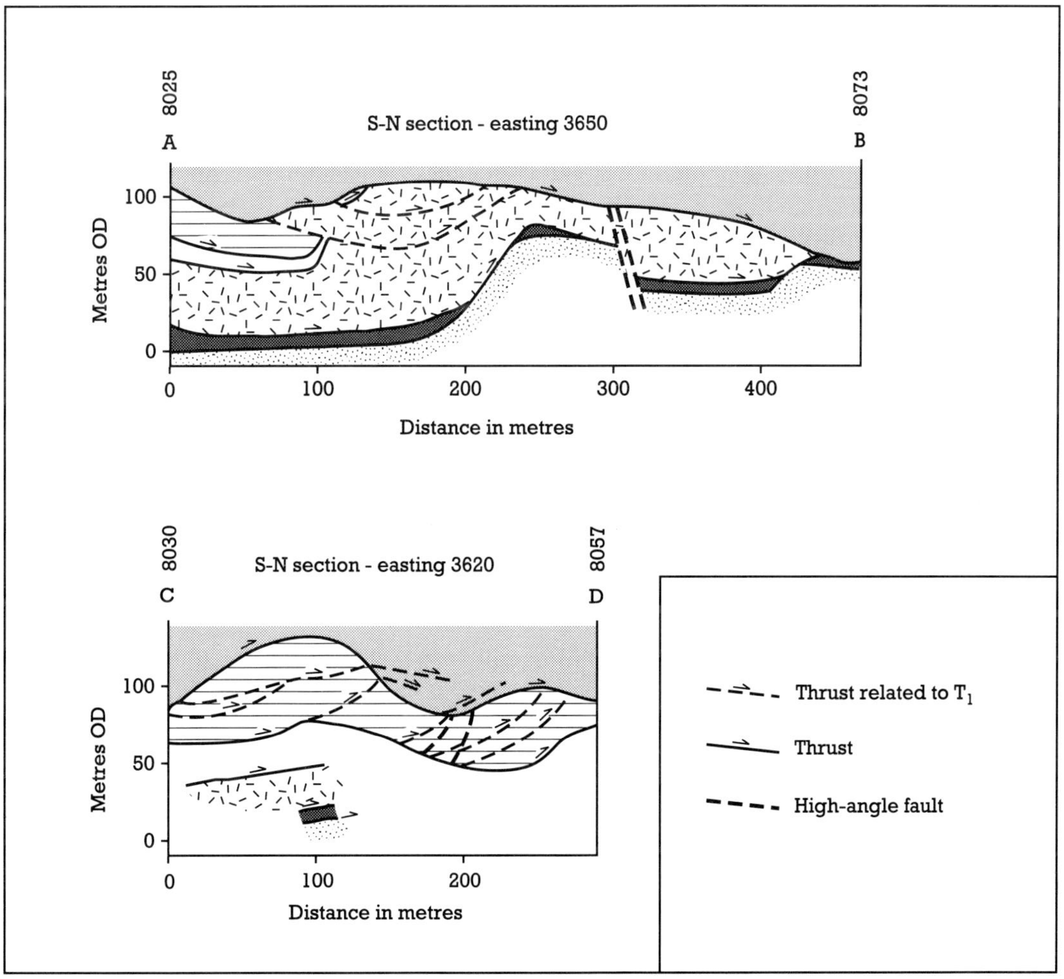

S-N section - easting 3650
A
B
8025
8073
Metres OD
Distance in metres
S-N section - easting 3620
C
D
8030
8057
Metres OD
Distance in metres
Thrust related to T_1
Thrust
High-angle fault

relations of the associated sediments have been interpreted as representing the development of a small, rifted basin, that is, one floored by basaltic ocean crust and with restricted pelagic sedimentation, but generated near to a neritic carbonate platform (Chandler and Isaac, 1982). This model was considered to be speculative by Selwood and Thomas (1986b), both in terms of facies interpretation and the significance of the volcanics which they consider were developed at the basin- slope margin. Limited chemical data on dolerites collected from the Greystone Quarry showed them to have a alkali-basalt lineage (rather than MOR-like tholeiites) and to be representative of an intraplate tectonic environment, in common with other intrusives in north Cornwall and Devon (Floyd, 1983).

The significance of this site lies in the close relationship between structural features (for example, the recognition of the Greystone Thrust), and emplacement of doleritic intrusives. Intimate contact features with the sediments not only illustrate the high-level emplacement of these volcanics into wet sediments, but that they were intruded during the commencement of deformation. Also, faunal dating of the sediments in the area has tied down the phase of magmatism to a 19-million-year period during the early Carboniferous; bracketing of magmatic events to such narrow limits is not readily achieved for other Variscan volcanic centres.

A debatable point, however, concerns the possibility that the intrusives and lavas represent basaltic ocean crust which floored a small, short-lived, rifted basin. This interpretation has implications for the tectonic regime of central southwest England during the early Carboniferous and the development of oceanic crust during this period. It has also been suggested that the volcanics and sediments were part of an ophiolite–flysch association initially underlain by the mafic–ultramafic Polyphant Complex, the latter representing the plutonic segment of the sequence (Chandler and Isaac, 1982). The presence of oceanic crust can only be considered as highly speculative since this interpretation of the variable chemistry of the volcanics is open to doubt (Chandler and Isaac, 1982); also there is the fact that the Polyphant rocks are dissimilar to ophiolitic cumulates. Limited chemical data from intrusives within the Greystone Quarry do not have a MORB signature, but are alkali basalts with intraplate chemical features characteristic of the north Cornwall–Devon magmatic province (Floyd, 1984). In tectonic terms, this indicates that the magmatic rocks were emplaced in an ensialic basin, rather than an oceanic-basin setting.

Conclusions

Here are exposed short sections of major, low-angle, fault-bounded structures which have brought together a pile of rock slices of very different age. Such thrusts may have transported the rock slices which they carry for tens of kilometres from the south. At Greystone, older Devonian rocks above one of the thrusts (the Greystone Thrust) have been carried over younger, Carboniferous rocks. The latter contained basalts and dolerites which had been injected into the clayey marine sediments, which in time lithified to become rock. They are dated as being around 350 million years old, on the basis of fossils of marine animals which the sediments contain. These date the penecontemporaneous igneous activity very precisely into a time bracket during the earliest part of the Carboniferous Period. The fact that some intrusions, which were apparently flat-lying sheet-like bodies (sills) have pillowed tops (see Gurnard's Head), suggests that they were injected as magma into marine sediments that still retained some sea-water, just as they were starting to be deformed by gravity-induced (slump) movements.

It has been suggested that both the sill-like intrusions and the lavas here have a chemical composition similar to basalts formed today at the mid-ocean ridges. This would imply that they had formed part of an early Carboniferous ocean in this region. However, this has been disputed, and it has been inferred, alternatively, that the basalts originated in a basin on the continental plate (an ensialic basin). This locality affords a unique opportunity to examine some of the major structural features of south-west England and their relationship to volcanism in the Carboniferous Period.

B17 PITTS CLEAVE QUARRY (SX 501761)

Highlights

The section here exposes a massive, uniquely columnar-jointed, multiple greenstone body

Pitts Cleave Quarry

showing an intrusive relationship to early Carboniferous sediments. The greenstone is a typical example of a mildly metamorphosed dolerite within the aureole of the Dartmoor Granite.

Introduction

The site includes all the levels and faces of the old, now disused, roadstone quarry about 3 km to the north-east of Tavistock. In all, a nearly continuous horizontal section of about 400 m is here exposed in a massive greenstone.

Within the early Carboniferous country rocks along the western margin of the Dartmoor Granite between Tavistock and Peter Tavy, many massive intrusive greenstones ranging in thickness from a few to one hundred metres can be found. As with many of the volcanics in central south-west England they are confined to the Greystone Nappe. They are predominantly lenticular sill-like intrusions emplaced in mainly Viséan (Lower Culm Measures) argillites and cherts. In the Survey Memoir (Reid et al., 1911) two types of greenstone were recognized depending on the degree of deformation suffered, such that 'albite diabases' still have their original ophitic texture preserved, whereas 'diabase schists' are foliated. Some of the intrusions have been subsequently contact metamorphosed to hornfelses by the adjacent Dartmoor Granite. The diabases are now recognized as variably degraded, low-grade regionally metamorphosed and deformed dolerites and basalts. Metamorphism within the allochthon of the Tavistock–Launceston area was generally syndeformational and low-grade, with temperatures of between 200–300°C and pressures in the range 0.3–1.0 kbar (Isaac, 1982).

The greenstone suite in the area around Peter Tavy differs in petrography and time of emplacement to similar intrusives found in the Meldon district to the north. The former suite (within which the site is situated) were intruded prior to the main deformational episode and they have been sheared and folded, whereas the Meldon group are post-deformational, but pre-granite in emplacement (Dearman and Butcher, 1959). Butcher (1958) considered that all the various greenstones were related via a single process of magmatic differentiation (such as fractional crystallization, although this view is no longer tenable on chemical grounds). However, little recent chemical work is available for the Petertavy suite of greenstones, although a few analyses from the Pitts Cleave greenstone (Parker, 1970) show it to be a moderately evolved Ti-rich alkali dolerite. The particularly high K_2O content probably reflects the effects of potassium metasomatism from the nearby Dartmoor Granite.

Description

Within the N–S-trending long face of the quarry can be seen the lenticular form of the sill, as well as large elongate rafts of hard, baked and spotted argillite (Figure 4.41) that divide the body into an upper and lower portion (Dearman and Butcher, 1959). The southern face shows spectacular curved columnar jointing in the upper sill, a feature rarely seen in the massive greenstones where most joints are more irregular (Figure 4.42). The change in attitude of the contact between the sediments and the sill from the southern end (nearly horizontal) towards the north (nearly vertical) suggests that a single limb of a recumbent fold may be exposed in the quarry (Dearman and Butcher, 1959).

The sill was originally a medium-grained, ophitic dolerite with finer-grained, often vesicular, margins. Owing to the effects of regional metamorphism it is strictly a metadolerite, with

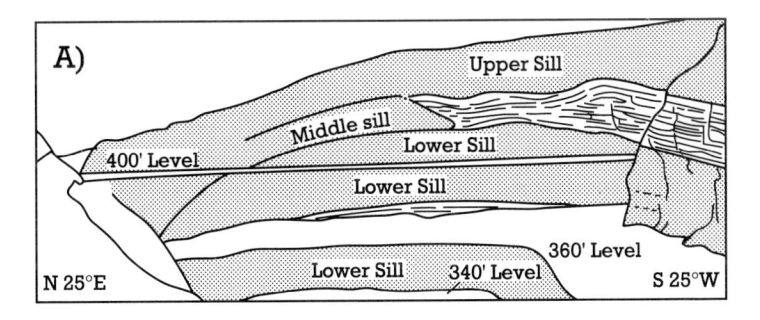

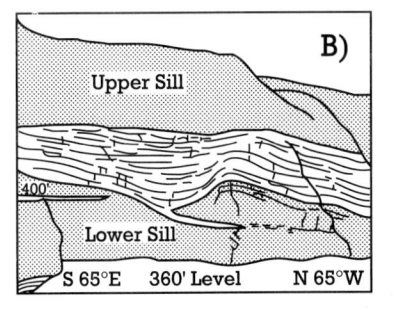

Figure 4.41 Diagrammatic sketch of intrusive dolerite bodies in the Pitts Cleave Quarry, near Tavistock (after Dearman and Butcher, 1959). A) main face (c. 230 m long) and B) southern face (c. 85 m long).

Pre-orogenic volcanics (Group B sites)

Figure 4.42 Well-developed columnar jointing in dolerite. Pitts Cleave Quarry, Tavistock, Devon. (Photo: P.A. Floyd.)

primary textures and minerals still discernible in the less-altered central part of the body. Typical alteration effects include a uralitic fringe of actinolite to primary, purple, titaniferous augite and the partial replacement of calcic plagioclase by albite, epidote and sometimes carbonate. Mafic minerals in contact zones can be completely pseudomorphed by chlorite and/or actinolite. Apatite and ilmenite are common accessories, although the latter may be skeletal and leucoxenized. The development of minor biotite after secondary chlorite is probably the only manifes-tation of the effects of contact metamorphism at a horizontal distance of c. 3 km from the granite. At the northern end of the quarry (upper 150 m level) have been found rare 'cognate inclusions' of clinopyroxene and brown amphibole megacrysts together with sporadic fragments of enclosed sediment (Butcher, 1982).

Interpretation

This site typifies a deep-level emplacement of

major greenstone bodies into the early Carboniferous of the Greystone Nappe of central south-west England. Relative to the chemically similar alkali dolerite sills intruded at high levels into late Devonian strata in north Cornwall, many of the bodies in this area were apparently emplaced at a greater depth and into consolidated sediments. Together with extrusive lavas and volcaniclastics, these sills represent one of the major effects of volcanism associated with basinal sedimentation. In the example here, intrusion appears to have been relatively deep in the sedimentary pile, with the massive dolerite sill enclosing rafts of consolidated argillites that were subsequently baked and thermally spotted. The site is unique among south-west England greenstones in exhibiting curved columnar jointing that may be related to a curved or undulating upper cooling surface.

Apart from typical secondary assemblages produced by regional metamorphism, the metadolerite also shows the initial effects of contact metamorphism by the Dartmoor Granite, with the development of biotite after regional chlorite.

Conclusions

The massive 'greenstone' intrusions seen here have been profoundly affected by later geological events. They were originally dolerite sheets that intruded muddy sediments at a deep level beneath a Carboniferous sea, around 340 million years ago. The sediments were consolidated and lithified before they were intruded and substantial fragments ('rafts') were disrupted and incorporated into the body of the penetrating magma. Sediments in direct contact with the hot magma were thermally baked and underwent localized recrystallization which produced a spotted texture. The dolerites were subsequently involved in the Variscan Orogeny, as well as being altered by the proximity of the later Dartmoor Granite that was emplaced towards the end of the Carboniferous. They were texturally and mineralogically changed by the development of new minerals (biotite, amphibole) superimposed on the original fabric in response to the hot granite body and fluids emanating from it. This site is a perfect example of an early Carboniferous mafic intrusion metamorphosed during the orogeny and then metasomatized by the Dartmoor Granite.

B18 TRUSHAM QUARRY (SX 846807)

Highlights

At this locality, a metamorphosed dolerite is seen within an unthrust unit of early Carboniferous sediments. The chemistry and mineralogy of the dolerite is distinctive, together with the effects of local attendant hydrothermal alteration.

Introduction

This site includes the old water-filled quarry (originally known as Crockham Quarry) and the southern faces of the adjacent working quarry, 1.5 km due east of Hennock.

To the east of Dartmoor, the Teign Valley sedimentary succession ranges in age from the late Devonian through to the late Carboniferous, with the stratigraphically lower units falling within the metamorphic aureole of the granite. The early Carboniferous, in particular, is characterized by the presence of numerous massive greenstone sills (within one of which the site is located), together with relatively minor basaltic lavas and various volcaniclastics. The rocks in this area are autochthonous basinal successions (Selwood and Thomas, 1986b), comprising black argillites, well-bedded cherts, dark limestones and intercalated volcanics (the Foundation Unit of Waters, 1970).

The petrography of the dominant greenstone intrusives in the Teign Valley has been described by Flett (*in* Ussher, 1913), who referred to them as ophitic diabases with local quartz-bearing and highly feldspathic variants. Rare olivine is preserved only as serpentine pseudomorphs within pyroxene, whereas plagioclase is invariably altered to albite, together with variable prehnite, sericite, chlorite and calcite. As with other south-west England greenstones, the intrusives are now low-grade metadolerites. The progressive contact metamorphic effects due to the granite were also recorded within the greenstones. Some chemical work has been done on the massive greenstones in this area by Chesher (1969) and Morton and Smith (1971), and two major-oxide analyses were presented by Ussher (1913). In general terms, the dolerites are incompatible-element-rich, and they exhibit variable, but often well-fractionated suites belonging to the alkali-basalt magma type. They also exhibit typical intraplate chemical features that are characteristic for this magmatic province of south-west England (Floyd, 1983).

Pre-orogenic volcanics (Group B sites)

Description

The Trusham Quarry site is situated within a massive metadolerite body that intrudes the basal early Carboniferous Combe Formation (Chesher, 1968). It is just outside the aureole of the Dartmoor Granite. The sill-like intrusive nature of the 65–70-m-thick dolerite can be seen, with both the upper and lower contacts exposed. The body dips at about 45°–50° to the south-east. The site, however, only includes the upper contact, which can be seen at the base of the southern wall of the old, water-filled, Crockham Quarry, but can then be projected into the working quarry and is exposed in the quarry face at the upper levels. Although the contact is undulating, it is generally concordant with the sedimentary bedding. The local sediments are cleaved, blue-black argillites and buff siltstones with occasional thick sandstone units. The argillaceous sediments adjacent to the sill have been thermally metamorphosed with the development of randomly orientated, pale spots, a few millimetres in diameter, and small chiastolite prisms. The Na-metasomatism (adinolization) of the contact sediments, often a common phenomenon of high-level sills intruded into wet sediments, has not apparently taken place.

The greenstone is a mildly metamorphosed alkali dolerite characterized by primary purple titanaugite, ilmenite and abundant large apatite crystals. Original plagioclase has invariably been replaced by secondary albite and minor chlorite (penninite) and epidote. Green granular pumpellyite was tentatively identified in some plagioclase laths. Subhedral prisms of pyroxene may also be partially replaced by chlorite, which gives the rock its greenish tinge. The dolerite is medium grained with a granular to intersertal texture, occasionally plagioclase-phyric and with pegmatitic patches developed towards the top contact. The proportion of mafic to felsic minerals may vary considerably, although a ratio of 40:60 is relatively common. At the contact with the sediments is a c. 2 m thick, chilled margin of green, chloritized basalt which is sometimes vesicular. Adjacent to oxidized joints and calcite veins, the normal green colour of the metadolerite is replaced by a pink coloration which is seen mainly in the plagioclase laths. This is due to the development of hematite produced by the oxidation of Fe in plagioclase by late hydrothermal solutions that passed through fractures in the rock.

During later tectonism, the intrusive mass suffered some internal deformation with the development of thrusts that enclose hydrothermally argillized sediment and a highly oxidized wedge of dolerite with foliated margins. Vertical shears cut, and thus post-date, the shallow thrusts.

Chemical data on the Crockham metadolerite (Chesher, 1969) show that it has an alkaline composition and is characterized by very high incompatible element contents, especially Ti, P, Y, K, Rb and Ba.

Interpretation

The dolerite of this site is an example of an early Carboniferous intrusive within the autochthon of south-west England, whereas many of the other basic bodies are restricted to the allochthon, especially within the Greystone Nappe. One of its main features is that both contacts are exposed, and it shows the local thermal effects of contact metamorphism by such intrusives. Unlike many other Variscan dolerites, the adjacent sediments are not adinolized, but have developed spots and andalusite within the baked argillites. The lack of features indicative of intrusion into wet sediments suggests that the sill was intruded at some depth below the water–sediment interface and that the sediments were relatively dry and consolidated. It is probably this feature of the sediments that restricted the development of adinoles, whose extensive development appears to be related to intrusion at a high level into waterlogged sediments, such as is seen adjacent to the Dinas Head greenstone (above – B10).

In other respects the dolerite here is not typical of intrusives within the early Carboniferous; it has a granular–intersertal texture rather than ophitic, is often highly felspathic and is relatively well fractionated and chemically evolved. Like many of the early Carboniferous intrusives of the Teign Valley it is characterized by large and abundant apatite crystals that reflect the generally high P content of these basic magmas. However, overall it still exhibits chemical features typical of the early Carboniferous alkali dolerites of north Cornwall and Devon. One further feature is the nature of late hydrothermal alteration within the dolerite that caused a fracture-related pink coloration of the plagioclase. Whether these fluids were related to the granite is unknown, as no contact-metamorphic effects are seen in the immediate vicinity.

Conclusions

At Trusham Quarry can be seen a 60–70 m thick (metamorphosed) dolerite sheet intruded into muddy sediments during the early Carboniferous. Unlike other large bodies of similar age and composition, the Trusham intrusion is here in its original setting, and has not suffered lateral transport on the back of major thrusts sometimes many kilometres from their original setting (see Greystone Quarry above). Part of the interest of this intrusion is that, on emplacement, it thermally baked the adjacent sediments, with the development of a new phase, andalusite (an anhydrous aluminium silicate). The original mineralogy of the dolerite has been partly replaced by new minerals in response to low-grade regional metamorphism and localized hydrothermal fluids during the Variscan deformation episode. Chemically, the intrusion is incompatible-element-rich and, like many north Cornwall and Devon intrusions, belongs to the alkali-basalt magma series.

B19 RYECROFT QUARRY (SX 843847)

Highlights

Ryecroft Quarry shows an intrusion which is representative of the massive alkaline greenstone sills intruded into autochthonous early Carboniferous successions. It illustrates internal lithological variation and cryptic mineral variation due to fractional crystallization.

Introduction

This site is situated in the disused and now heavily overgrown quarry set in the hillside of Ryecroft Copse opposite Ashton Mill weir and 0.6 km north-west of Lower Ashton. The quarry face has an horizontal extent of about 500 m.

The geological setting for the massive greenstone of this site is the same as for Trusham Quarry, about 4 km due south. It is characteristic of the larger intrusives found in the autochthonous basinal sediments of the early Carboniferous (Selwood and Thomas, 1986b) and the tectonic Foundation Unit of Waters (1970). In common with the other greenstones of the area, which were described by Flett (*in* Ussher, 1913) as diabases, it is now a low-grade metadolerite. The only detailed chemical work on the Ryecroft sill is that of Chesher (1969) and Morton and Smith (1971). Morton and Smith (1971) analysed various primary and secondary mineral phases; they also record that contact metasomatism is common at the contacts of the local sills, with the production of adinoles (up to 1 m wide) composed of quartz–albite–calcite assemblages and chlorite–sericite replacing andalusite.

Description

The Ryecroft Quarry greenstone is one of the thickest doleritic intrusives in this part of the early Carboniferous, varying from about 130 m to 150 m thick. It is generally sill-like in its attitude and, in the quarry, dips to the south-east, although over its full extent it has been gently folded. The lower contact is not exposed and only the uppermost one-third of the sill is readily seen in the quarry site.

The bulk of the exposed sill is a slightly metamorphosed dolerite with a melanocratic, cumulate lower portion that grades upwards through normal dolerite to quartz-bearing variants and minor, leucocratic syenitic segregations (Morton and Smith, 1971). This internal lithological variation is characteristic of large intrusive basic bodies that have undergone crystal fractionation with the production of late, acidic liquids. The primary assemblage of the dolerite was olivine, titanaugite, plagioclase, apatite, ilmenite and possibly K-feldspar, with quartz, albite and K-feldspar characterizing the late segregations. Large, sometimes cored, apatites are a characteristic feature. Common secondary minerals include chlorite (generally 20–30 modal percent), calcite, albite, prehnite and white mica. Biotite (0.5–3 modal percent), commonly nucleated on ilmenite, also appears to be secondary and may be partially replaced by chlorite. Low-temperature hydrothermal alteration is also indicated by the variable pinking of the plagioclase.

The vertical lithological differentiation exhibited by the sill is mirrored by minor cryptic variation in both primary and secondary phases (Morton and Smith, 1971). The clinopyroxenes are salitic augites with high contents of Ti, Na and Al which are characteristic of alkali basalts, and show a change in Mg/Fe ratio up through the sill

Pre-orogenic volcanics (Group B sites)

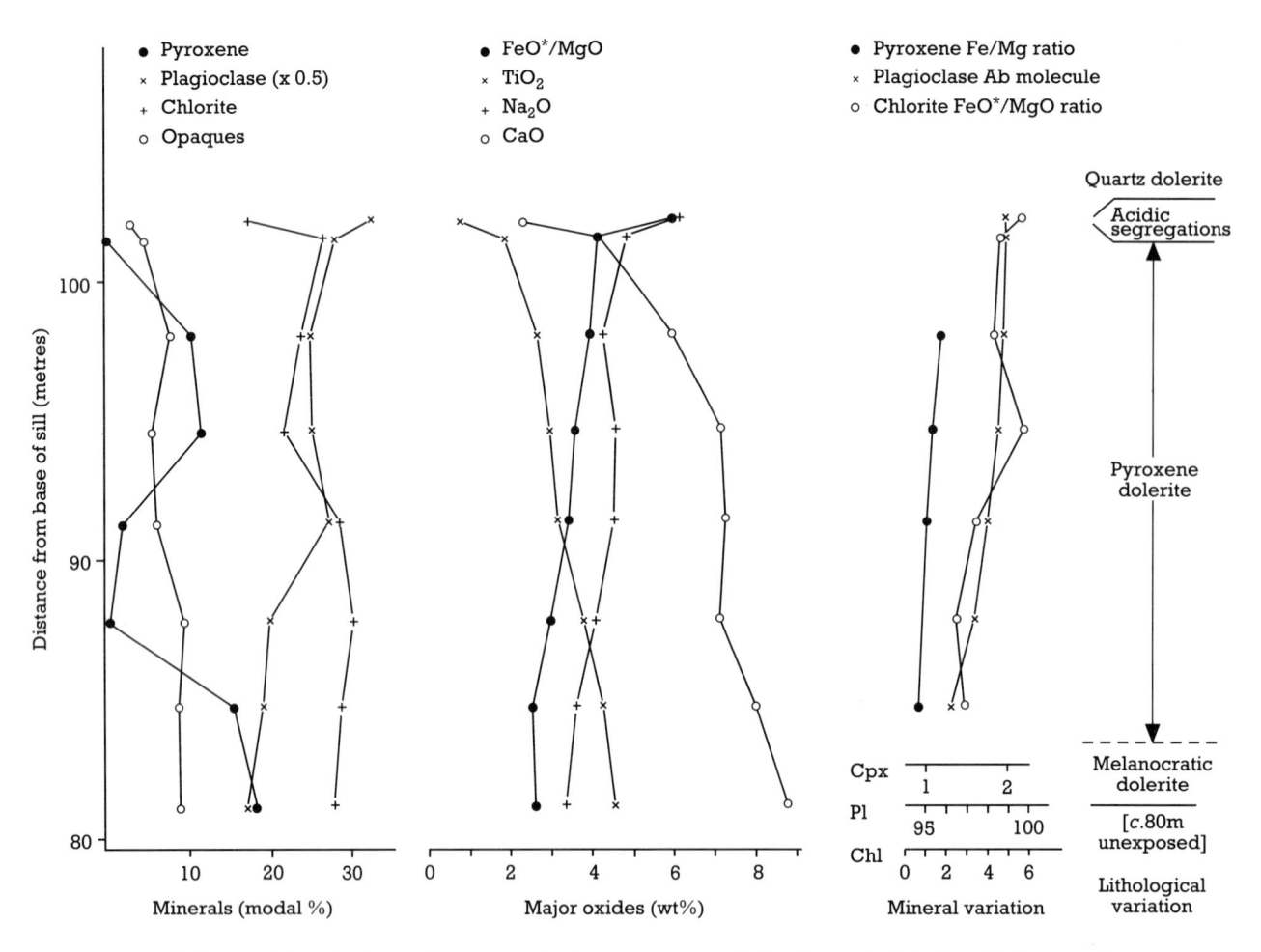

Figure 4.43 Modal and chemical variation in the upper part of the Ryecroft dolerite sill, Teign Valley, east Devon (data from Morton and Smith, 1971).

from 0.99 (near the base) to 0.84 (20 m from top). Most of the plagioclase is now almost pure albite, although the least differentiated lithologies still retain some An and Or components in their plagioclase. The chlorite exhibits a range of compositions from brunsvigite to pycnochlorite, which reflects the compositional variation of the primary host and position in the sill. The modal and chemical mineralogical variation is (shown in Figure 4.43) is characteristic of differentiated sills.

Chemically the Ryecroft sill is an alkali dolerite (Chesher, 1969; Morton and Smith, 1971), again characterized by high incompatible element abundances typical of the sodic alkali greenstone suite of north Cornwall and Devon. Major- and trace-element chemical variation mirrors the vertical lithological variation and modelling (Floyd, 1983) indicates that the sill evolved via a combination of *in situ* olivine, clinopyroxene and ilmenite fractionation.

Interpretation

The value of this site is that it is representative of some of the very thick alkaline dolerites intrusive into the early Carboniferous autochthonous succession of the Teign Valley. It has a relatively well-preserved primary texture and mineralogy, as well as exhibiting vertical lithological variation due to internal fractional crystallization during cooling. Unlike similar intrusives in north Cornwall, it generally shows a granular (rather than ophitic) texture indicating contemporaneous growth of plagioclase and clinopyroxene. It is also one of the few basic intrusive bodies in south-west England from which both primary and secondary mineral phases have been analysed. This work illustrated the control of both cryptic variation in primary minerals and the nature of the host phase in influencing the composition of secondary minerals. Thus, the composition of chlorite was seen to vary according to the degree of magmatic fractionation undergone by the replaced host phase. In general terms, this body is well evolved chemically with high incompatible-element abundances, but is characteristic of, and similar to, the Upper Devonian sodic alkali dolerites of north Cornwall and Devon that define that particular chemical province of south-west England (Chapter 3). All the secondary minerals suggest very low-grade regional metamorphic conditions (greenschist or lower) during the late Carboniferous, with the growth of biotite probably representing contact K-metasomatism by the Dartmoor Granite at a later stage. The site, however, is about 650 m outside the mapped aureole of the granite, and this indicates the distance travelled by migratory solutions beyond the extent of mineralogical reconstitution of the country rocks.

Conclusions

Here a massive, 130–150 m thick, concordant doleritic intrusion (a sill) has metasomatized the adjacent marine sediments into which it was injected. This process occurred some 340 million years ago during the early Carboniferous Period. The sill is remarkable for the evidence it presents for the operation of the fractional crystallization process during consolidation. This is now represented by progressive changes in mineral assemblages and compositions from the bottom to the top of the sill, with the most-evolved portion (near the top) being the most silica rich. The same mineral from different positions within the sill also shows chemical variation from the earliest to the latest formed. The preservation here of original and later superimposed metamorphic mineral phases, has made this site the subject of much mineralogical and chemical study.

Chapter 5

The Cornubian granite batholith (Group C sites)

List of sites

INTRODUCTION

The sites covered here are not only representative of the main megacrystic granites, but include many of the extreme variants typical of high-level, volatile-rich, calc-alkaline granites. They are shown in Figure 5.1 and are in approximately evolutionary order, the oldest granites first. Many, however exhibit late-stage or contact phenomena, so that this arrangement is fairly loose.

LIST OF SITES

Older granites, including some with xenoliths

C1 Haytor Rocks area (SX 758773)
C2 Birch Tor (SX 686814)
C3 De Lank Quarries (SX 101755)
C4 Luxulyan (Goldenpoint, Tregarden) Quarry (SW 054591)

Older granites in contact with metasediments

C5 Leusdon Common (SX 704729)
C6 Burrator Quarries (SX 549677)
C7 Rinsey Cove (Porthcew) (SW 593269)
C8 Cape Cornwall area (SW 352318)
C9 Porthmeor Cove (SW 425376)

Granite near contact between biotite- and Li-mica bearing varieties

C10 Wheal Martyn (SX 003556)
C11 Carn Grey Rock and Quarry (SX 033551)

Granites with Li-mica, topaz and fluorite

C12 Tregargus Quarry (SW 949541)

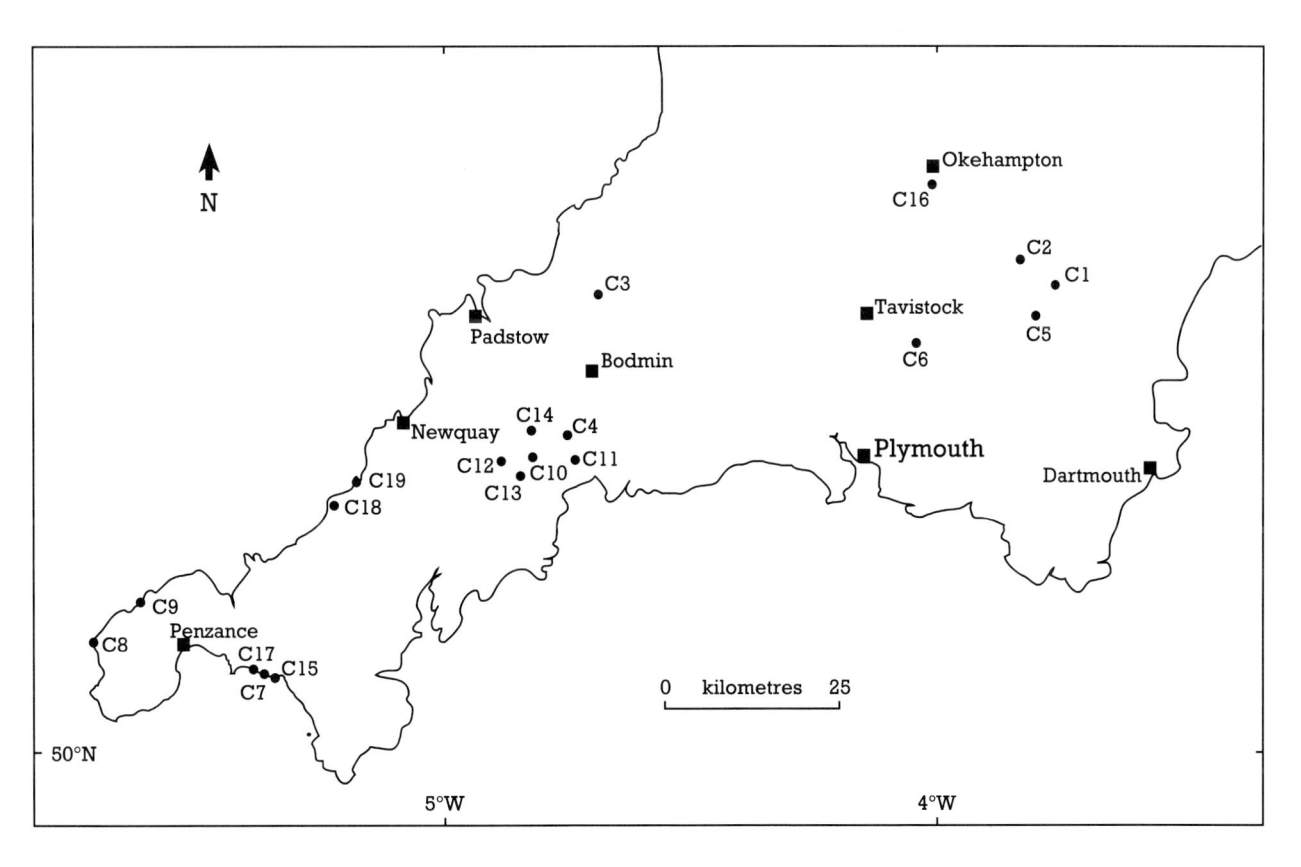

Figure 5.1 Outline map of south-west England showing the location of Group C sites.

Table 5.1 Petrographic summary of main granite types (based on Exley et al., 1983)

Type	Description	Texture	K-feldspar	Plagioclase	Quartz	Micas	Tourmaline	Other	Other names in literature
A	Basic microgranite	Medium to fine; ophitic to hypidiomorphic	(Amounts vary)	Oligoclase-andesine (amounts vary)	(Amounts vary)	Biotite predominant; some muscovite	Often present	Hornblende, apatite, zircon, ore, garnet	Basic segregations (Reid et al., 1912); Basic inclusions (Brammall and Harwood, 1923, 1926)
B	Coarse-grained megacrystic biotite granite	Medium to coarse; megacrysts 5-17 cm maximum, mean about 2 cm. Hypidiomorphic, granular	Euhedral to subhedral; microperthitic (32%)	Euhedral to subhedral. Often zoned: cores An_{25}-An_{30}, rims An_8-An_{15} (22%)	Irregular (34%)	Biotite, often in clusters (6%); muscovite (4%)	Euhedral to anhedral. Often zoned. 'Primary' (1%)	Zircon, ore, apatite, andalusite, etc. (total, 1%)	Includes: Giant or tor granite (Brammall, 1926; Brammall and Harwood, 1923, 1932) = big-feldspar granite (Edmonds et al., 1968), coarse megacrystic granite (Hawkes and Dangerfield, 1978). Also blue or quarry granite (Brammall, 1926; Brammall and Harwood, 1923, 1932) = poorly megacrystic granite (Edmonds et al., 1968), coarse megacrystic granite (mesocrystic type) (Hawkes and Dangerfield, 1978), coarse megacrystic granite (small megacryst variant) (Dangerfield and Hawkes, 1981). Also medium-grained granite (Hawkes and Dangerfield, 1978); medium granites with few megacrysts and megacrysts very rare (Dangerfield and Hawkes, 1981). Biotite-muscovite granite (Richardson, 1923; Exley, 1959). Biotite granite, equigranular biotite granite, and globular quartz granite (Hill and Manning, 1987).
C	Fine-grained biotite granite	Medium to fine, sometimes megacrystic; hypidiomorphic to aplitic	Subhedral to anhedral; sometimes microperthitic (30%)	Euhedral to subhedral. Often zoned: cores An_{10}-An_{15} (26%)	Irregular (33%)	Biotite 3%; muscovite (7%)	Euhedral to anhedral 'Primary' (1%)	Ore, andalusite, fluorite (total, <1%)	Fine granite, megacryst-rich and megacryst-poor types (Hawkes and Dangerfield, 1978; Dangerfield and Hawkes, 1981)
D	Megacrystic lithium-mica granite	Medium to coarse; megacrysts 1-8.5 cm, mean about 2 cm. Hypidiomorphic, granular	Euhedral to subhedral; microperthitic (27%)	Euhedral to subhedral. Unzoned, An_7 (26%)	Irregular; some aggregates (36%)	Lithium-mica (6%)	Euhedral to anhedral 'Primary' (4%)	Fluorite, ore, apatite, topaz (total, 0.5%)	Lithionite granite (Richardson, 1923). Early lithionite granite (Exley, 1959). Porphyritic lithionite granite (Exley and Stone, 1964). Megacrystic lithium-mica granite (Exley and Stone, 1982)
E	Equigranular lithium-mica granite	Medium-grained; hypidiomorphic, granular	Anhedral to interstitial; microperthitic (24%)	Euhedral. Unzoned, An_4 (32%)	Irregular; some aggregates (30%)	Lithium-mica (9%)	Euhedral to anhedral (1%)	Fluorite, apatite (total, 2%); topaz (3%)	Late lithionite granite (Exley, 1959). Non-porphyritic lithionite granite (Exley and Stone, 1964). Medium-grained, non-megacrystic lithium-mica granite (Hawkes and Dangerfield, 1978). Equigranular lithium-mica granite (Exley and Stone, 1982). Topaz granite (Hill and Manning, 1987)
F	Fluorite granite	Medium-grained; hypidiomorphic, granular	Sub-anhedral; microperthitic (27%)	Euhedral. Unzoned, An_4 (34%)	Irregular (30%)	Muscovite (6%)	Absent	Fluorite (2%), topaz (1%), apatite (<1%)	Gilbertite granite (Richardson, 1923)

Table 5.2 Average analyses of granites from the Cornubian batholith (after Exley et al., 1983)

	Type B Bodmin Moor	Type B Carnmenellis	Type B Geevor Mine	Type C Geevor Mine	Type C Bodmin Moor	Type D St Austell*	Type D Cligga Head	Type E Tregonning-Godolphin	Type F St Austell*	Granite porphyry Tregonning-Godolphin	Microgranite Meldon micro-granite dyke, NW Dartmoor†
	(N=10)	(N=12)	(N=7)	(N=1)	(N=3)	(N=6)	(N=2)	(N=10)	(N=5)	(N=2)	(N=1)
SiO_2	72.43	72.63	71.20	73.70	74.08	73.01	72.73	71.10	74.20	72.80	72.80
TiO_2	0.21	0.28	0.35	0.06	0.07	0.14	0.13	0.06	0.07	0.20	0.04
Al_2O_3	15.03	14.65	14.20	14.10	14.76	14.72	14.85	16.11	15.81	14.50	16.40
Fe_2O_3	0.32	0.50	0.80	0.60	0.19	0.47	0.34	0.35	0.08	1.85	0.84
FeO	1.48	1.24	1.38	0.44	0.86	0.74	0.94	0.81	0.17	1.21	-
MnO	0.04	0.05	0.03	0.03	0.03	0.03	0.03	0.07	0.01	0.05	0.09
MgO	0.44	0.48	0.60	0.05	0.18	0.14	0.33	0.09	0.08	0.26	0.05
CaO	0.84	1.12	1.12	0.56	0.44	0.44	0.41	0.59	1.31	0.28	1.28
Na_2O	3.11	3.11	2.82	2.86	2.74	3.42	3.21	3.73	4.06	0.12	2.77
K_2O	5.06	4.36	5.11	4.77	5.73	5.36	5.03	4.84	4.66	7.66	3.95
Li_2O	0.06	0.07	0.08	0.07	0.04	0.18	0.11	0.27	0.01	0.03	0.94
P_2O_5	0.25	0.18	0.24	0.32	0.25	0.33	0.15	0.50	0.46	0.26	0.48
B_2O_3	-	-	-	0.47	-	-	0.27	0.14	-	-	-
F	-	-	-	-	-	(0.38)	0.38	1.22	(1.36)	-	1.40
H_2O	1.01	-	0.73	1.38	0.88	-	1.13	-	-	-	-
Nb	-	17	30	40	-	57	-	93	81	21	67
Zr	121	137	185	40	34	(50)	65	46	(11)	94	38
Y	41	48	30	20	40	-	-	10	-	18	-
Sr	94	92	95	22	43	41	175	61	64	34	47
Rb	419	462	480	760	444	982	695	1218	615	814	2293
Ba	196	397	230	15	102	(83)	150	204	(43)	699	197
La	31	16	-	-	12	8	-	~3	-	14	15
Ce	38	-	-	-	2	34	95	36	19	68	27
U	-	-	-	-	-	-	-	19	-	20	24
Th	-	-	-	-	-	-	-	22	-	31	-
Pb	46	47	15	10	42	-	-	16	-	6	5
Ga	-	40	30	30	-	-	40	40	-	20	35
Zn	62	72	45	35	48	-	103	48	-	45	31
Ge	-	-	-	-	-	-	-	11	-	4	11
Sn	23	14	19	17	29	-	40	36	-	71	14
Cs	28	34	-	-	48	-	-	127	-	33	223
K/Rb	100	78	88	52	107	45	60	33	63	78	14

* Values in parentheses from the work of Exley (1959)
† Total Fe as Fe_2O_3
Oxide values in weight %
Trace element values in ppm

The Cornubian granite batholith (Group C sites)

Late differentiates

C13 St Mewan Beacon (SW 985534)
C14 Roche Rock (SW 991596)
C15 Megiliggar Rocks (SW 609266)
C16 Meldon Aplite Quarries (SX 567921)

Granite-porphyry (elvan) dyke

C17 Praa Sands (Folly Rocks) (SW 573280)

Mineralized granite

C18 Cameron (Beacon) Quarry (SW 704506)
C19 Cligga Head area (SW 738536)

LITHOLOGICAL AND CHEMICAL VARIATION

Summaries of the petrography and chemistry of the principal granite types of the batholith are given in Tables 5.1 and 5.2 (from Exley *et al.*, 1983) and descriptions of each type follow. Further details can be found in Exley and Stone (1982). Although widely used now, the classification of Streckeisen (1976) is not of much practical help in considering the Cornubian granites, because the differences which show their history and relationships consist of variation in the anorthite content of plagioclase, the kind of mica and the nature and amount of accessory minerals, none of which are parameters used by Streckeisen (1976). Still of limited value, but of more help, is the diagram of normative quartz–albite–orthoclase (Figure 5.2), while some important trace-element ratios for the various granite types are shown in Figure 5.3.

Type A: basic microgranite

This category is largely one of convenience as it embraces a range of types from diorite to granodiorite. They occur in nodules or rafts as enclaves within the present 'main' biotite granite to which they are precursors. Texturally, they are

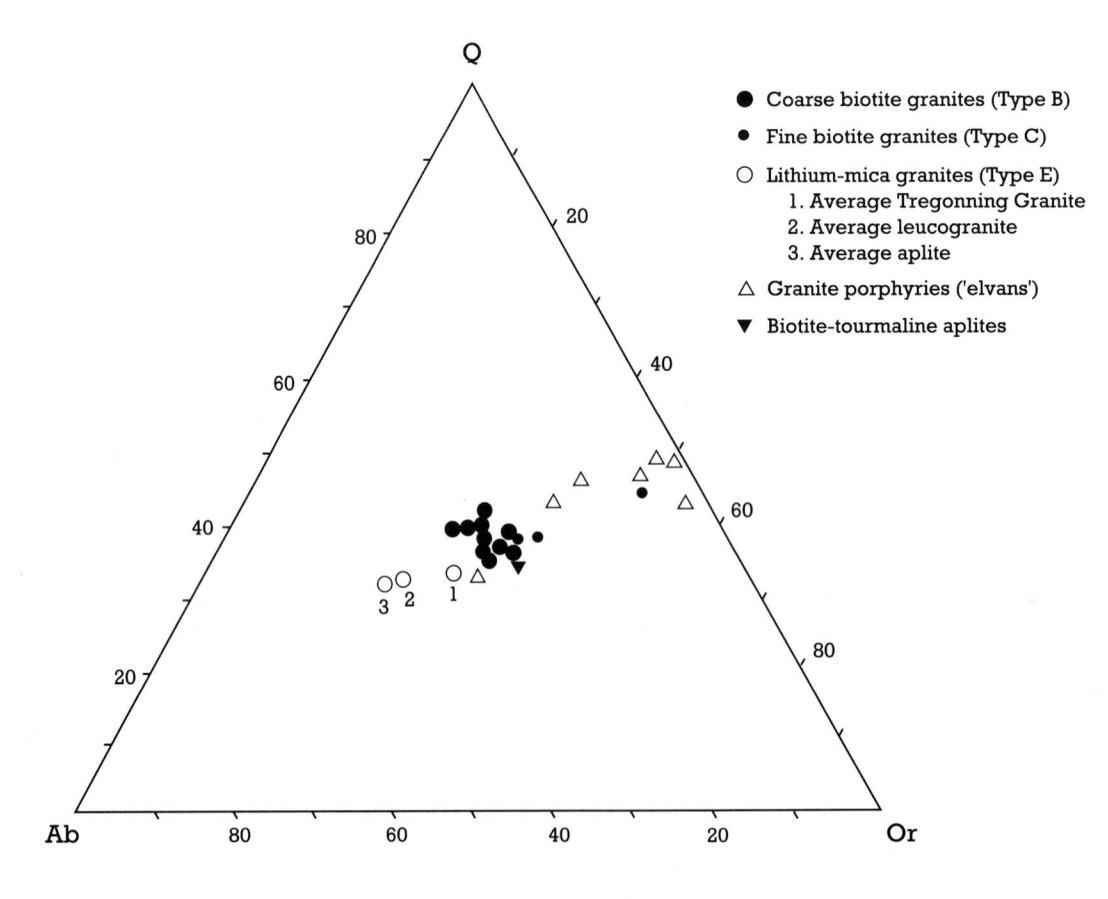

Figure 5.2 Normative quartz-albite-orthoclase (Q–Ab–Or) diagram (after Exley and Stone, 1982, Figure 23.2).

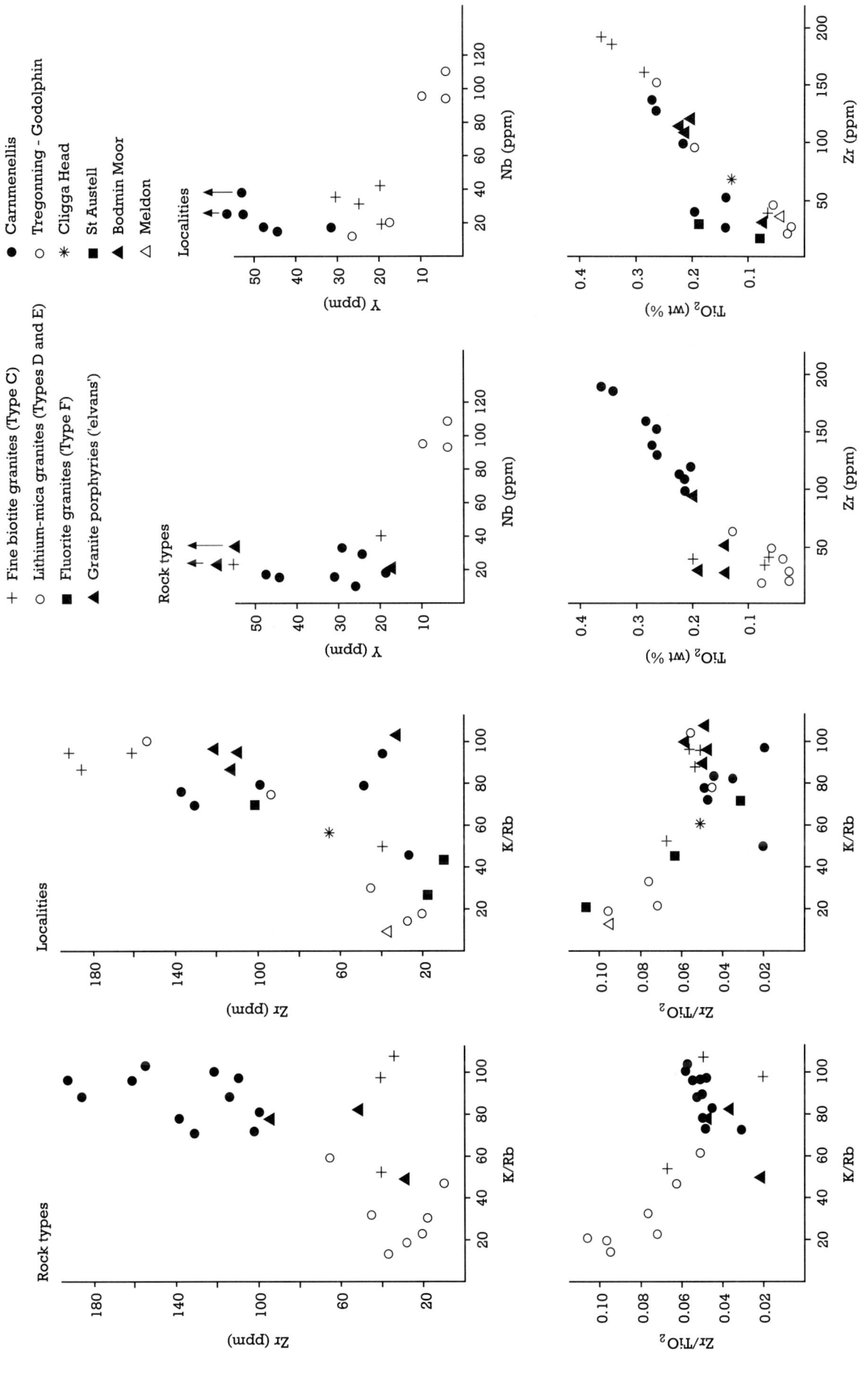

Figure 5.3 Variation diagrams for Zr–K/Rb, Zr/TiO$_2$–K/Rb, Nb–Y and Zr–TiO$_2$ in south-west England granite types and different plutons (after Exley *et al.*, 1983).

generally hypidiomorphic, although some are ophitic or subophitic, and their rather basic chemistry is reflected predominantly in calcic oligoclase or andesine, abundant biotite and sometimes hornblende. There are the replacement relations such as intergrowths and lamellae of secondary minerals, as well as aggregations and enlargement of crystals, and new generations of feldspar, mica and quartz which are the usual textural and mineral modifications resulting from degrees of feldspathization and granitization.

Type B: coarse, megacrystic biotite granite

Coarse megacrystic biotite granite is the 'main' granite of the batholith, appearing in all the outcrops, and it is estimated by Hawkes and Dangerfield (1978) to compose 90% of the present exposed area. In detail, there is a good deal of variation in both texture and composition, and Dangerfield and Hawkes (1981) have distinguished both a 'small megacryst variant' and a 'poorly megacrystic' type in addition to the coarsely megacrystic granite. Although these variants are present in all the major outcrops to some extent, the Bodmin Moor and Carnmenellis intrusions are notably different from the rest in being composed predominantly of the small megacryst variant.

There is also much variation in the size of the megacrysts, which are entirely microclinic, perthitic K-feldspar and which are usually aligned in a manner which suggests magmatic flow (see below, however). Potassium-feldspar occurs also in the groundmass, where it may be either in the monoclinic or triclinic state. In the case of Bodmin Moor, Edmondson (1970) has demonstrated a systematic regional distribution of the two forms in which the microcline is found in a wide zone almost surrounding an area in the central and southern parts of the outcrop from which it is absent. He argues that the secondary growth which produced megacrysts was contemporaneous with the exsolution of albite, and that both these processes and that of inversion from the monoclinic to triclinic state were catalysed by volatiles such as F and Cl. Apparently these late-stage structural readjustments were able to persist longer in the central and southern areas. Plagioclase is usually twinned on several laws, of which Carlsbad, albite and pericline are the most common, and is always strongly zoned, normally continuously as indicated in Table 5.1. It has long been known that the biotite of the Cornubian granites contains Li — for example Reid et al. (1910) quote 1.71% Li_2O in a biotite from the Land's End Granite, and Brammall and Harwood (1932) 0.32% in a biotite from Dartmoor — and recent work by Stone et al. (1988) has shown that the mineral is more properly a lithian siderophyllite and that it grades through ferroan zinnwaldite into true zinnwaldite. Muscovite, which is subordinate to biotite, is often secondary (Charoy, 1986; Jefferies, 1988), and the presence of tourmaline, at least some of which is magmatic, is noteworthy.

Points to be noted in the chemistry of the granites include enrichment in Li, P, B, F, Rb and light REE (Lees et al., 1978; Alderton et al., 1980; Charoy, 1986; Jefferies, 1985b, 1988), and depletion in Fe, Ca, Zr, Ba, etc. relative to average granite. These features are indicative of highly evolved, high-level granites in which incompatible elements are unusually enriched. Correlation coefficients for chemical elements in the main granites show two significant groupings (Table 5.3). Both biotite granites (Types B and C) belong to a chemical association characterized by strongly correlated Ti + Mg + Ca + Fe (Stone, 1975); Stone and Exley, 1978; Exley et al., 1983), whereas the association of Al + Mn + P is characteristic of Li-mica granites.

Detailed studies of the texture of the granite (Stone, 1979) suggest that the oldest minerals are andalusite (inclusions in biotite) and biotite followed by plagioclase, then quartz and fourthly K-feldspar, but there has been much of what Charoy (1986) describes as 're-equilibrium' resulting in recrystallization, as well as autometasomatism, which makes the textural history very complex. Thus the K-feldspar megacrysts contain earlier minerals, often in zones, which, among other evidence, such as the envelopment of later minerals, demonstrates a late metasomatic growth. The alignment noted above is, therefore, that of an earlier generation, usually of K-feldspar but sometimes of plagioclase, which acted as nuclei for the crystals now present. The development of the megacrysts, and their exsolution into perthite, probably took place as the magma itself cooled sufficiently to exsolve its water and it is usual to find a highly megacrystic facies near the walls and roofs of the plutons, the potassic aqueous phase having migrated to the cooler regions. In this respect it is of interest that the granite from more than 1800 m deep in the Carnmenellis pluton is

Table 5.3 Pearson product moment correlation coefficients for major and minor elements (after Exley and Stone, 1982, Table 23.1)

Ti	Al	Fe^{III}	Fe^{II}	Mn	Mg	Ca	Na	K	P	
-0.33	-0.61	-0.28	-0.36	-0.26	-0.29	-0.26	-0.05	+0.11	-0.45	Si
	-0.35	+0.40	+0.77*	-0.27	+0.90*	+0.75*	-0.21	+0.07	-0.48	Ti
		-0.16	-0.23	+0.43	-0.33	-0.24	+0.33	-0.28	+0.72*	Al
			+0.21	-0.16	+0.34	-0.13	-0.69*	+0.61*	-0.20	Fe^{III}
				-0.04	+0.76*	+0.60*	-0.01	-0.04	-0.09	Fe^{II}
					-0.34	-0.20	+0.23	-0.29	+0.61*	Mn
						+0.67*	-0.11	+0.02	-0.46	Mg
							+0.24	-0.37	-0.40	Ca
								-0.92	+0.33	Na
									-0.21	K

* Based upon 26 'average' analyses used and described in Stone and Exley (1978). Highly significant correlations have asterisks: these are values for which the Null hypothesis is rejected at the 0.01 significance level. Boxed values are those belonging to the femic element association.

equigranular (Bromley and Holl, 1986).

Some biotite and andalusite, as noted above, are regarded by Stone (1979) from the inclusion principle as 'restite' minerals derived from the assimilation of sediments. Jefferies (1984, 1985a, 1988), however, has argued that from the presence of radioactive mineral inclusions within it, much biotite must itself be magmatic. Indeed, he calculates that the total of assimilated material in the Carnmenellis Granite is only about 4%.

Type C: Fine biotite granite

Relatively small outcrops of this type occur in all the principal exposures, the largest being in the north of the Land's End outcrop (Dangerfield and Hawkes, 1981; Booth and Exley, 1987). Dangerfield and Hawkes (1981) have subdivided it, like Type B, into two textural groups, namely, megacryst rich and megacryst poor and, as with Type B, the Bodmin Moor and Carnmenellis plutons differ from the rest, this time in containing the megacryst-poor variant virtually to the exclusion of the megacryst-rich type.

In some cases, these granites do not crop out and their presence is deduced from abundant boulders on the surface. Equally, most of their contacts are not exposed. This lack of exposure makes their relationships with the surrounding rocks and their relative ages a matter of some speculation. In particular, it is unclear whether they are enclaves or separate intrusions. Unfortunately, because of the small areas involved, maps are not always useful.

The mineralogy is almost identical with that of Type B, but there is more quartz, the cores of the plagioclase crystals are less calcic and biotite is subordinate to muscovite.

Conspicuous among chemical differences from Type B are, on average, higher SiO_2, K_2O/Na_2O and Rb and lower FeO, MgO and CaO (corresponding with the changed mineralogy), Zr, Sr and Ba. In detail, however, there is considerable variation in the chemistry, which suggests that not all these rocks have the same origin, and some, in fact, have been interpreted as granitized sedimentary rafts (Tammemagi and Smith, 1975; Hawkes, 1982). Even those of unquestioned igneous origin do not seem to have been derived by any straightforward process from Type-B granite.

Type D: Megacrystic Li-mica granite

This variety is unique to the central and extreme western parts of the St Austell outcrop and, as can be seen from Tables 5.1 and 5.2, it occupies an intermediate position between Types B and E in both modal and chemical composition. Its texture is generally like that of Type B, that is, coarse-grained and megacrystic, although it has considerable variation in the size and frequency of megacrysts and the nature of the matrix. It is much the most heavily kaolinized area in Cornwall and for this reason it has only recently been realized that some of this textural variation reflects the presence of several distinct intrusions (Manning and Exley, 1984; Hill and Manning, 1987).

The chief changes in mineral content, relative to Type B, are the presence of less K-feldspar, more plagioclase (which is substantially more albitic), zinnwaldite in place of biotite, and the appearance of both topaz and fluorite in small amounts. These changes are the manifestation of a reduction in the K_2O/Na_2O ratio, a substantial increase in Li_2O, a reduction in MgO, redistribution of Fe and Al, and an increased importance of B and F, all brought about by the intrusion of the Type-E granite which Type D surrounds. This aspect is discussed further below and in the site descriptions which follow.

Type E: Equigranular Li-mica granite

Type E is the representative of the second chief intrusive phase, dated at about 270 Ma BP. It is exposed in the St Austell (and Castle-an-Dinas), Tregonning and St Michael's Mount outcrops, but it may underlie much of the exposed biotite granite (Manning and Exley, 1984; Bristow *et al.*, in press).

A characteristic feature of its contact with the rocks that it has intruded is the development of a roof complex of banded aplite, pegmatite and leucogranite; these are a consequence of the high volatile content of Type-E magma which is believed to have been derived at depth by a complex differentiation and reaction process. This is discussed more fully below and in the appropriate site descriptions.

Like Type D, it is distinguished at once by the assemblage albite–zinnwaldite–topaz and, as Table 5.1 shows, it has a much higher proportion of plagioclase to K-feldspar than Type B. The plagioclase is nearly pure albite, and this rock type has an average of 9% zinnwaldite, no other mica and 3% topaz. Its texture is aphyric/equigranular. This mineralogy is highly unusual among British granites, but is well known in tin-bearing granites elsewhere, such as the Krušné hory Mountains (Erzgebirge) of Bohemia in the Variscan Saxothuringian Zone (Štemprok, 1986).

Chemically, the rock is distinguished by a relatively low K_2O/Na_2O ratio, high Li_2O, P_2O_5 and Ba and very high Rb. Along with Types D and F, it is characterized by the strongly correlated association of Al + Mn + P which marks it as fundamentally different from the biotite granites (Stone, 1975; Stone and Exley, 1978; Exley *et al.*, 1983).

Type F: Fluorite granite

This rock, like Type D, has so far been found only in the west-central part of the St Austell pluton east of the Fal Valley, where it occurs in pockets several hundreds of metres across within Type E. Texturally, it is identical with Type E, that is non-megacrystic and equigranular, but its mineral content differs sharply, with no iron-bearing minerals such as zinnwaldite and tourmaline being found. Correspondingly, its total Fe content is much lower. On the other hand, the F content is high (Table 5.2) and this is shown not only in the presence of topaz and fluorapatite, but also in 2% of purple fluorite which gives the rock its easily recognized character in hand specimen.

Like Type D, fluorite granite is now thought to have originated from the reactions induced by the emplacement of Type E which are discussed in the relevant site descriptions.

Other varieties

Many minor (in volumetric terms) granitic rock types are also found in south-west England, ranging from the granite porphyry (elvan) dykes already mentioned, through microgranite sheets, quartz–topaz and quartz–tourmaline rocks, to aplites and pegmatites. Most of these are highly specialized and are significant indicators of the

Petrogenesis

evolution and chronology of the batholith. Details of their petrology are discussed in the various site descriptions.

PETROGENESIS

The granitic rock Geological Conservation Review sites described in this volume have been chosen to illustrate both the variety of types and their evolutionary history. The following section describes the various petrogenetic hypotheses developed over the years and how they relate to the rocks themselves.

Pre-1950 and the Dartmoor model

The questions of the linking of the granite outcrops at depth and the origins of the magmas from which they formed are inextricably entwined, but before the 1950s little appears to have been written about these matters. Two early and notable exceptions are De la Beche, the first Director of the Geological Survey, and W. A. E. Ussher. In his celebrated report of 1839, De la Beche was quite clear that a single granite mass underlay the individual 'protrusions' which he considered to be of the same age. Ussher (1892), in a fascinating paper full of detailed arguments about the relations between stratigraphy, structure and origins of both igneous rocks and sediments, took a similar view but also concluded that the granite '... resulted from the metamorphism of pre-existing rock of Pre-Devonian age ...'.

Despite these contributions and a considerable body of knowledge derived from mining activities, the petrologists of the Survey, who wrote most of the pre-1920s accounts, confined themselves chiefly to detailed petrographical descriptions and notes on the field relationships of the different granite facies. We have, for example, Barrow (*in* Reid *et al.*, 1910) observing that there occur within the Bodmin Moor outcrop masses of finer granite 'which are clearly intrusive', and Flett and Dewey (*in* Reid *et al.*, 1912) describing the 'basic segregations' in the Dartmoor Granite. There is, however, practically no discussion of the origins of the magmas or much about the derivation of different granite varieties: argument being limited to such remarks as 'the structure is rather characteristic of the residual material of granite magmas that is forced up as veins through a large coherent intrusive mass' (Flett and Dewey (*in* Reid *et al.*, 1912) referring to micropegmatite in fine-grained granite on Dartmoor). The same authors mention that the fine granites 'belong undoubtedly to the same magma' (as the coarse granites) and summarize the relationship by the comment that 'Though of one geological date, the intrusion is made up of various injections, some finer-grained than others, but in no case do the later veins and masses appear to have been intruded after the earlier mass had cooled. They therefore blend and pass into each other at the margins in a very characteristic way'. Basic segregations are ascribed by these authors to the 'crystallization of scattered lumps of small size, before the crystallization of the rest of the magma'. Elvan dykes were accepted by the Survey officers as late intrusions, but again their origins were not discussed beyond suggestions that they too, derived from the same magma (for example, Flett, *in* Ussher *et al.*, 1909).

There was much more interest in the volatile-rich varieties than in the main granite types and the processes of 'tourmalinization', 'greisening' and 'kaolinization' were speculated upon by most authors in these years. It was clearly recognized that they resulted from unusually high concentrations of B, F, Cl and OH in the magma and that, while some B was incorporated into 'primary' tourmaline, the vein tourmaline and other mineralizing effects resulted from the passage of the concentrated residual volatile phases through consolidated, but fractured, rock. 'Pneumatolytic' and 'hydrothermal' are adjectives commonly used in the literature. Subsequent studies have enlarged upon this early thinking as more data, including those from experimental work, have been acquired. The basic concepts have not suffered substantial revision, however.

An interest in the origins of granite magmas themselves was starting to show itself in Britain at about the time of World War 1 when petrology was becoming strongly influenced by the work of N. L. Bowen and his demonstrations of the production of acid magma by differentiation from crystallizing basic magma. Not all geologists found this process entirely satisfactory, however, and some, probably with continental (and especially Scandinavian) experience, preferred to argue for metamorphic changes in pre-existing rock, followed by mobilization. One of these was Holmes (1916) who, in reviewing Bowen's paper on 'The later stages of the evolution of igneous rocks', pointed out that only a limited amount of granite could be produced by differentiation from basalt, and that, while the earliest granites probably originated in this way, younger ones must have arisen by refusion, especially of sediments which were themselves derived from granite. In effect, Holmes supported Ussher's concept of 1892 as far as Cornubian granite is concerned. The dichotomy between supporters of differentiation and those of metamorphism (or 'anatexis' or 'palingenesis') led to the 'granite controversy' of the 1940s and 1950s, in which Holmes, together with H. H. Read and D. H. Reynolds, were the leaders of the latter school of thought in Britain. All made direct reference to Cornubian granites. Holmes (1932) took further his arguments for refusion, using Brammall and Harwood's evidence of assimilation in the Dartmoor Granite in support. Read (1949) regarded the granites as the end members of his Granite Series, 'almost dead when they arrived at their present positions'. And Reynolds (1946) illustrated her account of the chemistry of the granitization processes by reference, *inter alia*, to features in the Dartmoor granites.

Between the two World Wars, more sophisticated and detailed work accompanied the development of improved laboratory techniques. In consequence, much more chemical information became available, enabling petrologists to pursue more fundamental problems. In a series of papers between 1923 and 1932, Brammall and Harwood examined many aspects of the Dartmoor Granite and its minerals, and discerned in it stages of evolution, although they published no map to show the distribution of the different types.

In 1927, Ghosh published a paper about the eastern part of Bodmin Moor, describing two stages of coarse-grained granite in addition to the fine-grained variety, and followed this in 1934 with a similar study of Carnmenellis in which he distinguished three types of coarse granite. The granites of St Austell were considered in 1923 by Richardson, who, for the first time, separated eastern and western intrusions and clarified the importance of the Li-bearing micas there. He also emphasized the importance of 'mineralizers' in the later stages, pointing out how they modified the mineralogy, particularly by removing iron.

The Isles of Scilly granites were identified as having nine stages of development (the principal ones corresponding with those of Dartmoor) by Osman (1928). Nothing, however, appears to have been written about the Land's End intrusion in this period.

Of all these authors, Brammall and Harwood (1932) were the ones who really developed a petrogenetic scheme. They rejected a straightforward Bowen-type differentiation origin on the grounds that there is too much granite for the amount of basalt required in the area, and that such a magma would be deficient in such elements for instance, as Sn, W and B. They preferred a palingenetic starting point like that of Ussher (1892) and Holmes (1916), with the latter of whom Brammall evidently discussed the matter.

Their scheme envisaged a partial melting of pelitic metasediment (of the sort exemplified on Leusdon Common) to produce a rather sodic magma; evidence included the nature of xenoliths, presence of garnet, etc. This magma was 'basified' by assimilated country rock and by differentiation to produce more K- and Fe-rich varieties on the one hand and more siliceous varieties on the other. Evidence of these processes is provided by detailed textural descriptions, chemical analyses and calculations of the balances obtained in various reactions. Of particular importance is the sequence of analyses made across the Dartmoor Granite contact at Burrator. The magma was not intruded in one batch, however, and Brammall and Harwood interpreted the facies found at Haytor Rocks as indicating an older more 'basic' magma intruded by a younger, inner, more 'acid' magma. The reciprocal processes of 'basification' are those of 'granitization' and Brammall and Harwood describe changes in the textures and compositions of country-rock xenoliths of various types (such as those from Birch Tor) in varying degrees of assimilation by the granite magma. Although strictly based on and calculated for Dartmoor, this scheme was widely applied by others to the remaining Cornubian granites.

Brammall and Harwood were equally conscious of the importance of volatiles, especially B, F and OH, in facilitating the movement of elements, altering the physicochemical environment and making possible the generation of varieties such as those rich in tourmaline, although, since Dartmoor lacks the extremely Li- and F-rich granites found at St Austell, they did not find it necessary to suggest the involvement of these in the kinds of major evolutionary process Richardson (1923) had described.

The scheme devised by Brammall and Harwood is what is called the 'Dartmoor model' at the

Petrogenesis

beginning of this chapter. One of its remarkable features, reflecting the astuteness of its authors, is that from the start it contained the genetic concepts of both of the extremist schools involved in the granite controversy. It thus survived without serious criticism through the arguments of both 'magmatists' and 'transformists'.

As for alteration and mineralization, accounts published in these years start from the premise that the mobile constituents which brought them about, were initially present in the magma and were concentrated and released in the late stages of crystallization. The opinion of the Geological Survey is summarized by Reid and Flett (1907) in the Land's End Memoir (p. 54), where they say 'It is generally admitted that agents which occasioned these changes were vapours emanating from the granite at a time following its injection but anterior to its complete cooling and consolidation'. These vapours consisted mainly of water at a very high temperature. The occlusion of water in molten igneous magmas is a phenomenon of universal occurrence and needs no special explanation in this connection. But it is clear that the Cornish granite masses discharged not only steam, but other substances which have the power of profoundly modifying rocks when they penetrate them, especially at high temperature. Compounds of boron and fluorine were certainly present, as these elements are especially characteristic of the new minerals deposited (tourmaline, topaz, mica, fluorspar). Lithia and phosphoric acid are other substances which passed outward from the granite. Finally, most of the metalliferous ores may reasonably be ascribed to the same source. This is established beyond doubt for the tinstone, and is at least extremely probable for the uranium, tungsten, copper and iron ores. Perhaps the only ores in this area with a different origin seem to be those of silver–lead, zinc and some of the ironstones'.

Despite many papers on minerals and mining, there seems to have been little questioning of this general concept, or of where the magma acquired its relatively high concentration of B, halogen and metallic elements. At the same time, it is apparent that the importance of metasomatism was understood, although it was generally recognized only on a local scale, for example, in greisen-bordered veins, and not as a widespread, pervasive process.

One controversy did develop, however, and that concerned kaolinization which was recognized as a late, low-temperature process. The two schools of thought comprised those who believed, like Hickling (1908), that kaolinization was a consequence of deep weathering, and those like Collins (1878, 1887), Butler (1908) and the Geological Survey who saw it as a magmatic, hydrothermal process. Coon (1911, 1913) accepted both views, postulating an earlier, partial breakdown of feldspar by hydrothermal action, and a subsequent completion of the process by weathering. Given the confused field relations of altered and unaltered rock and the lack of sophisticated chemical techniques, these differences of opinion were bound to persist and in fact were not resolved for many years.

Post-1950, the St Austell model and minor rock types

Following a long quiescent period, research into the granites and their associated geology was resumed in the 1950s, by which time there had been great advances in geophysical and geochemical techniques, largely as a result of wartime developments.

In 1958, Bott, Day and Masson-Smith published the results of the first comprehensive geophysical survey of South-west England, establishing conclusively that the granite outcrops were protruberances from a batholith, as anticipated by De la Beche some 120 years earlier.

Although some modifications to the shape and extent of this have been made (for example, Bott and Scott, 1964; Bott et al., 1970; Holder and Bott, 1971; Tombs, 1977), Bott et al.'s concept has not been seriously challenged. Doubts have persisted about the floor of the batholith, however, the lower-crustal sediments having much the same density and seismic velocity as granite and continuing without substantial variation down to the Moho. As noted previously (Chapter 2), a seismic reflector has been found at depths of 10–15 m (Brooks et al., 1984) and it is now widely believed that this is caused by a southward-dipping thrust, an interpretation which accords with the nappe-and-thrust structure elucidated for the Cornubian Peninsula as a whole in recent years. Until the late 1970s, the main direction of granite research was towards classifying the relationships of the principal varieties, the question of the source of the early magma being no more soluble than it had been in the days of Brammall.

In the St Austell intrusion, Exley (1959) followed Richardson (1923) in separating eastern

The Cornubian granite batholith (Group C sites)

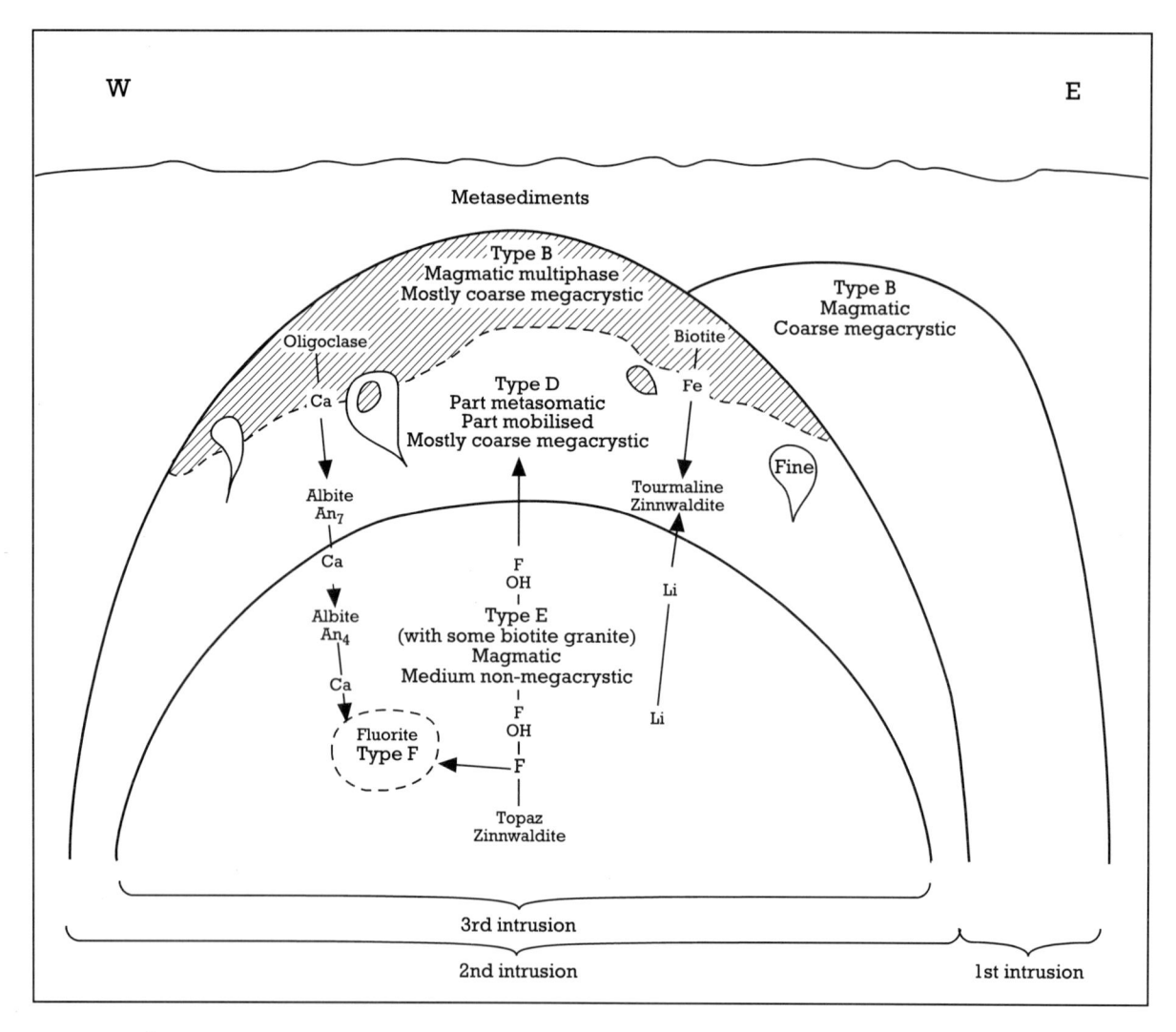

Figure 5.4 The St Austell model. Diagram showing the first intrusion of Type-B granite (Table 5.1) cut by multiphase second intrusion of biotite granite, with metasomatic aureole of Type D caused by intrusion of Type E.

and western intrusions and then went further by subdividing the western area into early and late Li-mica-bearing varieties and a fluorite-bearing variety. All these he considered to have been derived by differentiation from biotite granite magma. This view was later revised when it became clear that Exley's 'late lithionite granite' (now called the 'non-megacrystic Li-mica granite' (Type E) or, by Hill and Manning (1987) the 'topaz granite') was intrusive into biotite granite (Type B) which it had metasomatized to produce Li-mica (Table 5.1). The sequence has thus been reversed: much of the 'early lithionite granite' (Type D) now being recognized as late and referred to as the 'megacrystic Li-mica granite' and not magmatic in origin (Dangerfield *et al.*, 1980), although Bristow (in press) and Bristow *et al.* (in press) believe that a magmatic component is present. A further revision was made by Manning and Exley (1984) who ascribed the fluorite granite (Type F) to hydrothermal and metasomatic reactions accompanying the alteration of biotite granite to megacrystic Li-mica granite. The release of Ca from plagioclase, addition of F from topaz and redistribution of Li and Fe gave rise to fluorite in pockets of granite which were impoverished in mafic minerals (variations in nomenclature are listed in Table 5.1).

Owing to indifferent exposure and extensive alteration, the junction between the eastern and western intrusions and the extent of the metasomatic conversion of the latter into a Li-mica variety have never been precisely defined. Thus outcrops of biotite granite, although sometimes texturally different from that in the eastern intrusion, were known to occur in the western area. The confusion has been reduced to a large extent by Hill and Manning (1987), who have identified a number of granite types in the western area, indicating that, before being metasomatized, it was composed of a composite multiphase intrusion, the components of which themselves constitute an evolutionary, intrusive and metasomatic sequence. This consists of: biotite granite → equigranular biotite granite → globular quartz granite → tourmaline granite → aphyric granite → topaz granite. Exploration for, and expansion of, china clay workings have shown that substantial unmetasomatized bodies of biotite granite remain in the area. The magmatic history of the St Austell Granite, as it is now understood, is represented diagrammatically in Figure 5.4 and constitutes the 'St Austell model'.

The relationship between the biotite granite and the Li-mica granite magmas began to emerge when, shortly after Exley's early work on the St Austell granite, Stone carried out a detailed investigation of the small Tregonning–Godolphin Granite between the Carnmenellis and Land's End intrusions. Here, both biotite granite and non-megacrystic Li-mica granite occur together and Stone (1975) was able to establish that the latter had intruded the former. He deduced that the second magma had been derived at depth from biotite granite, the process envisaged consisting of the albitization of the plagioclase, the exchange of Li–Al for Mg–Fe to produce Li-mica from biotite, and the enrichment of the fluid phase (already widespread and F-rich) in K, leaving the rock richer in Na. At the depths and pressures considered probable, large volumes of melt could be generated in this way. Later experimental work by Manning (1979) on melts in the system Qz–Ab–Or with varying F contents, has shown that it is possible for magma of the required composition to be derived from volatile-rich bioite granite magma by differentiation.

The Li-mica granite of Tregonning is some 10 Ma younger than the neighbouring Carnmenellis Granite, just as the Castle-an-Dinas Li-mica granite near St Austell is 10 Ma younger than the nearby biotite granite. This fact, together with the widespread mineralization dated at about 280–270 Ma (Table 2.1), has led to the belief that the Li-mica granite intrusive phase was responsible for the main mineralization and that this type of granite is much more widespread than is indicated by the small outcrops in the St Austell and Tregonning areas. If this is correct, there seems to be no reason why both the evolutionary processes outlined above should not have operated, local conditions of temperature, pressure and volatile concentration being the determining factors.

None of the other granite outcrops show the range of varieties of the two described above, consisting predominantly of coarse-grained biotite granite (Type B) with inclusions (Type A) and small amounts of fine-grained granite (Type C), although Knox and Jackson (1990) have recently described an evolutionary suite of biotite granite intrusions from the southern marginal area of Dartmoor.

As far as the coarse-grained granite is concerned, suspicions that this was not composed of two or more varieties intruded separately began to emerge early in the post-war period. For example, Chayes (1955) demonstrated that the modes of Types 1 and 2 of Carnmenellis (described by Ghosh in 1934) were not statistically distinguishable, a conclusion reinforced by Al-Turki and Stone (1978) who also showed that some chemical differences were not significant. Ghosh's Type 3 was, however, significantly different from these. During the remapping of the Dartmoor Granite by the Geological Survey, Edmonds et al. (1968) became convinced that Brammall and Harwood's Giant and Blue granite varieties were not separate intrusions, but facies of the same rock. Likewise, Booth and Exley (1987) have noted variations in the coarse granite of Land's End, and it is the present author's opinion that the same feature is to be found on Bodmin Moor, although Ghosh (1927) regarded his Normal and Godaver varieties as having an intrusive relationship.

In all cases, the coarser, more striking megacrystic rock forms an envelope (usually incomplete) around a less obviously megacrystic core. Hawkes (*in* Edmonds *et al.*, 1968) refers to the two as the 'big feldspar granite' and the 'poorly megacrystic granite' and they formed the basis of the very useful field and mapping classification described by Hawkes and Dangerfield (1978). Owing to their mineral and chemical similarities, however, Exley and Stone (1982) preferred to regard them as variants of a single coarse-grained biotite granite, their Type B.

An explanation for the different textures as

variants was first suggested by Stone and Austin (1961) and further discussions are to be found in Exley and Stone (1964), Booth (1968), Hawkes (*in* Edmonds *et al.*, 1968), Stone and Exley (1968) and Stone (1979, 1984, 1987). Essentially it is agreed by these authors that the textures associated with the K-feldspar megacrysts show them to be secondary and developed by autometasomatism around either earlier K-feldspar or, in some cases, plagioclase. Their presence in the marginal regions of the intrusions was due to the exsolution of OH in the cooler parts, thereby providing the medium for the movement and recrystallization of alkalis. Stone (1979, 1984, 1987) relates this stage of evolution to the transition between magmatic (solidus) and post-magmatic (subsolidus) reactions and to feldspar unmixing. In his 1987 paper, he also draws attention to small but significant differences between the 'trace-alkali suite' of elements (Li, Rb, Cs, F and SiO_2) in Ghosh's Types 1 and 2 granites, suggesting that this variation is due to late-stage redistribution (which would have the textural consequences just noted) and that Type 2 represents 'relict areas not affected by these changes'. It should be noted that some authors, notably Webb *et al.* (1985), Vernon (1986), Leat *et al.* (1987) and Jackson *et al.* (1989), argue for a magmatic origin of the megacrysts. This is a view with which the present author does not concur in the light of his own experience of the textural relations and distribution of these megacrysts.

The other chief granite variety is the fine-grained biotite granite (Type C) and modern opinions about this include both those which regard it as a late, intrusive differentiate and those which regard it as granitized sediment, as was noted above (Exley and Stone, 1982; Exley *et al.*, 1983; Hawkes, 1968, 1982; Stone and Exley, 1986; Tammemagi and Smith, 1975). There has also been work during recent years on some of the less-abundant types, especially leucogranite, aplite, pegmatite, quartz–tourmaline rock, quartz–topaz rock, granite porphyry (elvan) and explosion breccia. Discussions about these are included in the relevant site descriptions and are briefly summarized below.

Leucogranite, aplite and pegmatite

Given the high concentrations of volatile constituents and such elements as Li, well-developed pegmatites are surprisingly uncommon in Devon and Cornwall, although they occur in small patches, veins and pods quite frequently and in many cases are obviously the result of pockets in the magma where volatiles have been concentrated; probably in some cases they represent globules of an immiscible phase.

One of the most characteristic developments of pegmatite is in conjunction with aplite and leucogranite in the 'roof complexes' of the Tregonning Granite, the related sheets near Megiliggar Rocks, and the non-megacrystic Li-mica granite intrusion in the St Austell mass, but they also occur on a large scale at Meldon near the north-western margin of the Dartmoor Granite.

The textures and compositions of these rocks suggest that they, like the metasomatic facies of the main granites described earlier, originated by reactions spanning the super-solidus/subsolidus stages of magmatic crystallization. Leucogranite is regarded as a direct magmatic descendant, owing its composition to fractionation of calcic plagioclase and dark minerals, while aplite represents the Na-rich liquid fraction left when K was preferentially partitioned into vapour phase. This, in turn, caused recrystallization into pegmatite.

Quartz–tourmaline and quartz–topaz rock

Quartz–tourmaline rock occurs in association with most of the main granites in areas where B concentrations have been particularly high, for example, at Roche Rock north of the St Austell Granite and along the western margin of the Land's End Granite at Porth Ledden. The evidence points to a magmatic origin through the separation of an immiscible liquid phase.

Quartz–topaz rock is rather rarer and mostly seen in the St Austell area, especially at St Mewan Beacon. Its origin is not yet fully explained, but it seems not to be staightforwardly magmatic and includes a significant hydrothermal component.

These two rock types also belong to the late-magmatic and immediately post-magmatic stages.

Elvan dykes

Granite porphyry dykes, sometimes consisting of single, but often of multiple, intrusions are to be found throughout the peninsula and follow the main, approximately E–W-trending joint direction for the most part. Some of these rocks are of straightforward magmatic derivation from biotite granite magma, but many (such as that at Praa

Petrogenesis

Sands), contain evidence of a solid component incorporated by fluidization. These rocks are the youngest of the magmatic intrusions.

Explosion breccias

Spectacular developments of breccias made up of both igneous rocks and metasediments, and usually heavily mineralized, occur in a number of places in Cornwall. Most are in the central and western parts, as at Venton Cove near Marazion, and at Wheal Remfry in the west of the St Austell Granite. They have resulted from violent reduction of pressure as fissure systems above magmatic fluid concentrations reached the surface allowing the rocks to implode.

Source material and the 1980s model

Until good techniques for the analysis of radio-isotopes and trace- and rare-earth elements became established, it was impossible to say much about possible source rocks for the early magmas, or about the conditions under which they formed. Thus, detailed though they were, discussions such as those by Brammall and Harwood (1932) remained unsupported by what is now considered to be essential chemical evidence. The general nature of more modern data and the tenor of arguments based on them have been outlined in Chapter 2, including the proposition that these are 'S-type' granites (Chappell and White, 1974) We now need to add further details.

1. Table 5.2 shows that a number of trace elements and their ratios are not appropriate to primitive granites resulting from deep-seated magmatic differentiation, e.g. Nb, Y and Zr are relatively low and Rb, Ba and Sn are high.
2. Table 2.1 shows that the initial ratios of $^{87}Sr/^{86}Sr$ range from 0.7095 to 0.7140, which are crustal values.
3. Hampton and Taylor (1983) record Pb isotope ratios of $^{206}Pb/^{204}Pb$ from 18.363 to 18.499, $^{207}Pb/^{204}Pb$ from 15.614 to 15.655 and $^{208}Pb/^{204}Pb$ from 35.261 to 38.508; again these are crustal not subcrustal values.
4. Both the concentrations of the radio-elements U, Th and Zr and the rare-earth elements, and the ratio of light to heavy REE (shown by the steepness of the slope of the chondrite-normalized values in Figure 5.5) indicate that the magma was derived from a source already well differentiated and not from a basic magma, even if it were contaminated as suggested by Thorpe et al. (1986), Thorpe (1987) and Leat et al. (1987). Darbyshire and Shepherd (1985) show REE patterns for Dartmoor, Land's End, Bodmin Moor and Carnmenellis (Figure 5.5) and suggest that differences between the first two and the last two indicate either different degrees of partial melting or different source rocks. They agree that all are indicative of a metasedimentary source, however. Most work on REE has been carried out on the Carnmenellis Granite (Jefferies, 1984, 1985a; Charoy, 1986; Stone, 1987) and this confirms that these elements are carried in the main by the accessory minerals: apatite, zircon, monazite, xenotime, ilmenite and uraninite. There is some uncertainty as to the extent to which these minerals (and the biotite which frequently encloses them) are magmatic or derived from their sedimentary host rock as 'restite'. Jefferies (1984) regards them as chiefly magmatic and Stone (1987) as restitic. The issue is, in any case, complicated by some fractionation and later metasomatic redistribution of these elements (for example, Alderton et al., 1980).
5. The highly aluminous nature of the granite and the occasional presence in it and its xenoliths of garnet, sillimanite, andalusite and cordierite, in addition to the chemical characters enumerated above, point clearly to a pelitic sedimentary source containing garnet, and Charoy (1986) has argued the case for material similar to Brioverian pelite.
6. The ammonium content of the granite ranges from 3–179 ppm with an overall average of 36 ppm and an average of 94 ppm for Bodmin Moor. These compare with a world's average of 27 ppm for granite and granodiorites and, since the source of ammonium is almost certainly organic matter in sediments, such high values are strongly indicative of contamination by sediment (Hall, 1988).

There remain, however, some anomalies, such as the sources of Sn, U, Cl and F which seems unlikely to have been available in any metasedimentary source in anything like the concentrations required to give rise to the quantities present in the granite. The only possible alternative source for these lies in the upper mantle where the heat

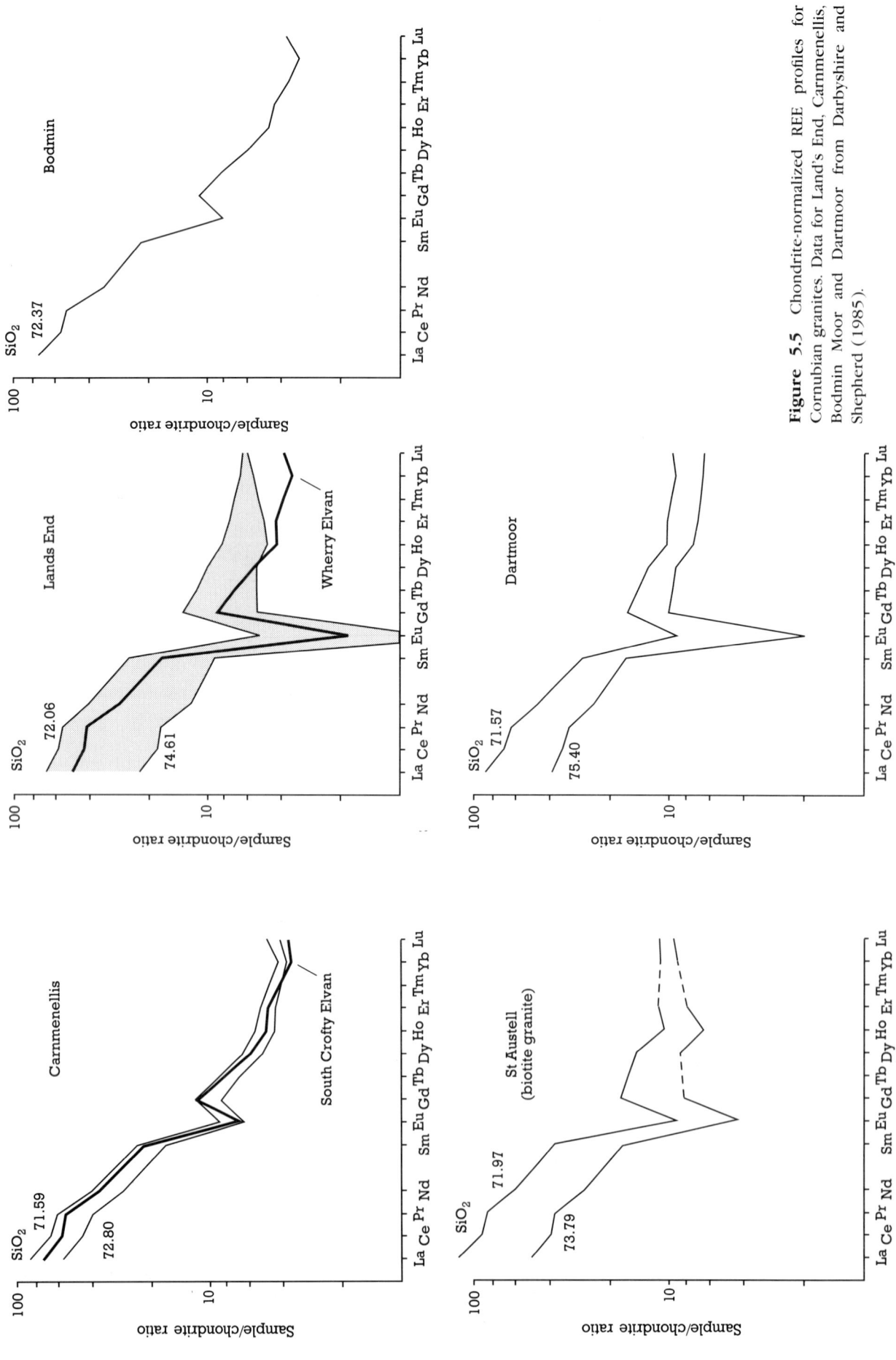

Figure 5.5 Chondrite-normalized REE profiles for Cornubian granites. Data for Land's End, Carnmenellis, Bodmin Moor and Dartmoor from Darbyshire and Shepherd (1985).

Petrogenesis

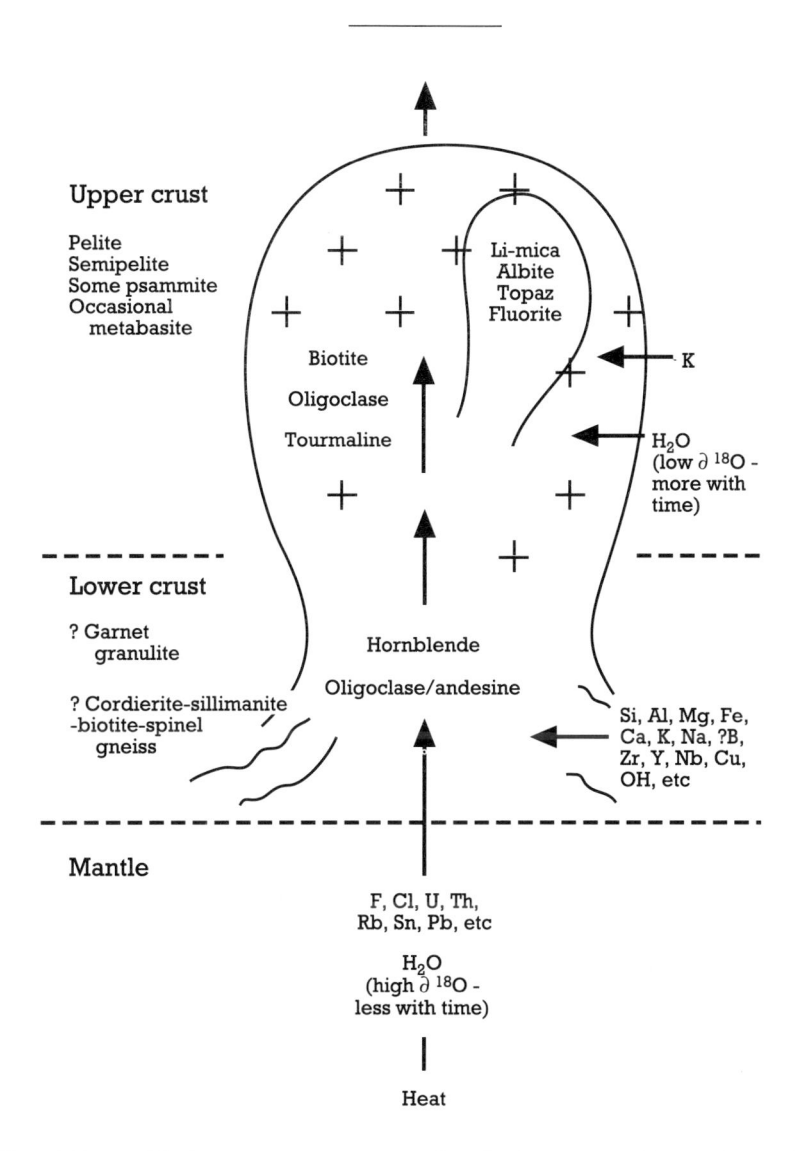

Figure 5.6 The 1980s model. Granitic magma generated in the lower crust (but with mantle components) and evolving both by assimilating upper-crustal constituents and differentiating Li-mica granite magma. Magma becomes increasingly hydrated by drawing in increasing quantities of meteoric water during ascent.

needed for partial melting must have also originated. The conclusion must be, therefore, that volatiles from high-level mantle rocks carried these components into the granite magma as it formed in the crust, and that their present distribution is a result of their partition into different phases as the magma evolved. Although aqueous, the amount of water in the magma at this stage must have been small, or the magma would not have been able to rise to high crustal levels; Charoy (1986) suggests something in the region of 3–5%. Most of the water which played so important a part in the later stages of evolution must have been acquired from the surrounding rocks.

The magmatic history accepted at present is demonstrated diagrammatically in Figure 5.6 and is put in the overall context in Table 2.2.

Mineralization and alteration – current views

As with magmatic origins and evolution, understanding of mineralization and alteration advanced little until it became possible to analyse trace elements and isotopes. In addition, however, the development of the heating/cooling microscope stage brought more useful results from the study of fluid inclusions than the crude and confusing technique of decrepitometry had done. As a result, in the late 1960s and 1970s experimental data started to appear which, through evidence from homogenization temperatures and salinities of inclusions, led to more exact interpretations of the mineralizing fluids themselves and their reactions. Details of many of the findings are summarized in Stone and Exley (1986), Bromley and Holl (1986), Willis-Richards and Jackson (1989) and Jackson *et al.* (1989), and site descriptions in this Chapter, but at this point it is necessary to emphasize two features of mineralization.

The first of these is that there were two principal stages of granite magma intrusion, and that although there was mineralization associated with both, the more important 'main' mineralization was related to the second, largely through the maintenence of a residual body of magma (Willis-Richards and Jackson, 1989). The earliest comprehensive work suggesting this seems to have been that published by Jackson *et al.* in 1982. These authors, concerned with the St Just area of the Land's End Granite, determined that there were several mineralizing events which included one at about 290–280 Ma and a second about 10 Ma later, and that the participating fluids changed from those of magmatic origin to later ones of meteoric type. Consideration of (*inter alia*) heat requirements for episodic circulation of fluids and the time intervals involved led Durrance *et al.* (1982) also to postulate a second magmatic event at about 10–20 Ma after the first. It was already known that the non-megacrystic Li-mica granite found in the intrusion in Gunheath clay pit near St Austell and in the Castle-an-Dinas Mine a short distance away had intruded biotite granite (Hawkes and Dangerfield, 1978; Dangerfield *et al.*, 1980), so that when the ages of these were found to be approximately 270 Ma BP (Darbyshire and Shepherd, 1985, 1987), it became clear that they represented the event in question. It is possible that the granite intrusions described by Knox and Jackson (1990) from southern Dartmoor also belong to this episode.

The second feature concerns the nature of the fluids which have been found to have a complex evolutionary history. The earliest stage in recognizing that they separated into different fractions followed the work of Jahns and Tuttle (1963) and Jahns and Burnham (1969) who described how partition of Na and K into liquid and vapour phases respectively gave rise to aplite and pegmatite formation, the process underlying the alkali metasomatism of the main granites, the formation of the leuogranite–aplite–pegmatite

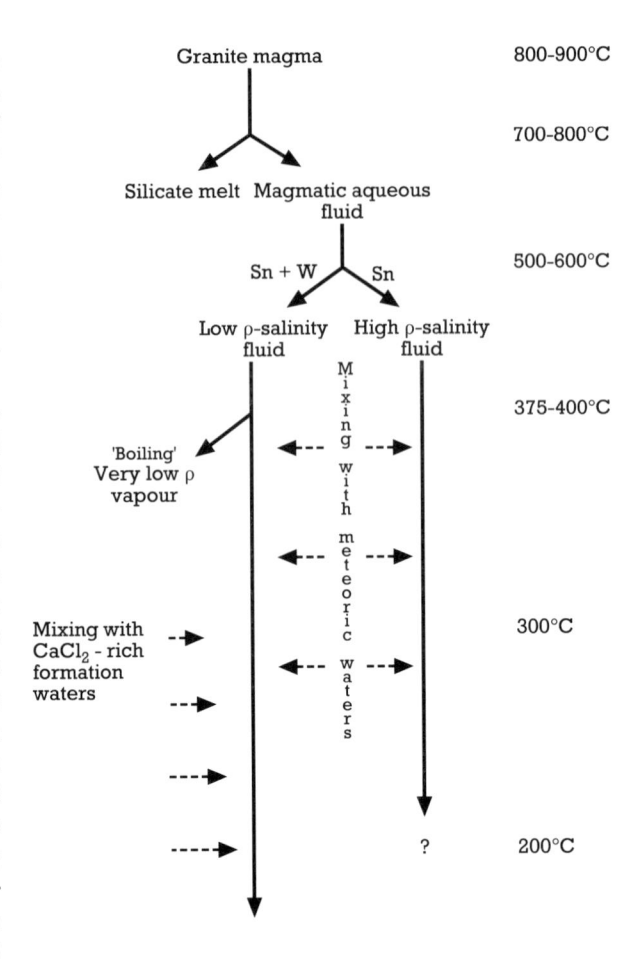

Figure 5.7 Schematic representation of fluid evolution in the eastern sector of the Cornubian metallogenic province showing the importance of 'immiscibility events' and mixing (after Shepherd *et al.*, 1985).

complexes and, at least partly, to pervasive greisening. Subsequently, however, Pichavant's experiments (1979) coupled with the field studies of Charoy (1979, 1981, 1982) demonstrated that, in B-rich magmas, an immiscible aqueous Si–K–B liquid could evolve to crystallize as rock types rich in tourmaline. Although it is possible that this phase could have carried several metals, Shepherd *et al.* (1985) argued that it too separated into two immiscible phases, one being of low density and low salinity, carrying some Sn and all the W, the other being of high density and high salinity and carrying most of the Sn. This is the pattern seen in the Dartmoor Granite mineral deposits; in the neighbouring country rocks immiscible liquids did not exsolve, but a meteoric water component became mixed with the magmatic component. This thinking is summarized in Figure 5.7.

Petrogenesis

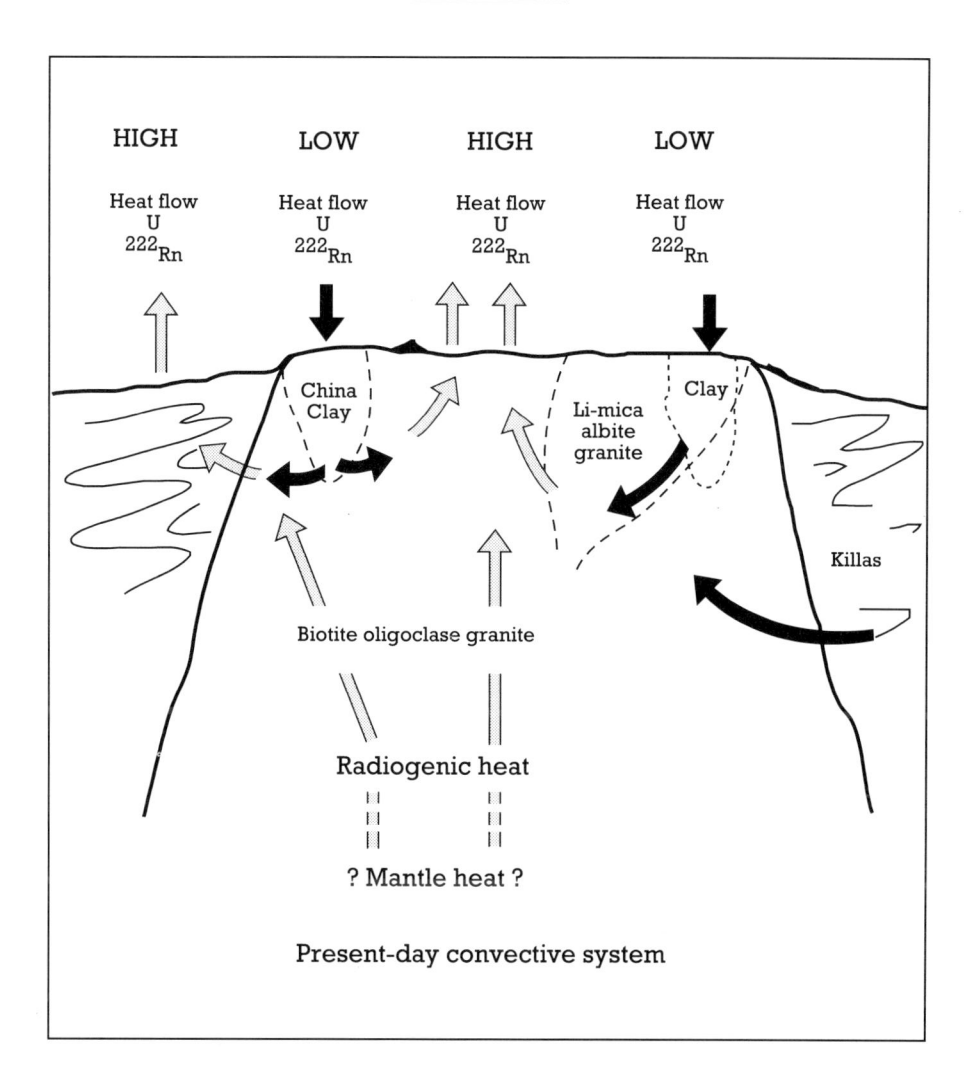

Figure 5.8 Diagrammatic representation of water circulation in Cornubian granite. Areas of low heat flow, U and ^{222}Rn concentration are associated with china clay and indicate draw-down; areas of high heat flow, U and ^{222}Rn concentration indicate uprise (based on Durrance et al., 1982).

The separation of the immiscible fluids resulted in greatly increased internal pressures which disrupted the outer parts of the granites in some places, e.g. Wheal Remfry, west of St Austell. The consequent sudden drop in pressure not only generated a far-travelling vapour phase which could deposit minerals either in the main fracture system or pervasively, but also caused the implosion of the surrounding rocks to give rise to breccias. This happened twice, first in minor fashion at the end of the 280 Ma magmatism, and again at the end of the 270 Ma magmatism when it initiated the main mineralization and also increased the permeability of the granites, thus facilitating later alteration.

The discovery of the importance of meteoric water as a constituent of the mineralizing and altering solutions was made by Sheppard (1977), using stable isotopes and was confirmed by Jackson et al. (1982). It has had a profound effect on the understanding of these processes.

Sheppard (1977) showed that mica in greisens contained both magmatic and meteoric water, but that kaolinite contained only the latter. He thus supports the weathering origin of kaolinite mentioned above, but the absence of a deeply weathered mantle overlying the Cornubian rocks and the close relationship between kaolinization and the joint system, often beneath unaltered rock, still obstructed its acceptance by advocates of magmatic hydrothermal action. The dilemma was resolved by the publication of a paper by Durrance et al. (1982), who, following the ideas of Fehn et al. (1978), established the concept of convective circulation systems associated with granites. These would draw in increasing amounts of meteoric water from the surrounding rocks as well as the granites, and this water, having passed through the pores and fractures in the granites, produced effects identical in appearance to those resulting from magmatic hydrothermal alteration (Figure 5.8). This is discussed in detail by Bristow *et al.* (in press), who also describe the continuation of kaolinization activity to the present day, including the superimposition of a Palaeogene tropical weathering period on the earlier alteration. What is now seen in china-clay pits and some quarries, such as that at Tregargus, is the result of all this. The overall sequence is shown in Table 2.2.

There have been two recent comprehensive accounts of the mineralization in south-west England in relation to the general geological setting and the batholith. That of Willis-Richards and Jackson (1989) describes how the Cornubian orefield contained pre-batholith deposits of Mn in sediments and Cu in basaltic rocks, how the syn- and post-batholith ores fall into eastern and western areas, the latter being richer in Sn, Cu and Zn than the former, and how the long-sustained heat flows have caused the continuation of convective systems. The second, by Jackson *et al.* (1989), concentrates particularly on ore-forming processes, discussing both their chronology and details of the morphology of the various types of deposit. The nature of both main-stage and epithermal mineralizing fluids is considered in terms of their temperatures, salinities and flow as modelled by computer. These papers support the brief description given above but add much important detail to it.

C1 HAYTOR ROCKS AREA (SX 758773)

Highlights

This classic site contains the best exposure of variants of the coarse, megacrystic granite of Dartmoor, together with a later, intrusive, fine-grained granite sheet. Its coarse-grained granites enclose a variety of genetically significant xenoliths. It also provides excellent evidence of tourmalinization.

Introduction

This site, which is centred on the fine summit tor of Haytor Rocks (Figure 5.9), is unique in containing the two major variants of the Dartmoor Granite whose relationships led to a view of the origins of Cornubian granites which held sway for 30 years or more. Not mentioned by the Geological Survey officers (Reid *et al.*, 1912), the two variants were crucially different in the eyes of Brammall (1926) and Brammall and Harwood (1923, 1932) who interpreted them as an upper, earlier, coarser variety intruded by a lower, finer, later variety, the supposed chilled contact between them being visible in the lower part of Haytor Rocks themselves. Subsequent research in all the Cornubian granites has shown that substantial variations in coarseness are usual among the main biotite-bearing type, although the rocks closer to the walls and roof of each pluton are generally coarser and more megacrystic than those further away. There is a gradational, not intrusive, relationship between coarser and finer, and the contact at Haytor Rocks is now recognized as being due to a separate fine-grained intrusive sheet.

Also to be seen in Haytor Rocks (and more rarely in the neighbouring quarries) is a variety of xenoliths in different stages of assimilation; not as interesting as those at Birch Tor but nevertheless instructive. The composition of xenoliths was used by Brammall and Harwood (1932) in calculations of the modification of the earliest Dartmoor magma into the first of the main magmas; and a cordierite hornfels from Haytor Rocks was described by them in 1923 and a highly granitized 'basic inclusion' from the Haytor Quarry in 1932. Later work suggests that xenoliths are less important than was supposed (Stone and Exley, 1986; Bromley and Holl, 1986). The authors of the Survey Memoir (Reid *et al.*, 1912) described the Dartmoor Granite as being less boron rich than average in south-west England, but the presence in the Haytor Quarries, and in the boulders in the area, of tourmaline-bearing veins and nodular masses of tourmaline called 'suns' indicates that the boron content of the magma was relatively high nevertheless. Brammall and Harwood (1923, 1925, 1932) discussed the significance of this and, especially in their 1925 paper, separated what are now regarded as magmatic, autometasomatic and post-magmatic tourmaline generations. Although later research has modified their views in some details, it has

Figure 5.9 Haytor Rocks, exposing the coarse megacrystic granite of Dartmoor. The megacrystic character of the granite is visible in the foreground exposure. (Photo: S. Campbell.)

been based on work done elsewhere than Haytor and is described within this volume.

Haytor Rocks were classified as an excellent and 'type' example of 'summit tor' by both Linton (1955) and Gerrard (1974) and figured in photographs in both Linton's (1955) and Palmer and Neilson's (1962) papers.

Description

The greater part of Haytor Rocks is composed of coarse, megacrystic biotite granite with alkali feldspar megacrysts averaging 40 to 50 mm in length. This variety, commonly seen all over Dartmoor, especially in tor outcrops, is like the main granites of other masses in the south-western peninsula and has been referred to their Type B by Exley and Stone (1982) and Exley *et al.*, (1983).

The lower parts of the western and north-western faces of Haytor Rocks, however, are made up of fine-grained granite, devoid of megacrysts and with pronounced vertical columnar jointing. The contact between the two granites is sharp, although irregular, and dips gently but unevenly towards the south-west; the marginal zone of the lower granite being especially fine grained. Taking into account the exposures in a nearby abandoned cutting, the lower granite must be at least 10 m thick.

Two hundred and fifty metres to the NNW is a small unnamed quarry, while 450 m to the NNE are the much larger eastern Haytor Quarries. In these exposures the rock is not as coarse or as abundantly megacrystic, and this led Brammall and Harwood to the conclusion that there were two granite types, a view that they supported by chemical analyses. They believed that the type forming the upper part of Haytor Rocks and other tors in the vicinity was earlier, sheet like and less potassic. It was known as 'tor' or 'giant' granite. The type found in the quarries was thought to constitute an underlying intrusion into the giant granite, with its chilled contact phenomena visible in the lower part of Haytor Rocks. This later variety was supposed to be derived from the earlier at depth, it was more potassic and was called 'blue' or 'quarry' granite. Other examples of apophyses and sheets of blue granite in contact with giant granite are cited by Brammall (1926) and by Brammall and Harwood (1923), but none is so large and significant as that at Haytor, and it was recognized that the provenance of these minor intrusions was not always clear.

The xenoliths in the area vary in size, but are mostly less than about 0.5 m across and are generally rounded. They range in composition from basic, through diorite and grandiorite, to metasedimentary; and give an indication of the variety of rocks penetrated by the granite magma. Tourmalinization, in evidence everywhere, especially in the quarries, is occasionally in the form of narrow quartz–tourmaline veins but more spectacularly as the nodular masses – sometimes granular, sometimes acicular – of quartz and tourmaline known as 'suns'.

Interpretation

The Haytor Rocks site demonstrates, better than any other single site, evidence which has been used to support hypotheses regarding the origins of the Cornubian granites. The most significant is the presence of the two major granite variants thought by Brammall and Harwood to be distinct and separate intrusions. However, it is now realized that most of the chemical and textural variations in the coarser Cornubian granites is gradational and does not allow the separation of types which can be shown to be different statistically (for example, Chayes, 1955; Al-Turki and Stone, 1978); thus both Haytor variants can be accommodated in Exley and Stone's Type B granite or the 'coarse megacrystic' and 'coarse poorly megacrystic' varieties of Dangerfield and Hawkes (1981). Moreover, it is generally accepted that the fine-grained intrusion at Haytor Rocks is a large sheet later than, and independent of, the coarse granites in both the tors and the quarries. It has a significantly different composition, and is thus considered to belong to the Type C of Exley and Stone (1982) or 'fine poorly megacrystic' type of Dangerfield and Hawkes (1981). Although both field relations and trace element ratios suggest that some Type-C granites were derived from the Type-B magma (Exley *et al.*, 1983), variations in their chemistry indicate that others were not and that the process was not straightforward (Exley *et al.*, 1983; Stone and Exley, 1986). For example, describing fine-grained granite from a few miles to the north, Hawkes (*in* Edmonds *et al.*, 1968) included both chilling and contamination among possible origins. The fine-grained sheets noted by Brammall and Harwood do not all have the same origin, therefore.

It was postulated by Brammall and Harwood (1932) that the earliest magma was 'sodipotassic' and that it was modified and made increasingly potassic by the assimilation of xenolith material, particularly metasedimentary fragments, from the rocks through which it ascended, and, while concentrating attention on material from the contact at Burrator, on the west of Dartmoor, they quote an analysis from a Haytor xenolith in their paper. According to Reynolds (1946), the chemical changes resulting from assimilation and granitization make suspect Brammall and Harwood's identification of some xenoliths as originally basic. Furthermore, the derivation of much of the biotite in the granite as a 'restite' mineral from the source rocks (Stone, 1979; Stone and Exley, 1986) and the physicochemical difficulties of assimilation by a magmatic system close to the 'granite' system ternary minimum of Tuttle and Bowen (1958), (Bromley and Holl, 1986) have reduced the significance of xenoliths in modern views about the petrogenesis. Nevertheless, the presence of a wide variety of xenoliths, many of which are metasedimentary, gives these granites the aspect of Chappell and White's (1974) S-type, although relatively high concentrations of some metallic and halogen elements suggest the addition of some mantle components (Exley *et al.*, 1983; Stone and Exley, 1986). The Dartmoor petrogenetic model is put into historical context in the earlier part of this chapter.

A relatively high boron concentration in the Cornubian magmas played a very important role in late- and post-magmatic activity of various kinds including mineralization. By comparison with granites further west, such activity was restricted in the Dartmoor area: quartz–tourmaline veins and nodules such as are well displayed in the rocks around Haytor, provide excellent examples of the early stages of these phenomena. Brammall and Harwood (1925) suggested that there was a reciprocal relationship between the proportions of biotite and tourmaline, a suggestion supported by a deficiency of biotite in the rock surrounding some of the tourmaline 'suns' found in the Haytor Quarries. They argued that Fe, Mg and Ti were distributed in biotite, tourmaline, rutile, anatase and brookite or zircon according to the temperature of the magma, concentrations of B and extent of post-magmatic alteration (Brammall and Harwood, 1923, 1925, 1927).

This site contains the exposures which led Brammall and Harwood (1923, 1932) to believe that the finer-grained Dartmoor Granite was intrusive into earlier, coarser-grained variety, a belief which is not now accepted. It also contains examples of the xenoliths whose assimilation these authors considered to have played a crucial role in modifying the composition of the initial magma. Lastly, it contains a variety of veins and nodules which illustrate the effects of tourmalinization resulting from the high boron concentration in the magma.

Conclusions

The Haytor Rocks site constitutes an ideal area in which to examine textural and compositional variations in the earlier, main suite of Dartmoor granites in particular and in the Cornubian granites in general. Additionally, there is evidence for the way in which the magma was modified both by the incorporation of xenolithic constituents and by reaction with boron. Geomorphologically, the site is dominated by a classic example of a summit tor.

C2 BIRCH TOR (SX 686814)

Highlights

This site has an important display of varied xenoliths illustrating material thought to have influenced the final composition of the main Dartmoor Granite magma.

Introduction

The site is situated 8 km south-west of Moretonhampstead. It encloses Birch Tor itself, which contains xenoliths and rafts of precursors of the Dartmoor Granite and is adjacent to the remains of the open-cast workings of the Birch Tor and Vitifer tin mine.

One of the cornerstones of Brammall and Harwood's hypothesis (discussed earlier), regarding the composition of the main Dartmoor Granite was that it had been modified from its original 'sodipotassic' nature by assimilation of country rocks and older granites which it had intruded. They noted several areas, for example Haytor Rocks and Bellever Tor, in which evidence for this can be found, but one of the most important is Birch Tor which shows a particularly good range of xenoliths contained within their 'giant' or 'tor' variety of granite. Particular reference is made to the dark rock at the base of the tor, in Brammall's paper of 1926 (in which there is a field photograph) and in joint papers of 1923 and 1932 (in which there are chemical analyses, with a photomicrograph in the latter). There has been no serious dispute about Brammall and Harwood's evaluation of the origin and composition of these xenoliths at high levels in the granites, and a similar range is described from a few miles to the north by Hawkes (*in* Edmonds *et al.*, 1968). Reynolds (1946) has, however, questioned the identification of some as originally basic as a result of her calculations of the changes resulting from granitization. Stone (1979) and Stone and Exley (1986) have noted that some of the biotite in the granite is of 'restite', not xenolithic, origin and Bromley and Holl (1986) have shown that on geophysical, as well as geochemical grounds, assimilation at depth was probably very limited.

Description

Birch Tor is a low, extended tor with two main outcrops and conspicuous subhorizontal jointing. It is composed of coarse, megacrystic, biotite granite and contains a variety of xenoliths of different shapes, sizes and compositions. Some inclusions are rounded and others flat and subangular, but the most conspicuous is a sheet- or raft-like mass, some 7 m long and at least 3.5 m thick at the base of the south face of the western summit. This was illustrated by Brammall (1926) who described it as 'dark, blue-grey microgranite', and has been figured and analysed by Brammall and Harwood (1923; 1932) who also gave an analysis of a smaller xenolith. Many of these xenoliths look similar and are microgranodioritic or microdioritic; evidently they have undergone recrystallization as a consequence of entrapment in the granite magma.

Interpretation

The Dartmoor Granite was the subject of intensive investigation, especially by Brammall and Harwood between 1923 and 1932. For the most part, it consists of coarse-grained, megacrystic biotite granite, but the size and concentration of megacrysts is rather variable and, partly on this basis, Brammall and Harwood separated it into two intrusions: an earlier, coarser 'giant' or 'tor' variety and a later, finer 'blue' or 'quarry' variety. It has been proved subsequently that these textural features are gradational and both varieties are now included in the coarse-grained megacrystic or poorly megacrystic categories (of Dangerfield and Hawkes, 1981) and Type B (of Exley and Stone, 1982 and Exley *et al.*, 1983). Other intrusive varieties also occur and can be found around Birch Tor, but these are present in comparatively small amounts, all are younger and are not relevant in the present context.

The granite here is believed to have originated by lower crustal anatexis, much of its biotite being 'restite' material carried over from the source rocks which were probably more granodioritic than those in the contemporary upper crust (Stone and Exley, 1986). It is therefore an 'S-type' granite essentially (Chappell and White, 1974), and, as noted in Chapter 2 and the 'Petrogenesis' section above, shows appropriate chemical and mineral features. In Brammall and Harwood's view (1932), the initial magma was both acid and 'sodipotassic' or even sodic, and it became more basic and potassic by assimilating country rocks of overall argillaceous composition but which also contained dolerite, dioritic and granodioritic units. They based this argument on examination of this section and on chemical analyses from xenoliths found in the main granite and from a series of analyses across the granite/country rocks contact at Burrator.

Although, as has been said, Brammall and Harwood believed that some xenoliths were of originally basic composition, Reynolds (1946) pointed out that their present compositions are inappropriate to such a derivation when desilication and other changes are calculated, and that the xenoliths were more probably metasedimentary in origin.

Brammall and Harwood's concept has also been modified in three respects by more recent work (Hawkes, *in* Edmonds *et al.*, 1968; Exley and Stone, 1983; Exley *et al.*, 1983; Stone and Exley, 1986; Bromley and Holl, 1986). First there is the recognition that biotite is a derived or 'restite' mineral (Stone, 1979), and this has important implications with respect to the water content of the magma. Secondly, is the recognition that the feldspar megacrysts are a secondary phase and have developed as a result of potash metasomatism, sometimes on nuclei provided by plagioclase crystals. Brammall and Harwood saw the 'potassification' of the magma by xenoliths taking place in such a way that feldspars grew from the liquid as primary crystals, whereas the later workers, following the lead of Stone and Austin (1961), attribute them to the action of a later and separate K-rich aqueous phase. Thirdly, assimilation of much country rock is precluded by the composition of the magma, now known to be close to the 'granite' system's ternary minimum melt composition, and by the rapid rate of settling of xenoliths. Moreover, the observed density increase with depth in the batholith is consistent with a concentration of sunken xenoliths.

Brammall and Harwood were strongly influenced by the work of N. L. Bowen, whose important research, culminating in the publication of *The Evolution of the Igneous Rocks* in 1928, was coincident with much of their own. Hence, among other things, they emphasized the importance of assimilation and differentiation in modifying magma, and concluded that the latter was responsible for their second magma fraction (giving 'blue' granite) and subsequent fractions. At the time, however, many physicochemical controls of crystallization were still unrecognized, including the nature of the partitioning of elements between solid, liquid and gas phases, and it is this later knowledge that has led to revisions of Brammall and Harwood's interpretations.

Conclusions

This site contains an outcrop of coarse, megacrystic biotite granite containing xenoliths of various rock types. These have been taken to illustrate the kinds of material whose incorporation into the magma substantially modified the chemistry and mineralogy of the Dartmoor Granite. It exemplifies a significant part of a concept of petrogenesis which was sustained for many years and was a foundation of modern thinking.

C3 DE LANK QUARRIES (SX 101755)

Highlights

These quarries contain fresh, coarse-grained, poorly megacrystic biotite granite, characteristic of the Bodmin Moor intrusion, strongly foliated and jointed, and containing pegmatitic patches, minor granitic veins and xenoliths. They also incorporate typical Cornubian, fine-grained, megacrystic biotite granite and granite porphyry dykes ('elvans').

Introduction

The De Lank Quarries provide a rare opportunity to see really fresh, coarse, Cornubian biotite granite of the type classified by Dangerfield and Hawkes (1981) as the 'small megacryst variant'. This is typical of Bodmin Moor, much of Carnmenellis and the Isles of Scilly, but uncommon elsewhere. This rock type is often foliated and this feature is particularly conspicuous at De Lank. Although the officers of the Geological Survey (Reid *et al.*, 1910; Reid *et al.*, 1911) noted these features, they did not classify the rocks as a separate type, and the 'Godaver' type of Ghosh (1927), found in the extreme east of the pluton, is not distinguished by these criteria. Indeed, the present author's research indicates that the Godaver Type is a minor variant of the main granite.

Although subhorizontal jointing is characteristic of surface exposures, its change in frequency with depth cannot usually be seen: the deep quarries at De Lank provide an invaluable demonstration of such jointing. Similarly, although surface exposures often display such phenomena as xenoliths, pegmatitic segregations and small veins of later, intrusive granite, these are usually weathered and of poor quality compared with fresh examples found in the quarry.

A major occurrence of fine-grained, megacrystic biotite granite is found a short distance to the north of the site, and an apophysis of this, exposed in the De Lank River in the north-east corner of the site, is one of the very few of this type in a fresh condition in Devon and Cornwall.

The remains of three substantial outcrops of dykes of granite porphyry ('elvan') are also present; although the bulk of the central parts of these dykes has been worked out, unusually good specimens are available in the ends of the cuttings and the contact facies remain excellently preserved.

Description

The De Lank Quarries are part of a group, few of which are now working, on the western margin of the Bodmin Moor Granite about 9 km NNE of Bodmin. This granite mass is one of the major cupolas on the Cornubian batholith. For the most part, it is composed of the small megacrystic biotite granite (Type B, Table 5.1; Exley and Stone, 1982) which is well seen in the quarries, but it also has four small areas of fine-grained granite (Type C, Table 5.1; Exley and Stone, 1982), one of which is just to the north of De Lank. The mass as a whole has been dated by the Rb/Sr method at 287 ± 2 Ma BP (Darbyshire and Shepherd, 1985).

Within the main quarries, the granite contains abundant megacrysts, mostly about 10–20 mm in length; these are of orthoclase microperthite, while the potash feldspars of the groundmass include microcline (Edmondson, 1970). There is a conspicuous, nearly vertical, foliation with an approximately north–south strike which is emphasized by the megacrysts. Although foliation is not rare in Cornubian granites, it is seldom as strongly developed as it is here.

In addition to the main rock-forming minerals, De Lank Granite contains about 1% tourmaline, contrary to the Geological Survey's assertion (Reid *et al.*, 1910), and it is thus similar to the rest of the mass.

The rock is well jointed in several directions, the chief subvertical orientations being about 075° and 340° and close to the mean for the northern part of the outcrop, with subordinate joints between these. Dip directions and amounts are variable. Subhorizontal joints are most prominent in the topmost 20 m where they undulate in approximate conformity with the land surface, but the granite becomes very massive at depth.

The rock is cut by aplite and microgranite veins and sheets up to 0.10 m thick, which have strikes parallel with those of the joints. The rock also encloses veins and pockets of quartzo-feldspathic pegmatite, as well as xenoliths which are sometimes stretched into *schlieren*.

The quarry area is limited on the north by an ENE–WSW fault zone dipping towards the south and with easterly dipping slickensides. This fault, the surfaces of which are coated with tourmaline, is a major structural feature, being one of a number which separate the outcrop of the Bodmin Moor mass into large blocks (Exley, 1965), and it controls the course of the adjacent De Lank River. It also separates the De Lank and Hantergantick quarries, the granites of which have perceptibly different compositions. Although similar block faulting almost certainly exists in other Cornubian granite masses, it is not as well demonstrated as it is on Bodmin Moor.

North of the De Lank Quarries, centred on Lower Penquite, is an area of fine-grained granite (Type C), which has an outcrop about 1 km in diameter and a long, narrow apophysis leading south. The latter is exposed in the De Lank River, close to the fault mentioned above, and is clearly intrusive, while the presence of the main outcrop is revealed by boulders in the fields. This variety is younger than, and intrusive into, the coarse granite.

Immediately south of the northern working quarry at De Lank, two granite-porphyry dykes ('elvans') striking ENE–WSW, about 10 m thick, are exposed in road cuttings and quarries on both sides of the river, and there is a third in a quarry on the south-west side. Much of the rock has been removed, but the chilled margins and faces at the ends are accessible and show the distinctive features of this rock which is fine-grained and often megacrystic. It is not clear whether the De Lank elvans are all single intrusions or multiple like that at Praa Sands.

Interpretation

This site provides a superb example of typical Cornubian, coarse, small-megacrystic biotite granite which is extensive elsewhere only at Carnmenellis. Here, however, it has a strong tectonic foliation. There is well-developed jointing and pegmatitic patches which, together with minor granitic veins, illustrate the effects of late magmatic fractions. The exposure of typical Cornubian fine-grained megacrystic biotite granite and granite-porphyry dykes indicate subsequent intrusive phases, while xenoliths provide examples of material incorporated by the magma during its ascent. Opportunities to see all these phenomena, and their relationships within such a small area and in such a fresh state, are rare.

As is usual in batholiths, the separate intrusions which comprise that in Devon and Cornwall vary somewhat in age (Table 2.1). The oldest is Carnmenellis (290 ± 2 Ma) and the youngest Dartmoor and Land's End (280 ± 1 and 265 ± 2 Ma, reset) with St Austell and Bodmin Moor between these at 285 ± 4 and 287 ± 2 Ma respectively (Darbyshire and Shepherd, 1985).

Textures also vary between individual plutons, and while the granites of Dartmoor, the eastern (oldest) part of St Austell and Land's End have relatively large megacrysts, those of Bodmin Moor and most of Carnmenellis are relatively small, although abundant (Dangerfield and Hawkes, 1981). Since the development of large megacrysts is a feature of the upper and outer regions of Cornubian plutons, it is possible that their absence from the Bodmin Granite indicates a deeper level of erosion, as is suggested also by the regular shape of their outcrops.

The foliation often seen locally in the Cornubian granites seldom extends for more than a few metres and is frequently curved, sometimes showing 'swirls' reminiscent of eddies in liquids. They have often been ascribed to magma movement, and some may have originated in this way, although some of the minerals, especially the feldspar megacrysts, are subsolidus and must therefore owe their alignment to pre-existing structures. The foliation at De Lank is quite different, and textural relations and extensive strain in the quartz, which is largely segregated into bands, show that it has a deformation, not igneous, origin. It is clearly different from that seen, for example at Haytor Rocks, Luxulyan Quarry or near Cape Cornwall, and is presumed to be associated with movement along the neighbouring St Teath–Portnadler Fault system (Dearman, 1963).

It has been argued, in the 'Petrogenesis' section above, that the granites are predominately 'S-type' (Chappell and White, 1974), and one of the pieces of evidence for this is the nature of the xenoliths. Excellent examples of these, often seen now as *schlieren*, are present at De Lank and are of 'restite' origin, comprising largely biotite and andalusite.

The field relations of the fine-grained granite apophysis in the De Lank River indicate that it is intrusive, and suggest that the larger mass to the north, to which the apophysis is presumed to extend, is intrusive also. Unfortunately, the latter is not seen *in situ*, and it is possible that it may represent 'granitized' sedimentary raft material as has been suggested for some fine-grained granite on Dartmoor (Edmonds *et al.*, 1968; Hawkes, 1982) and in the Land's End intrusion (Tammemagi and Smith, 1975).

The elvan dykes, which are some 10 Ma younger than the main granites, are believed to represent a differentiate from a deeper-seated biotite granite magma which underwent considerable modification by ion exchange and was emplaced as a fluidized system which included fragments of the granite through which it had passed (Stone, 1968; Goode, 1973; Henley, 1972; 1974).

Conclusions

This site provides a superb example of typical Cornubian coarse biotite granite of the small-megacryst type typical of Bodmin, Carmenellis and Scilly. It shows strong foliation and xenoliths (see 'Birch Tor' conclusion), as well as typical Cornubian fine-grained megacrystic biotite granite and granite-porphyry dykes. Opportunities to see all these phenomena and their relationships within such a small area and in such a fresh state are rare.

C4 LUXULYAN QUARRY (GOLDEN POINT, TREGARDEN) (SW 054591)

Highlights

Luxulyan Quarry contains the coarse, megacrystic biotite granite of typical Cornubian type which forms the earliest variety in the complex magmatic sequence at St Austell. Its fresh xenoliths of pelitic and semipelitic sediment provide evidence about the origin of the magma, and there is also evidence of post-magmatic activity in the form of luxullianite *in situ*.

Introduction

Luxulyan Quarry (which has also been called Goldenpoint and Tregarden) is situated in typical Cornish granite of the St Austell mass, described in detail by Ussher *et al.* (1909); these authors, however, did not realize that there was a sharp distinction between this eastern rock and the granite types seen a few kilometres to the west. The differences were first recognized by Richardson (1923) and later Exley (1959), who both concluded that the Luxulyan Granite represented an earlier, separate intrusion which has subsequently been interpreted as a boss about 9 km across and dated, by the Rb/Sr method, at 285 ± 4 Ma BP by Darbyshire and Shepherd (1985). The boss was emplaced by stoping and subsidence, and xenoliths of the country rock, found in the quarry, have been used by Lister (1984) as evidence bearing on the origin of the magma.

The rest of the St Austell outcrop consists of a second, slightly larger, intrusion which, having been emplaced to the west of the Luxulyan Granite, was itself intruded by a magma of entirely different composition and was altered by

The Cornubian granite batholith (Group C sites)

a complex interchange of elements in a volatile-rich environment. The changes and mechanisms are noted in general terms in Chapter 2 and the 'Petrogenesis' section above, and in detail in the appropriate site descriptions. It is sufficient here to note that, although widespread in the western area, they have affected the Luxulyan area to only a moderate degree (Figure 5.4).

Among the few volatile-induced modifications, however, was the formation of luxullianite. This rock, described by Bonney (1877a), Flett *in* Ussher *et al.* (1909) and by Wells (1946), was known only from boulders until found *in situ* by Lister (1978, 1979a, with a contribution by Alderton (1979)). Lister (1979b) also used material from this site in her study of quartz-cored tourmaline.

Description

The granite at Luxulyan (Figure 5.10) has been described (Dangerfield and Hawkes, 1981) as 'coarse megacrystic' and as Type B (Table 5.1) (Exley and Stone, 1982). It is characterized by biotite and zoned oligoclase (An_{25-30}), and contains abundant K-feldspar megacrysts between 20 and 100 mm in length. These are generally aligned and commonly include zones (containing quartz, plagioclase and biotite) indicative of growth following a potassium-rich metasomatism process. Other minerals include muscovite, tourmaline and trace quantities of apatite, topaz, andalusite, fluorite, zircon and iron ore. The rock is closely comparable with the other coarse biotite granites of Cornwall and Devon and its mineralogy and chemistry show that, despite its early arrival in the St Austell sequence, it is a highly evolved, high-level variety (Table 5.2).

Evidence bearing upon the emplacement of this granite is seen in the abundant, rounded xenoliths, which are usually about 20 mm across, but range up to about 130 mm. These are mainly composed of quartz and abundant biotite, but sometimes contain andalusite. The majority have lost any foliation that they might have had, but

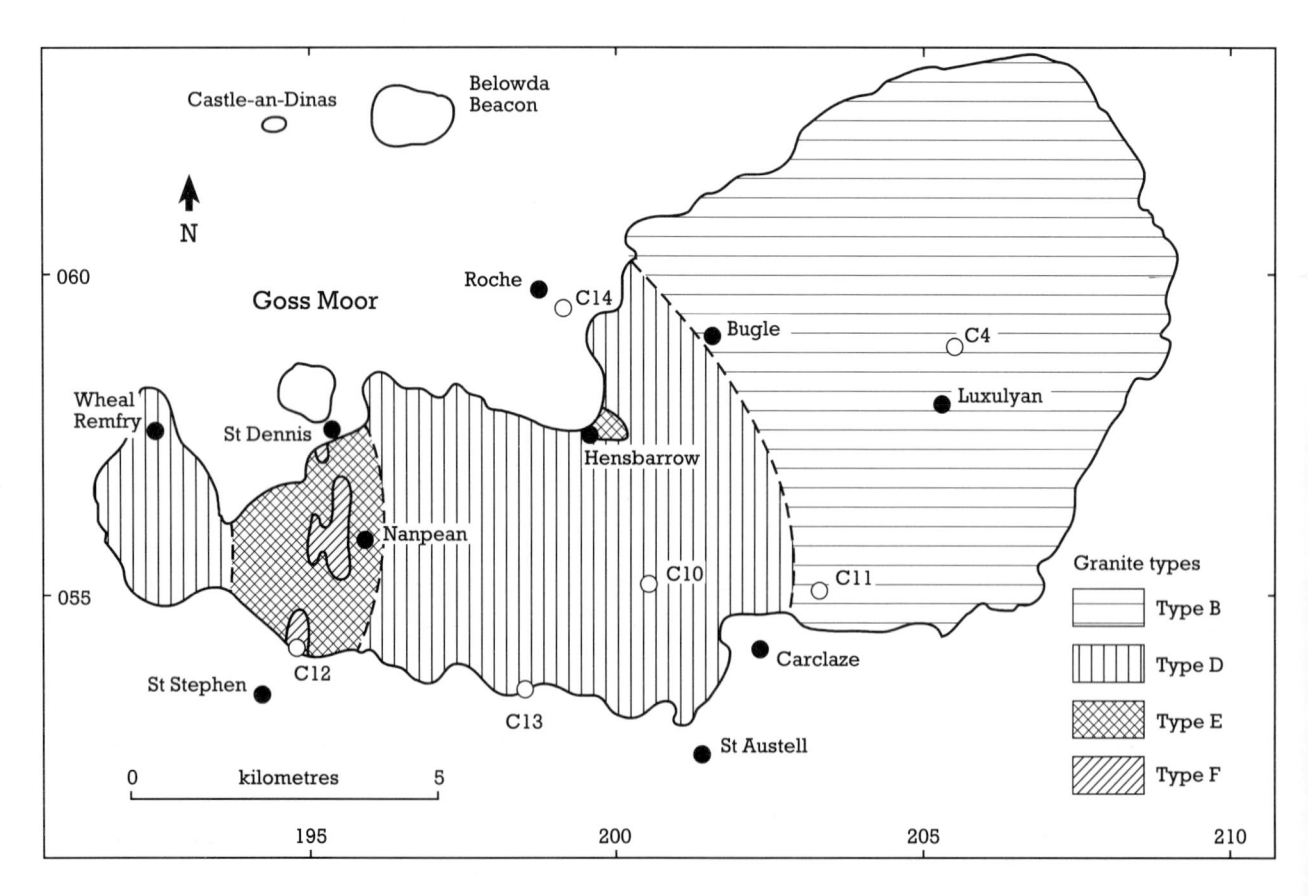

Figure 5.10 Map of the St Austell Granite outcrop, showing the chief granite types, localities mentioned in the text (filled circles) and the following sites: C4 = Luxulyan Quarry; C10 = Wheal Martyn; C11 = Carn Grey Rock; C12 = Tregargus Quarries; C13 = St Mewan Beacon; and C14 = Roche Rock.

have not yet been sufficiently 'granitized' to develop feldspar. They are clearly of pelitic or semipelitic origin and derived from the stoping of its walls by the magma. Fresh cordierite has been recorded from this quarry (Ussher et al., 1909).

Luxullianite, an attractive rock composed of red K-feldspar, acicular tourmaline and quartz, and formerly used as an ornamental stone, occurs in often discontinuous, near-vertical sheets which sometimes anastomose. These strike approximately ENE–WSW, and are up to a metre or two in thickness. The jointing is both more extensive and less regular than in many Cornubian exposures, and some joints show evidence of post-magmatic activity in the form of reddening and veining by quartz and tourmaline. Sometimes, such tourmaline has cores of quartz or feldspar (Lister, 1979b). There are also small pods, up to 0.5 m in diameter, of pegmatite consisting of the chief minerals of the granite but mostly deficient in plagioclase. A major joint, with a veneer of tourmaline striking 070°, serves as the quarry wall beneath the crushing plant.

Kaolinization, not due primarily to weathering, is confined to a zone striking N–S and tapering downwards in width from about 10 m. This separates the north-eastern quarry from the rest of the site.

Interpretation

The granite at Luxulyan is typical of the eastern part of the St Austell outcrop, which, having been recognized as significantly different from the granite in the western part (Richardson, 1923; Exley, 1959), was interpreted by the latter as the first member of a magmatic differentiation series. This interpretation followed from its relative enrichment in Ca and Fe (exhibited in oligoclase and biotite) and impoverished in Na (in albite), Li (in zinnwaldite), B (in tourmaline) and F (in topaz) (Tables 5.1 and 5.2). Later work has shown, however, that much of the western granite has a similar texture to that in the east and a composition intermediate between the medium-grained Li-mica–albite–topaz granites (Type E, Table 5.1) and the biotite granite. Hence the present interpretation is that the first member of the western intrusion was also a biotite granite, much of which was metasomatized by incoming Li-mica–albite–topaz granite derived at depth (see Chapter 3 and the 'Petrogenesis' section of this Chapter). The first importance of the Luxulyan site therefore lies in its exposure of the rock held to be the earliest in both hypotheses.

Theories about the derivation of the Cornubian granite magma agree that it resulted essentially from partial melting of a lower crustal source. However, the extraordinary enrichment of the batholith, relative to average granites, in such elements as Sn, W and Cu, Li, Sr and Ba, U and Th and B and F have led to speculation as to whether their provenance was middle or lower crustal or subcrustal (Simpson et al., 1976, 1979; Watson et al., 1984), and to what extent they were incorporated either from already enriched crustal material or from some subcrustal source. A study of xenolith material, some of which came from Luxulyan Quarry, has suggested that Sn, W, U and Ta were not derived from pelitic sediments, that V, Ba, Sr, Cu and Zn might have been, and that some elements which could easily have escaped from the magma (for instance, Li, Th and F) were in fact retained and concentrated in biotite-rich xenoliths. Those elements not derived from assimilated sediments must have been magmatic. Luxulyan xenoliths are thus of importance in the petrogenetic history of the Cornubian granites (Lister, 1984).

Granitic rocks generally similar to luxullianite have been found in various parts of south-west England, and it is agreed, from textural evidence, that they were formed by the post-magmatic alteration process described as 'tourmalinization'. Flett (in Ussher et al., 1909), however, when contrasting luxillianite with the tourmalinite of the Roche Rock 6–7 km to the WNW, observed that 'In luxullianite the process of metasomatic replacement has stopped at the half-way stage'. There has also been disagreement over the nature of the replacement and the original mineralogy. Thus, Bonney (1877a) thought that brown tourmaline had replaced biotite, but both Flett (1909) and Wells (1946) believed that biotite had never been present, brown tourmaline occurring instead. Again Bonney thought that acicular, blue, secondary tourmaline formed from feldspar, but Flett and Wells considered that it replaced both feldspar and brown primary tourmaline. Lister's (1978) examination of the first *in situ* luxullianite to be described makes it clear that biotite and primary tourmaline coexist, and that secondary tourmaline came from a hydrothermal generation and did not involve the breakdown of the primary crystals. However, the chemical changes between unaltered granite and luxullianite described by Lister (1978) differ from earlier suggestions, principally in showing a decrease in SiO_2 and an increase in K_2O and, following a discussion by Alderton (1979), she agreed that probably there had been a combined process of tourmalinization and K-feldspathization. It is worth noting as Charoy (1982) points out, that there is 'tourmalinization and tourmalinization', and that Lister and Alderton were not comparing like with like. As for tourmaline in veins, Lister (1979b) noted that some from this quarry contained cores of 'polycrystalline quartz and/or feldspar' and attributed this to skeletal growth resulting from undercooling of the tourmalinizing melt.

Regarding other alteration processes, Luxulyan is typical of the eastern St Austell area in showing only minor greisening and kaolinization, although it is interesting that the kaolinized zone in the quarry is of the wedge shape, described by Bristow (1977) as characteristic of such zones found throughout the Cornubian granites, although often on a much larger scale and at such a depth that this shape is revealed only when the zones are worked or from boreholes.

The biotite granite at Luxulyan is typical of Cornubian granites, and is enriched in elements such as Sn, W, Cu, Li, Sr, Ba, U, Th, B and F. It is uncertain to what extent these were contributed by crustal or mantle sources, but research on biotite in xenoliths from Luxulyan suggests that Sn, W and U (as well as Ta) did not come from sediments and were thus not crustal, that B, Sr and Cu (and also V and Zn) could have done so, and that magmatic Li, Th and F were trapped and thus concentrated in the biotite in xenoliths.

A similar type of granite formed the major part of a second intrusion to the west, but this was metasomatized to give the albite–Li-mica–topaz variety now present.

Luxullianite from the quarry has shown that there are two generations of tourmaline in these rocks, that tourmaline and biotite are not necessarily mutually exclusive and that the tourmalinization process may be accompanied by K-feldspathization.

Conclusions

Luxulyan Quarry provides an exceptional opportunity to examine the typical Cornubian biotite granite in a locality close to the succeeding lithium- and volatile-rich rocks of the St Austell complex of intrusions. The site shows xenoliths here consisting of metamorphosed, muddy sedimentary rocks. It is thought that the incorporation of these rock fragments has made a significant contribution to the final chemistry of the granite. The nearby village gives its name to the rock type luxullianite. This rock, made up of red feldspar, quartz and the dark mineral tourmaline (a complex boron-bearing aluminium silicate), was formed by alteration of the original St Austell Granite by hot fluids associated with the final phases of granite magmatic activity, which, flowing out from the solidifying granite, chemically altered and recrystallized the minerals which made up the granite and the rocks around it.

C5 LEUSDON COMMON (SX 704729)

Highlights

Migmatization is very rare in association with Cornubian granites, and Leusdon Common probably shows the best available example. Exposures here, which are mostly of boulders, show mixtures of granite and metasediment in intimate relationship resulting from mobilization and plastic flow. They come from the roof and the uppermost wall of the Dartmoor intrusion, which is composed of coarse biotite granite.

Introduction

Leusdon Common is a small area of gorse-covered ground some 6 km north-west of Ashburton. Exposed on its southern slopes are small, separated outcrops and boulders of the country rocks of the Dartmoor Granite, illustrating the nature of the uppermost wall or lower roof of the intrusion, together with a narrow apophysis of fine-grained granite. Brief accounts of the rocks are given by Reid *et al.* (1912) and Dearman (1962).

Description

The main body of the granite underlying the high ground at Leusdon is coarse-grained megacrystic biotite granite (Dangerfield and Hawkes, 1981; Type B, Table 5.1; Exley and Stone, 1982), of which the finer variety mentioned above is derivative. It too is megacrystic, at least in its uppermost 0.10 m, and it is *in situ.* It has given rise to spotted hornfelses at its contact, whereas the country/contact rocks found elsewhere on the site have a much more migmatitic aspect which is most unusual for Cornubian granite contacts. Cordierite is plentiful in the metamorphic rocks. Some boulders are composed of intermixed granite and sediment, and in other cases xenoliths in all stages of assimilation occur, some still showing original sedimentary banding and some being aligned. The granite in these cases is fine grained and varies from veins a centimetre or two thick to narrow, thread-like apophyses. These veinlets are often contorted and may penetrate into the metasediment, causing fragments to spall off. The metasediments also are often contorted.

Interpretation

These exposures exemplify some of the complex relationships between magma and country rocks involved in the emplacement by stoping of granite magmas. Although the mechanical effects are perhaps better shown at some Cornish contacts, for instance at Rinsey Cove, migmatization in such places is absent, as it is at two other Dartmoor roof contacts at Sharp Tor and Standon Hill. Neither is it prominent at the Burrator wall contact, although Brammall and Harwood (1932) referred to it in describing the changes brought about in Dartmoor Granite magma by the assimilation of Devonian metasediments.

The process of migmatite formation requires a combination of high temperature, high pressure and high volatile (especially water) content; it is, therefore, not commonly associated with high-level, relatively cool granites such as those in Cornubia, and there is no pattern in its occurrence here. The conclusion is that it took place in small, restricted regions near the upper parts of intrusions, where heat and water were locally concentrated, and that it may also have been connected with the potassium metasomatism which produced the megacrystic outer granites. Migmatization itself is of little consequence in the assimilation process discussed by Brammall and Harwood (1932), but the ramification of mobile granitic vein material, whether magmatic or migmatitic, was the chief stoping mechanism by which blocks of country rock were prised off to be engulfed by, and incorporated into, the granite magma.

Conclusions

There is no better site than Leusdon Common for the study of the complex relations between intrusion and country rock resulting from a combination of stoping and migmatization during emplacement of a Cornubian granite magma.

C6 BURRATOR QUARRIES (SX 549677)

Highlights

Burrator Quarries contain a rare contact between Dartmoor Granite and Upper Devonian sediments, showing slight mobilization, tourmalinization and vein intrusion. As the site of early investigation of compositional relationships between granite and country rocks, it is important in the development of hypotheses about the origins of Cornubian granites.

Introduction

This pair of small quarries lies 3 km east of Yelverton and 300 m south-west of the dam at Burrator Reservoir. It is one of the few places where the coarse, megacrystic biotite granite of Dartmoor can be seen in contact with the Devonian country rocks. This contact was a key feature in the evidence used by Brammall and Harwood (1932) in their petrogenetic study of the Dartmoor Granite, for they were able, by a series of analyses taken across it, to draw conclusions about the effects of assimilation on the granite magma. The significance of assimilation is now seen differently in the light of experimental work on the 'granite system', and more recent observations (Stone, 1979; Stone and Exley, 1986; Bromley and Holl, 1986; Bromley, 1989), but the changes discussed by Brammall and Harwood remain unchallenged.

Description

The country rocks on the west of the Dartmoor mass are essentially pelites and semipelites of Famennian age belonging to the Kate Brook Formation of the 'Kate Brook Tectonic Unit', the autochthonous component of the complex nappe

Figure 5.11 Contact between Dartmoor Granite and Devonian slates, re-exposed after face cleaning by the Nature Conservancy Council in 1980. (Photo: M.J. Harley.)

structure found between the Dartmoor and Bodmin Moor granites (Isaac *et al*., 1982; Isaac, 1985). These basinal sediments were first regionally metamorphosed to slaty rocks, and subsequently thermally metamorphosed by the granite. At Burrator they are spotted and banded cordierite- and andalusite-bearing hornfelses with a conspicuous flat-lying cleavage and extensive tourmalinization. Corundum, found in the Land's End contact aureole at Priest's Cove, has not been reported from here, and the tourmaline is chiefly a patchy yellow-brown and blue-green variety. The presence of much biotite probably indicates a degree of potassium metasomatism, and, certainly, the movement of potassium is recorded by the occurrence of perthitic feldspar close to the contact. Fine-grained granite and siliceous veins up to 0.3 m wide occur within the hornfelses, both concordantly and discordantly.

The contact itself, illustrated by Brammall and Harwood (1932), although very irregular, is sharp, and it shows evidence of some mobilization of the metasediments with attendant segregation of felsic and mafic constituents (Figure 5.11); Brammall and Harwood described this as a 'migmatic zone' although there is much less evidence of mobilization than at Leusdon Common (discussed above).

Tourmalinization has affected not only the contact hornsfelses but also the outer parts of the intrusion, and the granite in both quarries is considerably reddened by this process, especially adjacent to joints.

Interpretation

The Dartmoor Granite was emplaced at about 280 Ma BP, as dated by the Rb/Sr method (Darbyshire and Shepherd, 1985) making it the youngest Cornubian intrusion except for Land's End, and most of the rock is coarse grained and megacrystic, although a non- or poorly megacrystic variety also occurs. These were called 'giant' and 'blue' granite respectively by Brammall and Harwood (1923), who argued that the latter was intruded into the former: it has subsequently been recognized that the variation is gradational (Dangerfield and Hawkes, 1981; Hawkes, 1982), with both types corresponding with the Type B (Table 5.1) of Exley and Stone (1982) and Exley *et al.* (1983). It is often the case that the coarse and megacrystic nature of the granite persists right to its contact with the country rocks, as, for example, near Cape Cornwall, and this is seen in the Burrator Quarries, although the number of megacrysts is rather few and the crystals are only about 50 mm in length.

Brammall and Harwood (1932) published analyses from a sequence of 'shales' at Burrator as the contact was approached (over a distance of about 5 m) and used these, in conjunction with 'fresh' granite, to demonstrate both the chemical exchanges consequent upon intrusion and the effects upon the granite magma of the assimilation of xenoliths. In particular they concluded that the earliest of the Dartmoor magmas was 'acid' and was 'basified' by contamination at the same time as it was undergoing differentiation. The analyses remain valid and, in general terms, this view is still held, the earliest magma being envisaged as lower crustal and palingenetic in origin (see 'Petrogenesis' section above). It then rose through a pile of sediments and basic volcanics to its final high-level position where it consolidated in an essentially passive fashion, as the locally undeformed rocks show. However, it is now believed that some minerals, especially biotite, were derived from the source rocks, and that at low crustal levels the physicochemical condition of the magma prevented substantial assimilation (Stone, 1979; Bromley and Holl, 1986; Stone and Exley, 1986; Bromley, 1989) of the sort envisaged by Brammall and Harwood. The introduction of potassium from sediments was part of Brammall and Harwood's thesis, but it is now thought that late-magmatic potassium metasomatism was facilitated by an aqueous phase, and that the megacrystic texture was a consequence of this (Stone and Austin, 1961; Hawkes *in* Edmonds *et al.*, 1968; Exley and Stone, 1982; Exley *et al.*, 1983; Stone and Exley, 1986).

Although the assimilation of metasediment is not now regarded as being as important as Brammall and Harwood believed, the changes in the rocks at Burrator Quarries, and their interpretation by these authors, continue to be important to the understanding of the effects of Cornubian granite intrusion on the country rocks and the alterations which can result from assimilation.

Conclusions

This site is important both as a locality where contact phenomena of the Dartmoor (and, by implication, other Cornubian) Granite can be seen and as a place where classic work on

The Cornubian granite batholith (Group C sites)

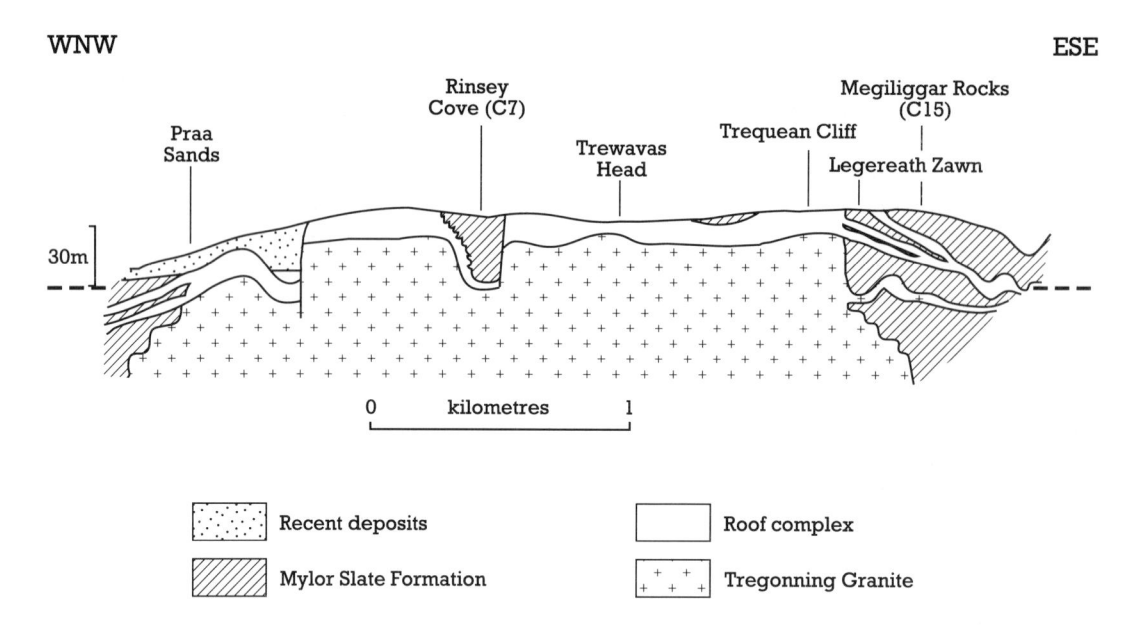

Figure 5.12 Diagrammatic section across the Tregonning Granite, based on coastal exposures, showing the location of sites at Rinsey Cove (C7) and Megiliggar Rocks (C15) (after Exley and Stone, 1982, figure 21.2).

Cornubian granite genesis was carried out. Although the conclusions from this work have been modified by subsequent findings, the data and principles still stand.

C7 RINSEY COVE (PORTHCEW) (SW 593269)

Highlights

This site has a unique section through a pelitic roof pendant in a granite pluton. Its late-stage and metasomatic minerals and textures in the granite and country rock, reflect the influence of magmatic volatile constituents.

Introduction

There are several places where contacts between Cornubian granites and their country rocks can be seen, but this section (Figure 5.12) through a roof pendant is unique. Not only do its margins show stepped contacts, xenoliths and granite apophyses, demonstrating the emplacement of the Tregonning Granite by stoping, but the granite itself, a lithium-mica-bearing variety, has developed a roof complex of leucogranite, pegmatite and aplite associated with a coarse-grained facies at the contact. Together with extensive tourmalinization, these features demonstrate the effects of volatiles such as OH, F and B during crystallization as they became progressively concentrated close to a nearby impermeable envelope.

A detailed field description and petrography was given by Hall (1930), general accounts of the petrogenesis are included in Exley and Stone (1982) and Exley *et al.* (1983), and details of the origin of the Li- and F-rich granite come from Stone (1975, 1979, 1984). The origin of the roof complex has been explained by Stone (1969), Bromley and Holl (1986) and Badham (1980).

Description

Rinsey Cove, or Porthcew, is on the south coast of Cornwall immediately to the east of Rinsey Head and about 4 km north-west of Porthleven. The Tregonning–Godolphin Granite pluton, which meets the coast here, is composed of two variants; the more northerly Godolphin facies consisting of fine-grained, megacrystic biotite granite (Type C, Table 5.1; Exley and Stone, 1982), while the southern Tregonning facies, exposed in the coast section, is made up of medium-grained, non-megacrystic lithium-mica–albite–topaz granite (Type E, Table 5.1; Stone, 1975; Stone and Exley, 1982). This has developed a local megacrystic facies and banding which, unlike that in western parts of the Bodmin Moor

176

Granite, is not tectonic in origin, and which is parallel with the contact with the roof, visible on the west side of Rinsey Head and in the cliffs on both sides of the cove. This contact is sharp and dips seaward at about 30°, the granite immediately beneath it being somewhat coarser than average and having a sheet complex of lithium-mica leucogranite–aplite–pegmatite just below.

Tourmalinization is common in xenoliths which are found in all stages of assimilation, and acicular tourmaline appears on the underside of the country rocks near the cliff top at the eastern side of Rinsey Cove.

The rocks which make up the roof and pendant, which occupy most of the Cove, are Mylor Slate Formation metasediments which, although predominantly dark and pelitic, contain semipelite and psammite and thus have a striped or banded appearance. The place of the Mylor Slate Formation in a wider context of stratigraphy and structure has been discussed by Leveridge et al. (1984) and Leveridge and Holder (1985). The rocks also contain numerous quartz veins, both contorted and cross-cutting, much of the silica for which seems to have been derived from the mobilization of quartz within the rocks by the compression and heat of metamorphism. The local structure, better seen here than at many contacts, consists of recumbent folds, on the limbs of which are minor folds resulting from an earlier deformation. There are two cleavages, of which the more striking (originally termed S_2 but now redesignated S_3 to correspond with the regional chronology) is subhorizontal and undisturbed by the granite (Stone and Lambert, 1956; Stone, 1966).

In addition to being folded, the metasediments have been thermally metamorphosed to spotted hornfelses, with the development of cordierite, andalusite (locally chiastolite) and, through metasomatism, tourmaline. Corundum, present in the Land's End metasediments at Priest's Cove, has not been found, and migmatization, like that associated with the Dartmoor Granite at Burrator and Leusdon Common, is absent.

On the west side of Rinsey Cove, the contact dips towards the Cove at 20–30°, cutting across the flat-lying cleavage, but is 'stepped' along joints in units from 0.5 to 4 m high, demonstrating emplacement of the granite by stoping and subsidence; disorientated xenoliths and apophyses of granite show how the magma penetrated the killas and prised blocks away. Coarse granite and aplite–pegmatite layers up to 3-m-thick are present. At the head of the cove, a cave has been eroded along the contact which here shows evidence of having been faulted and mineralized. The eastern contact is nearly vertical and may be either intrusive or faulted. The main body of the granite has a medium-grained texture almost up to the contact where there are 1-m-thick pegmatite and pegmatitic patches nearby. A smaller, elliptical pendant or large xenolith about 15 m across and also accompanied by pegmatitic patches is exposed between tide marks on the shore a few metres to the south.

Interpretation

It is considered that the magma producing the Tregonning Granite evolved from the Godolphin magma as a consequence of the deep-seated separation of a fluorine- and OH-rich fraction and ion exchange to enrich the Tregonning magma in Li, Na and Al at the expense of Fe, Ca and Mg. Biotite was thus replaced by zinnwaldite, and plagioclase became more albitic. This process is demonstrated elsewhere in the St Austell mass (discussed below, Tregargus Quarries). The sheets of the roof complex at Rinsey indicate the building up of a later, volatile-rich residuum under a relatively impermeable roof. The lithium-mica leucogranite represents a direct continuation of magmatic melt evolution, whereas the aplite represents a crystallized silicate liquid, and the pegmatite replaces aplite which was metasomatized by a K-rich aqueous fluid (Stone, 1969; Manning, 1982; Exley and Stone, 1982; Exley et al., 1983; Manning and Exley, 1984; Pichavant and Manning, 1984). The roof complexes of other granites, for instance Land's End (seen near Cape Cornwall and Porthmeor Cove), are much simpler because of the absence of the lithium-mica granite component, while the related sheets of Megiliggar Rocks, a short distance to the southeast, are both thicker and more complex. The difference between aplitic and pegmatitic textures could, however, equally be a result of pressure variations following successive openings of the rock initiated by the movement of foundering blocks, as suggested by Bromley and Holl (1986). Badham (1980) combined features of both these explanations, accepting the variations in fluid compositions but emphasizing the importance of physical conditions in determining textures. The lithium-bearing micas from the Tregonning Granite and its leucogranite sheets, whose exact compositions were uncertain for many years, have now been identified as zinnwaldite and lepidolite respectively (Stone *et al.*, 1988).

The only other exposures of this type of granite are found in the western part of the St Austell mass, but in this case they are intruded into granite. Moreover, the roof complex, seen only in china clay pits in the vicinity of Hensbarrow Beacon, has not developed as distinctive a display of leucogranite, aplite and pegmatite as that at Rinsey Cove.

Conclusions

This site shows outstanding sections through the upper part of one of the plutons of the Cornubian Granite mass and its junction with the older sedimentary rocks which it has intruded, baked and altered. Here may be seen evidence of how the granite was emplaced when still molten rock, by the undermining and dislodgement of blocks of the surrounding rocks by forcefully penetrating joints and fissures (a process called stoping), followed by subsidence of blocks into the magma. Blocks thus prised off from the country-rock walls and a possible hanging mass of baked sediments are seen to be surrounded by granite. In time, the molten granite would have assimilated the blocks, but here they have survived because the granite cooled and solidified before this could happen. The site has a well-developed roof complex, the upper-formed product of the granite magma. Although still of granite composition, the component pegmatite, aplite and leucogranite of the complex differ in detail from one another and from the earlier-crystallized granite as a result of losing some constituents of early crystals, increased concentration of volatiles, and interactions between the constituents.

C8 CAPE CORNWALL AREA (SW 352318)

Highlights

The Cape Cornwall area shows a stoped contact between metasediments and Land's End Granite, which shows evidence of being a two-stage intrusion. Greisen-bordered mineral veins, sapphire in thermally metamorphosed country rock and evidence of potassium metasomatism of granite and metamorphic aureole also occur here with late-stage veins, pipes and pods of pegmatite and quartz–tourmaline, and a quartz–tourmaline contact facies, all resulting from a concentration of a volatile-rich phase at the granite margin.

Introduction

Although the cliff and beaches immediately north and south of Cape Cornwall have become, in recent years, famous exposures for demonstrating both contact and late-stage igneous phenomena of the Land's End Granite mass, very little has been published on the area. The Land's End Granites as a whole were described in the Geological Survey Memoir by Reid and Flett (1907), with a revision, following publication of a new 1:50 000 map in 1984, by Goode and Taylor (1988); the latter, which contained little petrological discussion, is supplemented by an 'Open File' report (Goode *et al.*, 1987) in which there is substantially more detail. A specific study was made by Booth (1966), followed by a note on the granites' relations with the 'granite system' (Booth, 1967) and a paper by Booth and Exley (1987). Most of these contain passing references to the Cape Cornwall area, but van Marcke de Lummen (1986) used material from both Priest's Cove and Porth Ledden in his study of the crystallization sequence in the granites.

Charoy (1979, 1981, 1982) has quoted examples from Porth Ledden in his work on late-stage phenomena, especially tourmalinization, and Lister (1979b) has included specimens from both Priest's Cove and Porth Ledden in her investigations into quartz-cored tourmaline crystals. At least part of this tourmalinization, which Manning (1981) and Pichavant and Manning (1984) thought might have been due to a complex hydrothermal/metasomatic process, may be related to the 270 Ma BP intrusive phase noted in Chapter 2 (Bristow *et al.*, in press). Duddridge (1988) has argued that at least some tourmaline must be metasomatic.

Description

Cape Cornwall is situated on the west coast of the Penwith Peninsula about 2 km WNW of St Just; the cliffs and beach immediately to the north and south of the Cape exhibit a variety of

Figure 5.13 The headland of Cape Cornwall which exposes contacts between Land's End Granite and adjacent metasediments. (Photo: S. Campbell.)

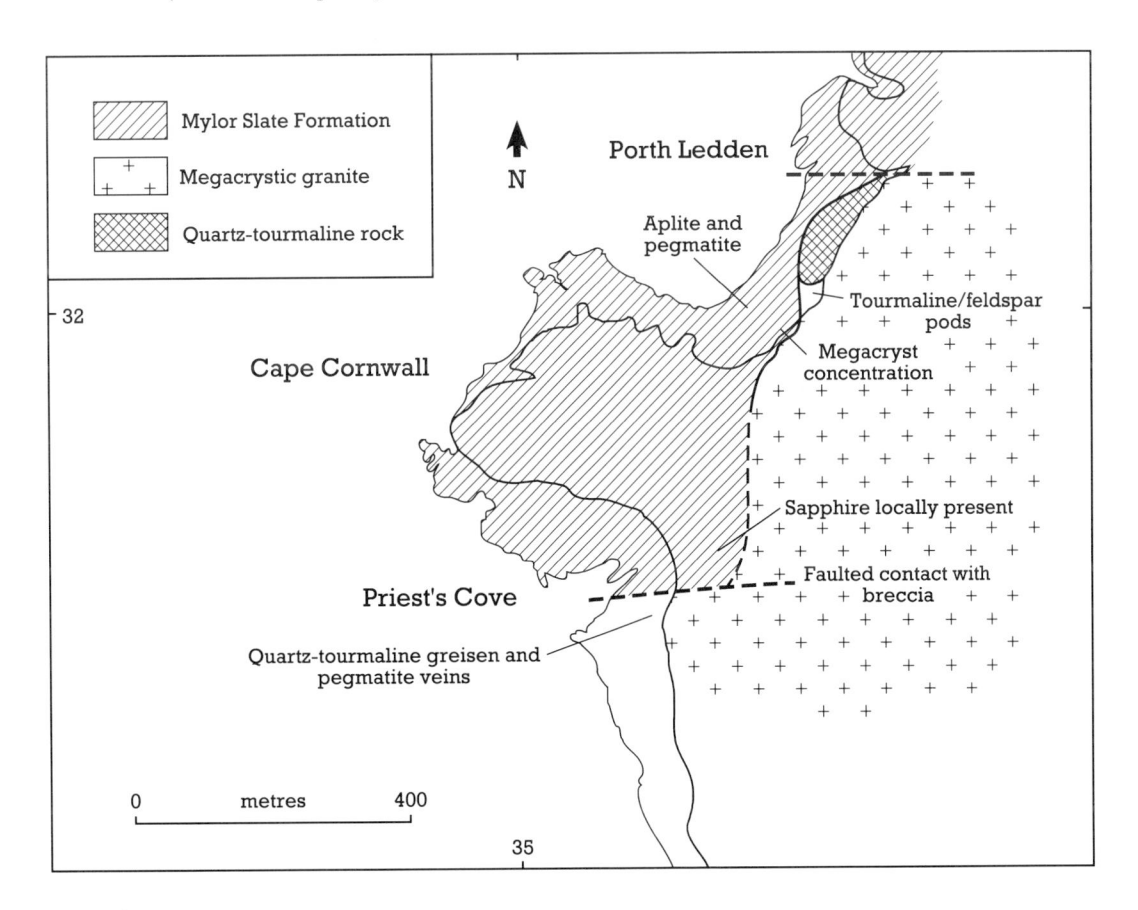

Figure 5.14 Geological sketch map of the Cape Cornwall area (site C8).

179

phenomena associated with the contact between the Land's End Granite and the Mylor Slate Formation metasediments ('killas') (Figure 5.13). Such phenomena are typical of Cornubian granites, but it is seldom that so many are displayed together in so small an area and can be related to one another so directly (Figure 5.14).

The country rocks are originally mudstones with subordinate sandstones, and these have been folded and metamorphosed into hard, splintery, grey, banded hornfelses, in which andalusite, chiastolite, cordierite, corundum and tourmaline have been identified. Corundum is of special interest, being unusual in Cornubian contact rocks and occurring as the variety sapphire in some exposures in the cliffs above Priest's Cove. Mobilization of the silica has produced extensive convoluted quartz bands and veins in the metasediments and, while these are typical of granite contacts throughout Cornwall and Devon, it is worth noting that there is no migmatization of the sort seen at Leusdon Common on Dartmoor (C5).

The contacts in Priest's Cove and near the stream at the northern end of Porth Leddon are faulted and hematized, and strike nearly east–west. The Priest's Cove contact is clearly exposed, vertical, and accompanied by a complex of quartz, quartz–tourmaline and pegmatite veins striking NW–SE, which is the trend of the tin-bearing veins formerly worked in the area. The breccia at this contact, formerly regarded as a result of faulting, is interpreted by Badham (1980) as 'a small breccia pipe formed during stoping'. Movement during and shortly after emplacement is usual in the Cornubian granites and, as here, often provided pathways for mineralizing solutions. Some of the quartz–tourmaline veins have dark, greisen borders a few centimetres wide.

The contact running down the cliff and across the beach in Porth Ledden is, in contrast, not faulted. Large K-feldspar megacrysts, both close to the contact itself and in isolated exposures presumed to be in an eroded roof pendant or large zenolith, provide examples of potassium metasomatism of the hornfels.

There is no evidence of more than slight disturbance of the country-rock structures as a result of intrusion, and granite emplacement must have taken place principally by stoping and subsidence; the step-like contacts which can result from this process, although discernible, are not well seen in this area and are better displayed at Rinsey Cove.

The Land's End Granite mass has been given an age of 268 ± 2 Ma BP (Rb/Sr method) (Darbyshire and Shepherd, 1985) but this is seriously at odds with mineral ages and the general age of the batholith and is thought to have been reset by mineralization (Table 5.1). Alternatively, it could relate to the later major intrusive phase at about 270 Ma. In the Cape Cornwall area, the granite contains many xenoliths of killas, and it is variable in both composition and texture. The main variety is a medium-grained and poorly megacrystic biotite granite (Dangerfield and Hawkes, 1981; Booth and Exley, 1987); this is a subdivision of the Type B of Exley and Stone (1982) shown in Table 5.1. It has megacrysts up to 40 mm long, but in some places it is almost aphyric, and elsewhere it contains veins and patches of a distinctly coarser variety. In Porth Ledden, a very coarsely megacrystic facies occurs close to the contact with the killas, where there is a pronounced parallelism between the megacrysts and the contact. This is quite different from the foliation seen, for example, in and about the De Lank Quarries on Bodmin Moor where there has been a strong tectonic influence. Good examples of aplite–pegmatite roof–complexes are also present, although these are not as well developed as those around Megiliggar Rocks near Porthleven, probably because the Land's End intrusion is devoid of a lithium-mica granite phase like that of Tregonning.

On the granite side of the Porth Ledden contact there is a lens of quartz–tourmaline rock which extends northwards from about halfway along the beach at the foot of the cliff. In some places this is subhorizontally banded and it passes downwards into aphyric, medium-grained granite. Within the main granite at the southern end, there are orbicular patches with tourmaline-rich rims and feldspathic cores which contrast with the pegmatitic pods, pipes and veins with tourmaline cores occurring in Priest's Cove. Some tourmaline is 'cored' (Lister, 1979b).

Many veins and apophyses of granite intrude the country rocks, their strikes often being controlled by the joint pattern; complexes of such veins are present in both Priest's Cove and Porth Ledden. The range of veins includes pegmatite (a notable one in Porth Ledden has curved feldspar crystals; *stockscheider*), medium-grained granite, aplogranite, microgranite and aplite. Some are relatively rich in tourmaline; some have a variety of textures including a development of pegmatite along their hanging-wall contacts.

Cape Cornwall area

Interpretation

The evidence of the veining and breaking up of medium-grained by coarse-grained granite suggests that in this area the Land's End magma was intruded in two pulses, the second following closely after the first had started to consolidate.

Together, these intrusions tilted the Mylor metasediments and thermally metamorphosed them to the equivalent of the anthophyllite–cordierite subfacies (examples of which are seen to the north of Porth Ledden at Kenidjack), with the development of sapphire locally as in Priest's Cove. Again, this is a rare phenomenon. The granites also sent out various veins and apophyses and incorporated fragments of country rock.

The remaining features, such as veins and pods of pegmatite and quartz–tourmaline, were dependent on the effects of volatile phases which became concentrated close to the granite–killas contact, as is usual in Cornubian rocks (although not so well displayed), but exceptional among British granites in general, presumably as a result of their lower volatile content. Thus the early concentration of water, particularly as it migrated to outer, cooler regions, would have accomplished the transfer of potassium within both granite and aureole, and led to the growth of megacrysts. At the same time, this migration facilitated the growth of feldspars in favoured localities, giving rise to pegmatitic bands, pipes and pods. Along with this aqueous phase was boron, which was concentrated enough in some places to form early, disseminated tourmaline within both granite and killas, although most of the tourmaline growth followed later, perhaps being linked with the 270 Ma BP intrusive episode (Chapter 2). In some places, such as at Wheal Remfry in the western lobe of the St Austell Granite mass, and at the Priest's Cove contact, the boron-rich fluid built up enough pressure to brecciate the confining rock (Allman-Ward *et al.*, 1982; Bromley and Holl, 1986; Halls, 1987; Bromley, 1989), but in others, for instance, Roche Rock and here at Porth Ledden, its effect was to cause pervasive tourmalinization (Bristow *et al.* in press). The fluid itself was probably a boron- and silica-rich component of the magma which separated either completely and formed areas of massive tourmalinite, or partially to form spots and patches (Badham, 1980; Charoy, 1979, 1981, 1982; and see 'Petrogenesis' section). The mechanism of separation was at least partly metasomatic, although complex (Manning, 1981; Pichavant and Manning, 1984): the precise process and effect in a particular case depending largely on physical conditions such as pressure, temperature and concentration gradients (Badham, 1980).

The cored-tourmaline crystals studied by Lister (1979b) were collected from near-vertical pegmatite veins and contained cores of polycrystalline quartz. Lister concluded from her examinations of the textures, chemical analyses and crystal growth mechanisms that they were formed by the supercooling of the tourmaline-bearing liquid. This would be a consequence of the sudden release of volatiles or of chilling, both of which could cause cooling below the normal crystallization temperature and thus instability at the 'crystal–fluid interface' and growth of unusual crystals. It is another effect of the high volatile content of Cornubian magmas. The greisen-bordered veins, some of which contained metallic minerals such as cassiterite and chalcopyrite, were emplaced last of all at about 270 Ma BP, that is, around 10 Ma after the intrusion of the main granite (Jackson *et al.*, 1982).

The exposures in the Cape Cornwall vicinity are outstanding, both as examples of the contact of the Land's End Granite mass with its metasedimentary country rock, and of the evolution of its late-stage facies. By analogy, they also provide an insight into the evolution of the other Cornubian granites and that of volatile-(especially boron-)rich granites in general.

The contact indicates that the granite was emplaced passively, by stoping, while K-metasomatism, the presence of pegmatites, extensive tourmalization and mineral veins show that the residual fluids of the magma were enriched in K (at least to begin with) and in B and a number of metallic elements. Depending on temperature and the extent of fracturing, these were able to penetrate both granite and killas and, because of good exposure, they illustrate how such processes must have operated in other granites of similar composition.

Conclusions

Both as examples of the contact of the Land's End Granite mass, its metamorphosed sedimentary country rocks, and of the evolution of its late-stage facies, the exposures in the Cape Cornwall vicinity are outstanding. These late-stage rocks include granitic rocks such as veins, pipes and pods of pegmatite and quartz–tourmaline rocks formed from volatile-rich solutions working their way to and crystallizing at the margins of the granite. Minerals also formed in the surrounding altered sedimentary rocks as a result of the influx of the late-magmatic fluids, and these include tourmaline and sapphire. By analogy, they also provide an insight into the evolution of the other Cornubian granites and that of volatile-(especially boron-)rich granites in general.

C9 PORTHMEOR COVE (SW 425376)

Highlights

This site uniquely exposes two satellite 'mini plutons' of the Land's End Granite showing development of a roof complex in biotite granite and a sequence of granitic dykes.

Introduction

This locality is about 3 km WSW of Zennor on the north-west coast of the Penwith Peninsula. It is one of several places where Land's End Granite can be seen in contact with its country rocks but is distinguished by the presence, on the eastern side of the Cove, of two small granite cupolas which have presumably arisen from the main intrusion. These are the only recorded examples in south-west England, of well-exposed, complete 'mini-plutons'.

Most references to the locality are very brief (Reid and Flett, 1907; Booth, 1966; Hall, 1974; Hall and Jackson, 1975; Exley and Stone, 1982; Booth and Exley, 1987; Goode and Taylor, 1988), the only comprehensive account thus far being that of Stone and Exley (1984), who describe the more accessible intrusion and its associated dykes; and also suggest a chronology, drawing attention to the evolution of its roof complex of interbanded leucogranite and pegmatite by differentiation. Bromley and Holl (1986) argue that the magma was emplaced after the foundering of a block of country rock, the overlying pegmatite/aplite complex developing through pressure changes as the block sank in stages.

Description

The country rocks at Porthmeor Cove are hard, grey, pelitic hornfelses of the banded Mylor Slate Formation which are sometimes spotted; these overlie a massive metadolerite sill some 20–25 m thick. The whole succession dips at about 20° to the north. The more northerly of the two cupolas on the eastern side of the cove, containing angular xenoliths and sending out apophyses, is visible in the cliffs but is inaccessible. The more southerly cupola, measuring about 19 m by 15 m, can easily be examined and displays several component rock types and a complex history (Figure 5.15).

It is an angular body, both in plan and in section, whose emplacement was controlled by the joints in the country rocks. All contacts are sharp and the roof is slightly domed. The main granite body is of megacrystic biotite granite (Dangerfield and Hawkes, 1981; Type B, Table 5.1; Exley and Stone, 1982; Exley *et al.*, 1983), but an aphyric, slightly banded granite occupies the central parts. A typical Cornubian leucogranite/pegmatite complex, 0.60–0.70 m thick, underlies the roof and this demonstrates the concentration of late-magmatic volatiles there and thus a considerable degree of differentiation *in situ*.

Three important dykes are associated with the cupola in addition to the vein complexes at the two seaward corners. The youngest dyke, composed of tourmaline microgranite, cuts through the cupola from north-west to south-east and then turns to the east-north-east, merging with a second dyke for a few metres before continuing into the hornfelses. A dyke of intermediate age, about 0.50 m thick and of megacrystic granite, arises directly from the cupola at its south-eastern corner and runs north-east into the country rocks. About 8 m from the pluton it cuts and displaces the oldest, 0.3 m thick, dyke which is of leucogranite (Figure 5.16). On the south side of its convergence with the youngest dyke, it forms a veneer on the steep cliff face and then joins an underlying leucogranite/aplite complex, only seen at low water. Bromley and Holl (1986) state that the top of a 'huge arrested xenolith' is visible at the lowest tides; this has not been confirmed by an examination at low-water spring tides, and rock relations appear to be more complicated than these authors suggest.

Interpretation

The complexity of rock relations at the base of the pluton does not invalidate Bromley and Holl's proposition that the subsidence of a large xenolith

Figure 5.15 Small granite cupola emplaced in pelitic hornfelses of the Mylor Slate Formation. Porthmeor Cove, Cornwall. (Photo: R.A. Cottle.)

Figure 5.16 Later dyke of megacrystic granite cutting and displacing an earlier leucogranite dyke. Porthmeor Cove, Cornwall. (Photo: R.A. Cottle.)

gave the opportunity for magma to move into the resulting cavity, decompression causing the volatiles to exsolve and the depleted magma to crystallize as aplite. Above this, the volatile-rich fluid would have cooled and solidified as pegmatite, and cycles of 'foundering, decompression and arrest' thus would have produced a banded complex.

The main Land's End Granite, which also has a roof complex of leucogranite, aplite and pegmatite over megacrystic biotite granite, has an exposed contact towards the head of the cove, but it is separated from the cupola by a fault. Roof complexes are well known in south-west England, and another occurs at Porth Ledden, but they become even more spectacular where lithium-mica granite is associated with them as in the sheets adjacent to the Tregonning intrusion and seen at Megiliggar Rocks (Stone, 1969, 1975).

It is significant that despite a degree of differentiation sufficient to produce pegmatite, there is no development of lithium–mica–albite–topaz granite at Porthmeor and Porth Ledden, suggesting that this variety (Type E, Table 5.1, found only at Tregonning–Godolphin and St Austell) does not evolve directly, and at high levels, from biotite granite magma (Exley *et al.*, 1983).

It appears that the main Land's End Granite pluton was intruded, tilting its envelope of sediments and sills to the north, and developed a roof complex through volatile concentration. Either this complex, or a closely related one, then gave rise to the earliest dyke, which is of leucogranite, at Porthmeor. The Porthmeor cupola, composed of biotite granite like the main Land's End intrusion, was subsequently passively emplaced by stoping, sent out the second dyke, also of megacrystic biotite granite, and developed its own roof complex. Finally, the third dyke, of tourmaline microgranite whose source is not seen, was intruded through all the older rocks (Hall and Jackson, 1975; Stone and Exley, 1984). The small cupola therefore repeats the cycle in the main granite in which the build-up of volatiles, principally B, F and OH, during crystallization caused the formation of rocks impoverished in Fe, Ca and Na but enriched in minerals, such as tourmaline, containing these volatiles. The partitioning of some elements between residual silicate magma and OH-rich vapour determined the formation of leucogranite, usually fine grained, or pegmatite. Where jointing provided lines of weakness, the magma from the cupola penetrated to form the second dyke; the last dyke, again following the jointing but younger than the others, represents a last stage in the granite evolution in the area.

Conclusions

Here there occur two domed granite masses (cupolas), one of which is a composite intrusion, made up of an outer (older) coarse-grained granite and a younger component forming the core of the intrusion. The top of the cupola is formed of a capping mass of rock, less than one metre thick, of two types – one coarsely crystalline (pegmatite) and the other finer grained and pale coloured (aplite/leucogranite) above the main mass of the biotite granite. Such rocks have been associated with the final volatile-rich (boron- and water-rich) granite magma, differentiation, cooling and solidification in place at the top of the granite mass. Pegmatite/aplite roof complexes are usually developed from Cornubian Li-mica–albite–topaz granite magmas, but the Porthmeor Cove example is a rare and instructive example of development from biotite granite. This exposure also shows the unusual emplacement of a satellite pluton after the main intrusion, and a clear sequence of related granitic dykes.

C10 WHEAL MARTYN (SX 003556)

Highlights

This site is exceptional in that it contains relatively fresh Li-mica granite, generated by metasomatic alteration of biotite granite, which forms the greater part of the western intrusion at St Austell and provides the main source of china clay.

Introduction

As described in the 'Petrogenesis' section of this chapter, the St Austell Granite comprises two main intrusions. That in the east is centred on Luxulyan and consists of coarse-grained biotite granite (Type B, Table 5.1), while that in the west is chiefly made up of megacrystic granite of variable grain size (mostly coarse) and characterized by the Li-mica zinnwaldite (Type D, Table

5.1). Formerly thought to have been a magmatic differentiate, this latter granite is now considered to have resulted from the metasomatic alteration of large parts of an earlier biotite granite by an intruding Li-mica granite magma (Richardson, 1923; Exley, 1959; Hawkes and Dangerfield, 1978; Dangerfield *et al.*, 1980). Other varieties within the western intrusion include the Li-mica granite product of the crystallization of this magma and fluorite granite derived from it (respectively Types E and F, Table 5.1; Manning and Exley, 1984; Hawkes *et al.*, 1987; and Figure 5.4). There are also small remnant patches of biotite granite, suggesting that the western intrusion was multiphase, and these have been interpreted as constituting a sequence of fractions of magma representing stages in the cooling history (Hill and Manning, 1987; Bristow *et al.*, in press).

The western intrusion is the area in which kaolinization has been most intense and from which the great majority of china clay is extracted, so that fresh examples of the original rocks are hard to find; the Wheal Martyn site provides these rare examples. The nearby china-clay pits of Greensplat, seen from the China Clay Museum's viewing point, demonstrate the severity of the alteration and show the manner of working china-clay rock.

Description

The Wheal Martyn site is roughly in the middle of the St Austell Granite outcrop, about 3 km north of St Austell. It is towards the eastern side of the second, western intrusion which has a diameter of about 11 km and is centred just to the north-east of St Dennis. Its northern part underlies the Lower Devonian metasediments and the superficial deposits of Goss Moor (Figure 5.10).

The excavation at Wheal Martyn contains relatively fresh Type-D granite (Table 5.1). In this, there are rather few megacrysts of subhedral feldspar, 10–15 mm long; rounded composite quartz grains 5–7 mm in diameter; and tourmaline needles up to 5 mm long, all set in a groundmass with an average grain size of 1–2 mm. Flakes of pale-brown Li-mica, identified as zinnwaldite by Stone *et al.* (1988) are also visible. It is probable that this rock would be classified as 'globular quartz granite' by Hill and Manning (1987). There is also a small patch of fine-grained biotite granite (Type C, Table 5.1), suggesting that in this area some fine-grained granite was emplaced relatively early.

Interpretation

Following the intrusion of a biotite granite (Type B, Table 5.1) boss in the east, a series of biotite-bearing granites, variable in texture, was emplaced on its western side. Owing to its coarse, megacrystic texture and its composition (Type D, Table 5.1), which includes albite (An_7) and Li-mica, the bulk of this rock was thought to be a magmatic differentiate from the main biotite granite magma (Exley, 1959). However, the realization that the non-megacrystic Li-mica granite (Type E, Table 5.1) was intrusive into megacrystic rocks led Dangerfield *et al.* (1980) to conclude that the surrounding Type-D granite had originally been biotite granite (Type B) and that it had been metasomatized. This conclusion is now widely accepted, although Bristow *et al.* (in press) consider that some is of magmatic origin. The metasomatism was achieved in the presence of an aqueous fluorine-rich fluid, by introducing lithium which replaced iron to form zinnwaldite, and sodium which replaced calcium in oligoclase to give albite: adjustments in other constituents balancing these changes. Surplus calcium was combined with fluorine into fluorite, and surplus iron, together with boron, crystallized as tourmaline. The introduction of the lithium, sodium and some of the fluorine and water, was effected by the intrusion of the Type-E magma which had separated at deeper levels from biotite granite magma with which it had undergone some ion exchange. The rock, formed directly from this magma, is now seen as a non-megacrystic Li-mica granite (Type E), containing patches of fluorite granite (Type F) in the Nanpean area and near Hensbarrow (Stone, 1975; Dangerfield *et al.*, 1980; Exley and Stone, 1982; Manning, 1982; Exley *et al.*, 1983; Manning and Exley, 1984).

An early form of alteration resulting from the presence of OH- and F-rich fluids was a pervasive greisening which attacked the K-feldspar, replacing it by a secondary white mica and quartz. The extent of this is very variable and the granite at this site shows it to a limited degree.

The whole area surrounding the site has been affected by intensive kaolinization. This was a consequence of Cretaceous or Palaeogene sub-aerial weathering initiated by hot water circulating through the joint system in the granite (Exley,

1959, 1964; Sheppard, 1977; Durrance et al., 1982; Exley and Stone, 1982; Exley et al., 1983; Manning and Exley, 1984; Bristow et al., in press). That closely spaced joints produced pervasive kaolinization can be seen in many clay pits, including those visible from the Museum viewing area, as can also the relationship between this process and tourmaline veins and greisen-bordered veins. The latter provide evidence for the movement of late fluids through the solidified granite and for the way in which these replaced the feldspar in the joint walls with aggregates of mica and quartz – an identical process to that seen in the body of the rock and noted above. Where they occur, ore minerals are usually found in greisen-bordered veins of this type. (See also the 'Petrogenesis' section of this chapter and the description of Tregargus Quarries.)

Conclusions

This site provides a rare exposure of relatively fresh, metasomatically originated, megacrystic Li-mica granite, which is one of the main varieties of the St Austell outcrop.

C11 CARN GREY ROCK AND QUARRY (SX 033551)

Highlights

This site is one of the few showing granite intermediate in character between that of the two main St Austell intrusions.

Introduction

Carn Grey Rock and its adjacent quarry lie 3.5 km to the north-east of St Austell, beside the road to Trethurgy. They lie in the contact zone between the first and second intrusions of the St Austell mass (Figure 5.10). The more western of these, much of which is now characterized by Li-mica, cuts the eastern biotite granite along a zone extending roughly between Carclaze and Bugle and in the direction of Roche. It has been thought, variously, that this rock is of direct magmatic origin or to have been metasomatically produced from biotite granite (Richardson, 1923; Exley, 1959; Hawkes and Dangerfield, 1978; Dangerfield et al., 1980; Manning and Exley, 1984; Hawkes et al., 1987; and also the 'Petrogenesis' section above). The western intrusion does contain some biotite granite, however, and is variable in texture, suggesting that its origins are not simple (Hill and Manning, 1987; Bristow et al., in press). Some biotite granite is found close to the presumed intrusive contact, indicating a degree of complexity there (discussed below). Although situated on the eastern side of this contact, Carn Grey Rock resembles the western types of (Li-mica) granite in respect of its texture, sodic plagioclase, high tourmaline content and low biotite content.

Description

The granite at Carn Grey Rock is medium- to coarse-grained and rather poorly megacrystic, with megacrysts up to 40 mm in length, and quartz in rounded aggregate grains. In thin section it shows many features indicative of re-crystallization, such as strain, zoning and intergrowth in minerals. Its composition is that of biotite granite from the eastern intrusion, except that it has less biotite and calcic plagioclase and more tourmaline. The biotite is very pale, and Richardson (1923) believed that both biotite and 'lithionite' (zinnwaldite) were present; the optical properties of these micas are similar, however, and they are unlikely to coexist as discrete phases. Indeed, Leech (1929) disagreed with Richardson and considered the Carn Grey Granite to be a distinct type, comparing it with that of Merrivale on Dartmoor. Carn Grey Rock is a rather 'flat' tor, about 4 m high, with well-developed subhorizontal jointing which is also seen in the quarry below, where it shows an antiformal structure. It is believed that this site was the source of many standing stones and menhirs in the St Austell district.

Interpretation

The first of the St Austell intrusions, which has a centre near Luxulyan and a diameter of about 9 km, is made up of coarse-grained granite with biotite, zoned oligoclase (An_{25-30}) and potash feldspar megacrysts (Type B, Table 5.1; Exley and Stone, 1982). The second, which was intruded across the western edge of the first, is centred near St Dennis, and is about 11 km in diameter.

Most of it contains zinnwaldite (Stone *et al.*, 1988), albite (An$_7$) and potassium feldspar megacrysts, but is generally not as strikingly megacrystic as the first intrusion and has a variable texture which includes some fine-grained rock and, in addition, pockets of biotite granite (Manning and Exley, 1984; Hill and Manning, 1987; Bristow, in press; Bristow *et al.*, in press). The zinnwaldite-bearing rock is Type-D granite (Table 5.1; Exley and Stone, 1982). Originally thought to be a member of a differentiated magmatic sequence (Richardson, 1923; Exley, 1959), much of this rock is now considered to be the result of metasomatism of biotite granite by the intrusion of a late-magmatic differentiate from biotite granite magma. This brought in lithium and sodium, and the resultant rock can be seen in its solid state in the Nanpean area and near Hensbarrow Beacon as a non-megacrystic Li-mica granite with albite (An$_{0-4}$) and topaz (Type E, Table 5.1). Its origins are discussed in the 'Petrogenesis' section above and in relation to Tregargus Quarries.

In most of the western area (see Figures 5.4 and 5.10), an aqueous, F-rich fluid exchanged Li and Na for Fe in biotite and Ca in oligoclase. The Fe not used in the resulting zinnwaldite, combined with B to form tourmaline, and Ca not retained in albite combined with B to form fluorite. Fluorite granite (Type F, Table 5.1) occurs in pockets within the Type-E granite in the Nanpean area (Manning, 1982; Exley and Stone, 1982; Exley *et al.*, 1983; Manning and Exley, 1984; Hawkes *et al.*, 1987).

The contact between the two main intrusions is not exposed, and severe kaolinization makes field relations difficult to interpret, but biotite granite has been reported from several localities in the western area (Richardson, 1923; Bray, 1980; Allman-Ward *et al.*, 1982; Hill and Manning, 1987), and the evidence seen so far suggests that the contact is an irregular zone rather than a plane. Although the texture of the Carn Grey Rock is typical of the western intrusion, its composition is intermediate between the eastern (Type B) and the western (Type-D) granites, and it seems to represent the easternmost point to which the metasomatism penetrated and a case where the changes were not complete.

Carn Grey is an important site, providing one of the fresh exposures in the south-west of the eastern St Austell intrusion. It has textural and compositional characteristics and a geographical position which suggest that it provides a link between the main original rock types of the eastern and western intrusions, where partial alteration by Li, Na and F, brought in by the youngest intrusion, can be seen. Successive intrusions and subsequent Li metasomatism do not occur in any of the other Cornubian granite masses.

Conclusions

Carn Grey is an important site, providing one of the rare fresh exposures in the south-west of the eastern St Austell granite intrusion. The site lies in the contact area of the first and second granite intrusions which make up the St Austell mass. Texturally, the granite here has the characteristics of the medium- to coarse-grained megacrystic (with larger crystals, to 40 mm) western granite, but the chemical/mineral composition approaches that of the eastern granite. It therefore has characteristics and a position which suggest that it might provide a link between the chief rocks of the eastern and western intrusions.

C12 TREGARGUS QUARRIES (SW 949541)

Highlights

Tregargus Quarries are unusual in that they contain the two late-magmatic variants (non-megacrystic lithium–mica–topaz granite and fluorite granite) of the St Austell sequence; examples of late- and post-magmatic mineralization and alteration are developed.

Introduction

The Tregargus Quarries (Figure 5.10) are sited in fluorite granite which contains patches of non-megacrystic Li-mica–albite–topaz granite (Types E and F, Table 5.1). These facies, which are surrounded by megacrystic Li-mica–albite–topaz granite (Type D, Table 5.1), are part of the western intrusion of the St Austell mass recognized by Richardson (1923), Exley (1959), and others subsequently. This intrusion was emplaced alongside an earlier biotite granite boss centred on Luxulyan (Figure 5.10), a distinction not identified by the early Geological Survey work (Ussher *et al.*, 1909). The western intrusion was originally biotite granite also, but it was largely metasomatically altered by the incoming Li-mica–albite–topaz granite magma (Hawkes and Dangerfield, 1978; Dangerfield *et al.*, 1980; Hawkes *et al.*, 1987), which had itself been generated from biotite granite at depth (Stone, 1975; Manning and Exley, 1984; Pichavant and Manning, 1984).

Fluorite granite ('china stone' or 'Cornish stone') such as that at Tregargus, was thought by Ussher *et al.* (1909) and Howe (1914) to be an intermediate stage in the breakdown of granite to china clay, but Coon (1913a, 1913b) regarded it as a late variety modified by circulating water. Exley (1959) believed it to be the youngest member in a differentiation sequence, but the current view (Manning and Exley, 1984; Pichavant and Manning, 1984) is that it is a by-product of the complex metasomatic processes by which the Li-mica–albite–topaz granite altered its biotite granite host.

Post-magmatic greisening and tourmalinization are seen in the veins of the quarries, and there is considerable early kaolinization. The last process has given rise to much controversy. One school of thought, exemplified by Hickling (1908) and Sheppard (1977), argued that it resulted from extensive deep weathering, but a second (for example, Collins, 1878, 1887, 1909; Butler, 1908; Ussher *et al.*, 1909; Howe, 1914; Exley, 1959, 1964, 1976) held that it was due to hydrothermal fluids of magmatic origin. A third, currently popular thought, attributes kaolinization to a combination of processes of both kinds (Coon, 1911, 1913a, 1913b; Tomkeieff, discussion in Exley, 1959; Bristow, 1977; Durrance *et al.*, 1982; Bristow *et al.*, in press).

All of these evolutionary concepts are put into historical context in the 'Petrogenesis' section of this chapter.

Description

The Tregargus Quarries are just under 1 km north-west of St Stephen, a village 7 km west of St Austell, and are towards the southern limit of an elliptical area, roughly 3.5 km from north to south and 1.25 km from east to west, which, for convenience, is here called the Nanpean area (Figure 5.10). Within it occur two of the four main varieties of St Austell Granite (Exley, 1959; Exley and Stone, 1982; Exley *et al.*, 1983; Manning and Exley, 1984) which have been extensively worked for china stone, the latter formerly being used as a flux in pottery manufacture.

In the Nanpean area most of the rock contains lithium-mica (zinnwaldite – Stone *et al.*, 1988), albite and topaz. Its texture is not megacrystic, and it encloses irregular patches, up to 400 m across and at least 70 m deep, of fluorite granite. The latter have rapidly transitional, non-intrusive contacts with the former, which the fluorite granite resembles in texture and composition apart from having no iron-bearing minerals (the mica is muscovite) and an average of 2% fluorite (Table 5.1). The two most important areas are immediately west of Nanpean and around Tregargus.

The rock in the Nanpean area escaped the potassium metasomatism which gave rise to large feldspar megacrysts elsewhere in the St Austell mass, but they do show, in varying degrees, greisening, tourmalinization and kaolinization. The first is both pervasive through the rock, but restricted to the replacement of some of the K-feldspar, and localized, where it occurs as true greisen borders a few centimetres thick alongside quartz and quartz–tourmaline veins. Tourmalinization, although pervasive in other places, having produced much (4%) tourmaline within the megacrystic Li-mica granite (Type D, Table 5.1), is localized in veins at Tregargus.

Kaolinization, and its relationship to joints and faults, is best seen in the non-megacrystic Li-mica granite, the fluorite granite largely remaining fairly fresh, although, where it has been affected, it used to be graded from fresh 'hard purple' to altered 'soft white' by quarrymen before being sold for pottery making (Exley, 1959; Keeling, 1961).

The predominant rock at the Tregargus Quarries is fluorite granite, and all grades of china stone occur, although not much 'hard purple'. The non-megacrystic Li-mica granite (Type E, Table 5.1; locally known as shellstone) is to be found in irregular areas, no larger than a few square metres, but in such a way that the relationships between the two types may be seen. All the rock has been altered and veined in the manner described above. Among less-usual minerals found in the quarries are autunite, gilbertite and löllingite.

The Cornubian granite batholith (Group C sites)

Interpretation

As has been indicated in the 'Petrogenesis' section, the St Austell Granite, which is unique in Britain, has had a more complicated evolution than the other Cornubian granite masses, starting with the intrusion of a boss of coarse-grained megacrystic granite characterized by biotite and zoned oligoclase (An_{25-30}) (Type B, Table 5.1; Exley and Stone, 1982), centred on Luxulyan. Subsequently, a second intrusion was emplaced to its west, cutting across both the earlier granite and its metamorphic aureole. The second intrusion also consisted of biotite granite, and it is thought to have included several facies (Hawkes and Dangerfield, 1978; Manning and Exley, 1984; Hill and Manning, 1987). The later intrusion, however, did not rise as far as the first so that a large, but thin, area of its thermally metamorphosed Lower Devonian roof still remains north of St Dennis, penetrated by granite in several places (Figure 5.10).

Much of the biotite in the second intrusion has been replaced by lithium-mica (now known to be zinnwaldite; Stone *et al.*, 1988), and the four main varieties of St Austell Granite were thought at one time to constitute a magmatic differentiation series of biotite granite (Type B) → megacrystic Li-mica granite (Type D) → non-megacrystic Li-mica granite (Type E) → fluorite granite (Type F) (Richardson, 1923; Exley, 1959). However, subsequent work has shown that this is not the case. It is now believed that, as a consequence of fluorine and water concentration, a residual 'magma' separated at depth from the biotite granite magma and that this contained relatively high concentrations of Li and Na rather than Fe and Ca, similar to the state of affairs in the Tregonning–Godolphin area in west Cornwall (Stone, 1975; Manning, 1982; Pichavant and Manning, 1984). This residual magma intruded the second St Austell boss to give the non-megacrystic Li-mica–albite–topaz granite now seen in the Nanpean area (and also near Hensbarrow Beacon; see Figure 5.10). Under the influence of highly volatile fluids (particularly of fluorine and water) emanating from this intrusion, biotite and oligoclase of the earlier rocks were altered. Iron was replaced by Li to give zinnwaldite, and Ca was replaced by Na to give albite (An_7); other adjustments, especially of Al, took place concurrently and gave rise, *inter alia*, to topaz. Excess Ca combined with F to produce fluorite, and hence local patches of fluorite granite, and excess Fe combined with B to form tourmaline (Manning and Exley, 1984; Pichavant and Manning, 1984). The megacrystic Li-mica granite (Type D, Table 5.1) in most of the western intrusion is thus interpreted now as a metasomatic aureole of the Nanpean (Type-E) Li-mica granite.

Some mineralization was associated with the intrusion of the coarse biotite granites in Devon and Cornwall, but the main period followed about 10 Ma later, at about 270 Ma BP (Jackson *et al.*, 1982). It is believed, on field and other evidence, to have been consequent on the intrusion of the Li-mica–albite–topaz granite (Manning and Exley, 1984; Darbyshire and Shepherd, 1987; Bristow *et al.*, in press). The existence of both pervasive and vein-concentrated greisening and tourmalinization throughout the western intrusion attest to the circulation of volatiles, especially water, F and B, and to a sequence of metasomatism, alteration and replacement continuing through the late-magmatic and into the post-magmatic stages (Figure 5.4).

The fact that kaolinization is often found beneath unaltered rocks and to great depths effectively seemed to have ruled out weathering as a cause, until Sheppard (1977) described stable isotope ratios which indicated clearly that meteoric water was of the greatest importance. This led to a revival of interest (Bristow, 1977) in a combined magmatic/hydrothermal and weathering mechanism such as had been proposed by Hickling (1908), Coon (1911, 1913a, 1913b) and, more positively, by Tomkeieff (discussion *in* Exley, 1959). These authors envisaged an initial hydrothermal alteration, breaking down feldspar to mica and increasing porosity, followed by weathering to complete the process. The recognition of convection cells, circulating meteoric water driven by radiogenic heat (Figure 5.8), and the probability of such activity during the Mesozoic and Cenozoic (Durrance *et al.*, 1982), has enabled a detailed chronology of the intrusion–mineralization–kaolinization continuum to be built up (Bristow, 1987; Bristow *et al.*, in press; Table 2.2). During this, the original magmatic water was progressively replaced by meteoric water from the country rocks, bringing elements which, at appropriate temperatures, formed the various mineral deposits and finally brought about low-temperature processes such as kaolinization which have continued through to the present. Evidence of all the chronological stages of Table 2.2, except stages 1 and 2, is present at Tregargus.

St Mewan Beacon

Conclusions

This site contains the only examples now readily accessible of the two youngest varieties of the chief members of the St Austell suite. The granite here originally contained the dark mica biotite, but because of concentrations of certain elements in the final granite magma and its associated solutions, reactions were set up which converted the original mineral composition and produced a different granite enriched in new elements, including lithium and fluorine. Formerly variously attributed to hydrothermal alteration and magmatic differentiation, the Li-mica–albite–topaz granite is now thought to have been derived from a late-stage magma originating at depth, and the fluorite granite to have been the result of complex metasomatic exchanges with earlier biotite granite. Also to be seen are greisen and quartz–tourmaline veins and kaolinized rock. These have spatial relationships with both the fracture system and fresh rock which show that they have resulted from solutions, now known to be largely of meteoric origin, which have travelled up from deeper levels and penetrated the walls of the fractures.

C13 ST MEWAN BEACON (SW 985534)

Highlights

This site displays a rare exposure of quartz–topaz–tourmaline rock of hydrothermal origin, formed immediately under the metamorphic rocks of the granite roof.

Introduction

St Mewan Beacon is situated on the southern margin of the St Austell Granite, 3 km WNW of St Austell and just outside the Blackpool china-clay pit (Figure 5.10).

The St Austell Granite was emplaced in three episodes, the second of which cuts across the first, near St Dennis (Figure 5.4). Both the first and the second consist of megacrystic biotite granite of typical Cornubian type (Type B, Table 5.1; Exley and Stone, 1982). A third intrusion of Li-mica–albite–topaz granite (Type E, Table 5.1) was emplaced within the second boss, and this is now exposed between St Dennis and St Stephen and near Hensbarrow Beacon. It is believed to have been derived from biotite granite at depth (see 'Petrogenesis' section and site descriptions) and upon emplacement to have metasomatized much of the second intrusion, albitizing the oligoclase, converting biotite to zinnwaldite and introducing topaz. It is this type of granite (Type D, Table 5.1) which is adjacent to St Mewan Beacon. Accompanying and following these intrusions, the introduction of boron gave rise to extensive tourmalinization which preceded greisening, metalliferous mineralization and kaolinization (Manning and Exley, 1984).

Field relations, textures and composition have, in the past, been used to suggest either a 'pneumatolytic' (Ussher *et al.*, 1909) or 'magmatic' (Collins and Coon, 1914) origin for the rocks of the Beacon, but Manning (1981) and Pichavant and Manning (1984) have concluded, from fluid-inclusion and other experimental data, that the rock was formed by complex hydrothermal processes.

Description

The rocks exposed at St Mewan make up a line of low crags along the south-facing slope. Storage tanks now occupy a small quarry at the western end, from which rock was formerly taken to pave grinding mills for china stone.

For the most part, the rocks are equigranular, fine- to medium-grained and made up of quartz and topaz with subordinate tourmaline, but banded quartz–tourmaline rock occurs in the southern side of the quarry, the banding dipping at about 40° to the south. The suite forms a contact facies between the main part of the granite, which is very kaolinized here, and its country rock consisting of tourmalinized pelites, semipelites and psammites of the Lower Devonian Meadfoot Group (Collins and Coon, 1914).

Interpretation

In addition to the quartz, topaz and tourmaline, the rocks of the Beacon contain accessory muscovite (sometimes as a replacement for topaz), apatite and opaque ore. The proportions of the main minerals vary to give rocks which may be very quartz- or tourmaline-rich, especially near the margins of the outcrop, but the average composition is about 60% quartz, 25% topaz and 15% tourmaline. They therefore fit into the St

Austell sequence after the main intrusions and metasomatism, and before the main post-magmatic tourmalinization (between Stage III and Stage IIIb of Table 2.2). However, not only are these unusual rocks very hard, but experiments on melting relations and fluid-inclusion composition suggest that they are too refractory to have been produced from a straightforward magmatic melt, although they could have crystallized in equilibrium with saline hydrothermal fluid at about 620°C (Manning, 1981). The latter would link them to the hydrothermal (i.e. high-temperature, low-pressure) mineralization stage; occasional exposures of comparable rocks are found in clay workings and mines, although these are seldom long-lived enough and accessible enough to be examined. Manning (1981) concludes that 'multistage and complex processes' were involved, but that more work is required on stability relations in highly saline systems containing B, F and OH in order to advance knowledge of these systems and processes further. The commencement of such work is reported in Pichavant and Manning (1984), where it is concluded that in the H_2O-saturated system Q_z–Ab–Or–H_2O added B partitions into the vapour phase while added F partitions into the melt, and that added B effects little change in the minimum melt compositions.

St Mewan Beacon provides a rare chance to see an unusual topaz-rich rock of high-temperature hydrothermal origin arising in the change from late- to post-magmatic conditions and providing a link in the evolutionary continuum. It is unlikely to have crystallized from an ordinary melt, and is probably the result of interaction between magma and a volatile phase rich in F, B and OH.

Conclusions

St Mewan Beacon consists of an igneous rock made up predominantly of the minerals: quartz, tourmaline and topaz, believed to have formed within the topmost portion (roof) of part of the St Austell granite intrusion. It has been suggested that it formed by the modification of solidifying granite magma by hot (hydrothermal) solutions containing fluorine, boron and water, through the alteration and reorganization of the chemistry and mineral content of the crystallizing granite. The fluids were a legacy of the waning igneous activity which formed the Cornubian granites. The site provides a rare chance to see an unusual rock of high-temperature origin arising in the change from late- to post-magmatic conditions and providing a link in the evolutionary continuum.

C14 ROCHE ROCK (SW 991596)

Highlights

Roche Rock is a unique site within an intrusion of quartz–tourmaline rock which was formed either from a boron-rich differentiated magmatic fluid or from a complex late-magmatic hydrothermal process.

Introduction

Roche Rock (Figure 5.10) is a lenticular, craggy outcrop, rising to a height of 20 m above its lower slopes and surmounted by a ruined chapel (Figure 5.17). It is on the southern outskirts of the village of Roche, about 8 km NNW of St Austell. The rock itself is entirely made up of quartz and black tourmaline (schorl) in varying proportions, is generally rather friable and sometimes distinctly vuggy. Its origin has been ascribed variously to magmatic (Power, 1968; Charoy, 1981, 1982) and hydrothermal/metasomatic (Flett, *in* Ussher *et al.*, 1909; Wells, 1946) processes. Manning (1981) regarded it as probably being due to the latter, but as a result of much more complex reactions than previously recognized. It is likely that these are related to those responsible for breccia formation and mineralization in other places and they are associated with the 270 Ma BP intrusive phase (Chapter 2; Allman-Ward *et al.*, 1982; Bromley and Holl, 1986; Halls, 1987; Bromley, 1989; Bristow *et al.*, in press).

Description

There is nothing to add to the simple field observations noted above, beyond saying that in hand specimen most of the tourmaline is in short prisms, although some is acicular, with needles up to 40 mm long, and radiating 'suns' are not uncommon. Although the total outcrop has been mapped as an elliptical area about 600 m by 300 m, an area of about 250 m by 100 m is fully exposed at this site.

The average grain size is seen to be 1–2 mm and both quartz and tourmaline are usually sub-

Roche Rock

Figure 5.17 The craggy outcrop of Roche Rock consists of quartz–tourmaline (schorl) rock. Roche Rock, Cornwall. (Photo: R.A. Cottle.)

to anhedral, with sutured or indented margins, and the tourmaline is yellow brown with patches and zones of blue.

Interpretation

The school of thought favouring hydrothermal origin, based its conclusions primarily on textural evidence (Flett, *in* Ussher *et al*., 1909; Wells, 1946), apparent replacement of one mineral by another being common and tourmaline often being found in veins and other obviously post-magmatic situations. Hence the Roche Rock tourmalinite came to be regarded as the tourmalinized end-member in a sequence, starting with granite and in which luxullianite was an intermediate stage. Power (1968), however, in a general study of south-west England tourmaline, discovered clear differences between blue-green acicular tourmaline of the kind usually found in veins, and regarded as hydrothermal, and the brown prismatic tourmaline found in the body of the granites and considered to be primary and, at latest, late magmatic. This established the Roche Rock tourmalinite as magmatic.

Quartz–tourmaline rocks and tourmalinites are not rare in Cornwall; in addition to the Roche Rock (which is outstanding), there is an example at Porth Ledden, and such rocks are often uncovered in china-clay workings and mines. Charoy (1981, 1982), noting the experimental work of Pichavant (1979), ascribed all these to the separation of immiscible boron- and fluorine-rich magmatic fluid, accompanied by alkalis and silicon, in the late stages of Cornubian granite crystallization (see 'Petrogenesis' section above). Bearing in mind the results of other laboratory experiments on relatively simple granite systems incorporating boron and fluorine, Manning (1981) suspected that the process was more complex than this, especially in view of a tendency towards highly saline fluid inclusions in the minerals, and that the process may be multistage. Like quartz–topaz–tourmaline rocks such as that at St Mewan Beacon, with which he associated them, Manning thought that tourmalinite might have followed closely on magmatic crystallization, but saw a need for more experimental work, particularly on stability relations in systems with high salinity and containing B, F and OH. Some preliminary results of such work are reported by Pichavant and Manning (1984); these suggest that B and F can depress the magmatic freezing temperature of the granite system to 650°C at 1 kbar, and that the effect of F is greater than that of B in altering minimum melt compositions.

The second intrusive phase of the batholith at about 270 Ma BP (see Chapter 2) brought with it much boron and resulting tourmalinization. Frequently, this produced both explosive and implosive reactions which caused extensive brecciation by building up high internal pressure, which was suddenly released by fracturing to the surface, as at Wheal Remfry in the west of the St Austell mass (Allman-Ward *et al.*, 1982; Bromley and Holl, 1986; Halls, 1987; Bromley, 1989). It seems probable, however, that where violent rupturing of the rocks did not occur, intense tourmalinization, with both primary and secondary generations of tourmaline, took place in areas of high boron concentration. Roche Rock could well be an example of this (Bristow *et al.*, in press).

Conclusions

Roche Rock is probably the largest and certainly the best exposed quartz–tourmaline rock in Cornwall and perhaps in the country. It affords a unique site for the examination of this late granitic facies *in situ*. The origin of this distinctive rock has been variously attributed to its being the last product of the magma crystallized from a residual granite melt, particularly rich in boron to form so much of the mineral tourmaline, or, alternatively, as the product of changes wrought by hot solutions charged with boron left over from the magmatic phase during which the granites of Cornwall were formed (see St Mewan Beacon).

C15 MEGILIGGAR ROCKS (SW 609266)

Highlights

This is the only well-exposed site which shows contacts between lithium-mica granite and pelitic hornfels, sheets of leucogranite, aplite and pegmatite developed from a late-stage granitic roof complex, and unusual minerals.

Introduction

This site (Figure 5.12) includes both cliffs and foreshore over a length of about 1 km below Tremearne Farm, just over 2 km north-west of Porthleven. Within it are the eastern contact of the Tregonning part of the Tregonning–Godolphin Granite with its country rocks and a series of granitic sheets developed from the intrusion as an extension of its roof complex.

The mixed sediments of the Mylor Slate Formation are strongly folded, cleaved and veined in a manner typical of the Cornish 'killas'; the local structures have been discussed by Stone and Lambert (1956) and Stone (1966, 1975) and the regional setting of the Mylor Slate Formation by Holder and Leveridge (1986). The sediments have been thermally metamorphosed into cordierite- and andalusite-hornfelses near the granite.

The granite (Type E, Table 5.1; Exley and Stone, 1982) is thought to have originated at depth by a complex exchange process. As described earlier ('Petrogenesis' section), this was facilitated by an F-rich phase and involved the replacement of Fe–Mg in biotite by Li–Al to produce zinnwaldite, and the albitization of plagioclase. In the opinion of Stone (1975 *et seq.*), this occurred in solid biotite granite, which was then remobilized as Type-E magma, while Manning (1982) considers that it could have been part of a process of magmatic differentiation (Stone, 1975, 1984; Manning, 1982; Exley and Stone, 1982; Exley *et al.*, 1983; Manning and Exley, 1984; Pichavant and Manning, 1984). Either way, this variety of granite is exposed elsewhere only in the St Austell outcrop.

From the eastern contact, a series of sheets of pegmatite, aplite and leucogranite, changing from one to another both vertically and laterally, cuts through the cliffs and along the beach (Figure 5.18). Stone (1969, 1975) and Exley and Stone (1982) have ascribed the varied rock types to the contrasted partitioning of elements between discrete liquid and vapour phases, but Bromley and Holl (1986) have suggested that these phases themselves arose from pressure variations triggered by the opening of cavities by subsidence of country rock. Badham (1980) argued that temperature variation controlled the development of the various rock types. The sheets are character-

Megiliggar Rocks

istically lithium rich, and Stone (1984) and Stone *et al.* (1988) believe that some of the lepidolite in these late rocks may have been magmatic, rather than metasomatic like most of the Li-mica in the region. Many unusual minerals occur here, among them amblygonite (Stone and George, 1979) and triplite (George *et al.*, 1981).

Description

The country rocks at Megiliggar Rocks are banded, light and dark grey and buff psammites, semipelites and pelites of the Mylor Slate Formation which have been deformed twice (first into minor upright and overturned folds and then into major recumbent folds) and cleaved, as has been explained in the account of the nearby Rinsey Cove site. These features are typical of the killas of Cornwall and are particularly well seen in these cliffs. Close to the granite, the metasediments have been baked and are now spotted hornfelses with cordierite and andalusite. Corundum, however, has not been reported as it has at Priest's Cove.

The neighbouring granite intrusion consists of a northern, fine-grained, megacrystic biotite granite (Dangerfield and Hawkes, 1981; Type C, Table 5.1, Exley and Stone, 1982), which is called the Godolphin Granite (after Godolphin Hill), and a southern, medium-grained, non-megacrystic lithium-mica- and topaz-bearing granite (Type E, Table 5.1) named after Tregonning Hill. The latter component is exposed in the cliffs.

Its eastern contact with the Mylor Slate Formations can be seen at the eastern end of Trequean Cliff, especially on the shore, and across the head of Legereath Zawn. In both localities it is sharp and without significant marginal change in the texture of the granite. It is almost vertical, although tourmalinization streaks (*'schlieren'*) and a thin pegmatite occur within the granite at the first locality and there are sheets and veins of granites of different types at the base of the cliffs and in stacks at the second.

Presumably owing to the steepness of the contact, a sheeted roof complex is not well-developed here, but from Legereath Zawn eastwards there is a series of granitic sheets of similar compositions to those found in the roof complex above the eastern end of Rinsey Cove. These sheets dip gently towards the south-east, vary in

Figure 5.18 Pegmatite–aplite–granite sheets cutting Mylor Slate Formation metasediments in the cliffs at Legereath Zawn, near Tremearne Par. Megiliggar Rocks, Cornwall. (Photo: C.S. Exley.)

The Cornubian granite batholith (Group C sites)

Figure 5.19 Pegmatite–aplite–granite layering in one of the granitic sheets. Megiliggar Rocks, Cornwall. (Photo: C.S. Exley.)

thickness from 0.10 m to 3 m approximately, and both coalesce and split on occasion. They cut across the cleavage in the Mylor metasediments.

The chief rock types in the sheets are pegmatites (Figure 5.19), aplites and lithium-mica leucogranites, and these types change from one to another with distance from the main granite. Pegmatites are especially well developed under pelite 'roofs', even where these are provided by xenoliths, of which there are many in various stages of detachment, disorientation and digestion. The most-complicated single outcrop is that forming Megiliggar Rocks and, here and nearby, all variations of texture are present, from narrowly banded 'line rocks' to pegmatites 0.10–0.15 m thick, together with a wide range of compositions. Some rocks are 'graded' in respect of grain size and/or tourmaline content, and have uneven bases. Many quasi-sedimentary features occur (Figure 5.20).

The occurrence at Megiliggar of blue-green apatite crystals up to 30 mm long is well known (although they are far from common); that of other unusual phosphate minerals such as amblygonite (Stone and George, 1979) and triplite (George *et al.*, 1981) less so.

Megiliggar Rocks

Figure 5.20 Pegmatite–aplite–granite boulder on Tremearne Beach, demonstrating the quasi-sedimentary character of the igneous layering. Megiliggar Rocks, Cornwall. (Photo: C.S. Exley.)

Interpretation

The Tregonning Granite is believed to have derived from biotite granite at depth, by the separation of a fluorine- and water-rich fraction of the magma as described earlier for Rinsey Cove and in the 'Petrogenesis' section. The origin of the pegmatite–aplite–leucogranite sheets lay in the high volatile concentrations in this new magmatic fraction, which contained high concentrations of such elements as Mn, P, Sn, Ga and Ge along with the OH, F and B. The Li-mica leucogranite would seem to have been the direct continuation of the magmatic process, but the eventual partitioning of Na into the silicate melt and K into aqueous vapour, gave rise to aplite, and by reaction with the aplite, metasomatic pegmatite respectively (Stone, 1969; Exley and Stone, 1982). The development of the vapour phase could have resulted from sudden pressure changes as blocks of country rock subsided to make way for the magma; repetition of such movements would produce alternating aplite and pegmatite (Bromley and Holl, 1986). Badham (1980) did not distinguish between vapour and liquid phases, but argued that the separation and crystallization of leucogranite at higher temperatures was followed by the separation and crystallization of distinct aplite and pegmatite fluids at lower temperatures and subsequent diffusive alteration.

The rock types and mineralogies of the sheets in this complex are much more varied than those found at other places, such as Porthmeor Cove, because of the presence of Li, P and F in the Li-mica granite magma.

The Megiliggar Rocks section exhibits the only well-exposed series of leucogranite–aplite–pegmatite sheets developed from a lithium-mica granite, and allows detailed examination of these unusual late-stage facies and their relationships with each other, their parent granite and their host metasediments. They have developed as a result of the concentration of volatile constituents under an impermeable roof of metasediment and the partitioning of elements from residual magma between liquid and vapour phases. In the only other exposures of Type-E granite, the contacts are with other granites and the roof complexes are relatively poorly developed.

The Cornubian granite batholith (Group C sites)

Conclusions

The Megilligar Rocks section exhibits the only well-exposed series of sheets of aplite, pegmatite and leucogranite developed from a lithium-mica granite. They comprise sheet-like offshoot intrusions from the main granite mass, are the last representatives of igneous activity locally, and are among the last products of the declining igneous activity that had formed the massive granite masses of Devon and Cornwall. The emplacement of this main body of magma had already folded and baked the surrounding (older) sedimentary rocks. The site allows detailed examination of these unusual late-stage granitic facies and their relationships with each other, their parent granite and their host, metamorphosed, sedimentary rocks.

C16 MELDON APLITE QUARRIES (SX 567921)

Highlights

Meldon Quarries is a classic site which contains a unique, late-stage aplitic dyke with a variable texture and composition, and a very wide range of unusual minerals both within the dyke and in associated veins. There has been metasomatism of semipelitic and calcareous sediments by elements introduced in association with the dyke.

Introduction

This site is about 4 km south-west of Okehampton, and comprises two small quarries in the Meldon 'Aplite' Dyke and adjacent metasediments, together with much older quarries in nearby calcareous metasediments. It is famous for the variety of lithium-, beryllium-, boron-, fluorine-, sulphur- and rare-element-bearing minerals occurring in both the dyke, which has a unique composition (Edmonds *et al.*, 1968), and the metasediments (Worth, 1920; Dearman and Butcher, 1959; Kingsbury, 1961, 1964, 1970; Mackenzie, 1972; Chaudry and Howie, 1973, 1976; Hawkes, 1982; von Knorring and Condliffe, 1984).

The dyke was intruded into the Carboniferous Lower Culm sediments lying within the metamorphic aureole of the Dartmoor Granite mass. These form steep, overturned folds (Dearman and Butcher, 1959; Edmonds *et al.*, 1968), which are now thought to be linked with the northward thrusting of a nappe from the south (Selwood and Thomas, 1984). The mineralogy of these metasediments varies not only with their original composition and its modification by thermal metamorphism, but also with the degree to which they suffered metasomatism associated with the intrusion of the dyke.

Description

The Meldon Dyke is probably at least 3 km long, running north-east from Sourton Tors to the Meldon Railway Quarry (Worth, 1920; Hawkes, 1982), but outcrops are intermittent, and thickness is variable both along the strike and vertically. The dyke often splits into smaller veins and apophyses and does not always reach the present surface. In the larger of the two quarries on the site, south of the Red-a-Ven Brook, the dyke is about 20 m wide at the lowest level, but it separates into several branches at higher levels, where the thickest member is some 10 m thick. The dip is about 50° to the south-east. In the smaller quarry, on the north side of the brook, the dyke is made up of several branches, the thickest being only about 4 m. In the Railway Quarry, 500 to 600 m away, only thin stringers of aplite are found (these thicknesses refer to those in the former working faces).

The dyke is parallel with and about 1.5 km distant from the margin of the main Dartmoor Granite intrusion, and it is of the same age, i.e. about 280 Ma (Darbyshire and Shepherd, 1985). It is not quite parallel with the strike of the aureole metasediments, and cuts obliquely across the junction between the Meldon Chert Formation and the underlying Meldon Shales-and-Quartzite Formation, both of which are of Lower Culm Measures (Namurian) age (Dearman, 1959; Dearman and Butcher, 1959). The structure of this area is complex, consisting of a pair of antiforms overturned against the granite to the south-east and separated by a shallow, asymmetrical synform. The antiforms occur in the Meldon Shales-and-Quartzite Formation, the Meldon Chert Formation forming the core of the intervening synform. Formerly thought to be an overturned fold (Dearman, 1959; Dearman and Butcher, 1959; Edmonds *et al.*, 1968), the structure has been reinterpreted from evidence along strike by Selwood and Thomas (1984), who believe the Meldon Shales-and-Quartzite Formation and part of the overlying Crackington Formation to be allochthonous and to have been thrust into their present positions as a nappe from the south, before the intrusion of the dyke. The remaining Crackington Formation was subsequently thrust up from the north. All of the beds have north-westerly dips at various angles, however, and the aplite is situated on the south-easterly limb of the synform where a fault has developed and the dip of the metasediments is about 30°.

The Meldon Shales-and-Quartzite Formation was originally composed of interbedded mudstones and siltstones with, towards the top, a volcanic sequence of spilitic and keratophyric agglomerates and tuffs, and an uppermost group of shales and quartzites. The sedimentary nature of the pyroclastic deposits is emphasized by Dearman and Butcher (1959). As a consequence of thermal metamorphism by the granite, these rocks have been hornfelsed, the semipelitic varieties being characterized by biotite, chiastolite and andalusite, and the metavolcanic rocks by amphibole, pyroxene and garnet.

The Meldon Chert Formation originated as a series of bedded limestones and cherts (hence the old term 'Meldon Calcareous Group') whose relationships are well seen in the bank of the West Okement River in the north-west corner of the site. Depending on composition, they have been thermally metamorphosed into more-or-less pure marbles and 'calc-flintas', the latter predominating, with the development of wollastonite and grossularite. However, it is not possible entirely to separate purely thermal from metasomatic effects in the vicinity of the dyke and many other minerals occur in the metasediments. They include axinite, andradite, idocrase, hedenbergite, hornblende and tourmaline.

The aplite itself is a predominantly fine grained, equigranular rock with euhedral to subhedral laths of albite, anhedral orthoclase, quartz and a small amount of lepidolite. It is considerably more sodic, less potassic and less siliceous than the average Dartmoor Granite and these features, combined with the presence of lithium, were said by Edmonds et al. (1968) to make it unique in Britain. The chief accessory minerals, amounting to about 5%, are topaz and elbaite. The last mineral is usually pale green, but often has pink zones suggestive of the rubellite which is present in later hydrothermal veins and their metasomatized aureoles. The dyke is mostly white and consists of albite, quartz and pinkish lepidolite. There are, however, bluish marginal facies containing quartz, albite and accessory orthoclase, tourmaline, apatite and colourless lepidolite. There are also patches of a coarser brown rock, with an increased quartz and pink or brown lepidolite content and tourmaline, topaz, fluorite and apatite. It often shows mineral banding (Chaudry and Howie, 1976) and also includes streaks and pockets of pegmatite in which orthoclase is a significant phase and, in addition to the rock-forming minerals already named, such species as petalite, fluorite and apatite occur.

Chaudry and Howie (1973) have separated the pegmatites into two types, one containing albite and elbaite, and the other containing topaz and petalite. The lepidolite in the pegmatites is of a lower-temperature variety than that in the aplite proper (Chaudry and Howie, 1973), and there are resemblances to lepidolites found at Megiliggar Rocks (Stone et al., 1988). Many other minerals occur, either sporadically as accessories or in a more concentrated fashion along joints, in veins or in drusy cavities. Axinite is particularly well developed in cross-cutting 'reaction veins' (Mackenzie, 1972). Lithium-bearing varieties include, in addition to the micas, amblygonite and spondumene; and beryllium-bearing species include not only beryl, but also beryllonite, chrysoberyl, eudidymite, milarite, rhodizite and bavenite. There are also the zeolites heulandite and stilbite, the clay montmorillonite, the boron-bearing species datolite, and others such as pollucite (caesium), palygorskite, prehnite and columbite (niobium). Moreover, the late sulphide mineralization, which was extensively developed in the area of the nearby Red-a-Ven Mine, has given rise to the presence (especially in the metasediments) of pyrite, chalcopyrite, pyrrhotite and arsenopyrite.

Interpretation

This site illustrates both the normal development of an aplitic dyke associated with a granite in the metamorphic aureole, and the unusual concentration of minor elements and volatile constituents characteristic of the Cornubian batholith. As with the aplites of the Tregonning Granite roof complex, so well seen at Rinsey Cove and Megiliggar Rocks, the Meldon Aplite formed at a late stage when Li, Be, Rb, B, F and OH were highly concentrated, and it provided a route by which S compounds could travel out of the parent granite at a later stage. While no direct link between the dyke and the Dartmoor Granite has yet been identified, clearly there was one; it may still exist at depth or it might have been broken by tectonic activity.

The relations between aplite and pegmatite and their dependence on the partitioning of K into an aqueous phase and Na into a silicate liquid phase, have been noted in other site descriptions, for example, Megiliggar Rocks. These are all roof complexes, however, whereas the Meldon Dyke, showing similar phenomena, has developed away from the granite body.

There have been virtually no recorded discussions as to why and how the aplite magma, with its remarkable composition and range of unusual mineral phases, should have developed so locally and asymmetrically with respect to the main granite. It seems certain that it evolved in a cusp on the side of the batholith where volatiles were concentrated. It was noted in the 'Petrogenesis' section that mineralization and lithium enrichment are thought to be associated with a second major intrusive event at about 270 Ma and the resemblance between the Meldon Aplite and the Rinsey Cove and Megiliggar aplites and pegmatites strongly suggests that it belongs to this episode. Although no conclusive evidence of the 270 Ma event has been found on Dartmoor, and, as stated above, the Meldon Dyke has been dated at 280 Ma, Knox and Johnson (1990) have described post-main granite intrusions including fine- to medium-grained leucocratic topaz- and tourmaline-granite from the Lee Moor area.

The site demonstrates the metasomatic development of minerals carrying the various elements noted above and the sequential crystallization of some of them in various rock types, both igneous and metamorphic. Minerals with such elements as Be tend to precede those with Li, Rb, B and F, while those with much OH, such as zeolites and clays, are the youngest. Although groups of these minerals are to be found commonly elsewhere in the mineralized zone of the batholith, nowhere else is there so wide a variety in so small an area.

Conclusions

Here a steeply dipping igneous intrusion, the Meldon Aplite Dyke, 20 m wide at its maximum, cuts through the baked sedimentary rocks which surround the Dartmoor Granite. The aplite, a fine-grained, pale-coloured granitic rock consisting normally of the minerals quartz and feldspar, was intruded at the end of the last major phase of igneous activity in this part of Britain, between 280 and 270 million years ago. Like the main Dartmoor Granite mass, the aplite baked and chemically altered the surrounding rocks into which it was injected: thus the chemical changes wrought by the granite were overprinted by further metamorphism imposed by the dyke. The aplite, which was one of the final products of the magmas which had formed the granites of the south-west peninsula, was responsible for a uniquely diverse suite of minerals both within itself, and through the reactions of the penetrative hot solutions which flowed outward from it, with the surrounding metamorphosed sediments.

C17 PRAA SANDS (FOLLY ROCKS) (SW 573280)

Highlights

This site contains fresh exposures of an unusually large granite-porphyry dyke with evidence of multiple intrusion, chilled margins and flow banding. Microscopic features provide evidence for a complex emplacement mechanism involving solid, liquid and gas phases.

Introduction

Praa Sands is a 1.5 km long beach between Marazion and Porthleven. The site (Figure 5.12) is at its western end (Folly Rocks) and consists of a major granite-porphyry ('elvan') dyke, and its country rocks which are Mylor Slate Formation metasediments. The structure of the dyke, the main body of which has a narrow, banded, chilled margin outside an even narrower non-banded rock, has led Stone (1968) to postulate that it was a multiple intrusion, and that its textures and chemistry showed it to have contained liquid magma, solid fragments and a gas phase, all of which interacted. This explanation of the mechanism, although more complex than the simple magmatic intrusion suggested by Reid and Scrivenor (1906), was supported by Goode (1973). Henley (1972, 1974), following a study of other elvan dykes, pointed out their potassium enrichment, and Exley and Stone (1982), Exley et al. (1983) and Stone and Exley (1986) gave accounts of their possible derivation. They have been dated at about 269 Ma BP by the Rb/Sr method (Hawkes et al., 1975).

Praa Sands (Folly Rocks)

Description

The country rocks at this site are folded and cleaved black and grey Mylor Slate Formation pelites with numerous contorted quartz veins, and they have been hardened by contact metamorphism for a few centimetres adjacent to the dyke. This dips at about 80° to the north-east and, by analogy with other dykes in Cornwall, has an Rb/Sr age of 269 ± 8 Ma (Hawkes *et al.*, 1975). The dyke is about 18 m wide overall and strikes approximately 128°, which is parallel with the local mineral veins but divergent from the general trend seen inland. Such large, multiple intrusions are unusual in south-west England.

Three rock types are discernible, the first being a banded, very fine-grained felsitic variety which forms an outer selvedge some 0.30 m wide. This rock contains rounded megacrysts of quartz and alkali feldspar up to about 1 mm in diameter. Within this selvedge, on both sides of the dyke, is a zone about 0.15–0.20 m wide in which the rock is again fine grained but not noticeably banded. This also includes rounded quartz and feldspar megacrysts but these are rather sparsely distributed. Finally, separated from the second variety by a very narrow zone in which the grain size increases markedly but gradationally, there is the main, central body of the dyke which, although relatively coarse, is still microcrystalline. Here, quartz and alkali feldspar megacrysts are common, the latter being mostly subhedral, sometimes zoned, and aligned subparallel with the contacts. There are also rather rounded, fine-grained, poorly megacrystic xenoliths. In thin section, the rock is seen to contain fragments of granite and compound crystals of quartz and alkali feldspar, as well as broken xenocrysts of quartz and feldspar.

Interpretation

Stone (1968) believed that the outer, felsitic rock constituted the first phase of a two-stage intrusion, the main central part being a second phase with either fine-grained chilled margins or a progressively intruded magma filling a slowly opening fissure, as suggested earlier by Reid and Scrivenor (1906), and giving a rock of increasing coarseness. The textures of all these granite-porphyry types show evidence of reaction and recrystallization of early-formed mineral phases and xenocrystic and probably xenolithic fragments as well.

Chemically, the Praa Sands elvan is both potassium rich and has a high K:Na ratio; this led Stone (1968) to propose that there had been extensive ion exchange among the components of the fluid phase which must have been present at the time of its emplacement to effect both alteration and recrystallization of minerals. Moreover, such a fluid phase probably provided the medium for transportation of both solid and magmatic particles in a fluidized system. Goode (1973) supports this view, adding that, in general, Cornish elvans also show evidence of having drilled channels for themselves by their own hot gases (gas coring) and having brecciated the adjacent rocks during intrusion.

Exley and Stone (1982), Exley *et al.* (1983) and Stone and Exley (1986) argue, from detailed chemical and other evidence, that the granite-porphyry magma might have been derived from biotite granite magma of the type which supplied the main batholith rocks. This evidence includes enrichment in K and Rb, impoverishment in Na, and a statistical clustering with granite types B and C.

A number of elvan dykes in the Perranporth area were examined by Henley (1972, 1974), but these are not as well exposed as that at Praa Sands. Nevertheless, Henley was able to conclude, from both textural and chemical features, that elvans solidified from a magma that was already enriched in potassium as a consequence of leaching of sodium and silicon from pre-existing solid granite by late-magmatic fluids. Such aqueous residual fluids, rising rapidly from deep-seated reservoirs because of fracturing, might be stable only with K-feldspar and mica, and would incorporate fragments of granite and corroded crystals from it. An interesting feature of this concept is its dependence on a sufficient time interval between the main period of granite crystallization and dyke emplacement to permit erosion of the granite and stress relief, thus allowing deep fracturing. On the evidence then available, Henley put this at 55–75 Ma, but it is now thought to be nearer 20 Ma (Table 2.1).

The Praa Sands elvan is a very important, fresh exposure of a granite-porphyry dyke intruded in stages by a combination of magmatic intrusion and fluidization processes and deriving from both differentiated magma, developed at depth and K-rich, and incorporating solid granite broken from the walls of the fissures through which the gas-charged material passed.

Intrusions of this type are most commonly

The Cornubian granite batholith (Group C sites)

associated with volcanic activity and are known in many parts of the world. In the case of Cornubia, they are either very late-tectonic or even post-tectonic and may indicate links with volcanoes of which there are now no traces, except, perhaps, the rhyolites of the Withnoe–Kingsand area of south-east Cornwall (Cosgrove and Elliott, 1976; see Chapter 6).

Conclusions

Here is exposed a remarkable example of one of the last products of the major igneous phase which affected south-west England around 270 million years ago. After the major granite masses of Dartmoor, St Austell, etc., smaller vertical or steeply dipping sheets (dykes) of granite porphyry ('elvans') were emplaced. The elvan dyke at Praa Sands is 18 m across, and is remarkable in that it was formed by the injection of more than one magma into an opening fracture. Its outer portion, which baked the surrounding rocks, shows banding indicating the movement of the molten rock through the fissure. The central part of the dyke is coarse grained and contains fragments of older rocks (xenoliths) broken off from rocks at deeper levels by the rising magma. It has been suggested that the fissure which hosts the dyke continued to open to make this two-phase injection of magma possible. Between the central and outer rock types is a zone made up of a third rock produced by the reaction between the gas-charged later magma and the earlier outer melt. The Praa Sands Dyke is an important one for the study of the processes of igneous intrusion and reactions between magmas with different chemical and physical natures.

C18 CAMERON (BEACON) QUARRY (SW 704506)

Highlights

This quarry contains the only surface exposure of the St Agnes Granite contact, with rare pervasive greisening, which is possibly unique in Britain. There is also replacement and telescoped ore mineral paragenesis.

Introduction

Cameron (or Beacon) Quarry is situated about 800 m WNW of St Agnes Beacon on the north coast of Cornwall, and is the only place where the St Agnes Granite contact, with the country rock of Upper Devonian hornfels, is exposed at the surface. Together with the Cligga Head Granite, 5 km to the north-east, this granite forms a small cusp on a northerly prolongation of the Cornubian batholith (Bott *et al.*, 1958; Tombs, 1977).

At the contact between the sedimentary rocks and the biotite granite, the pelites have been metamorphosed to spotted hornfelses, and the granite has a fine-grained margin with some pegmatitic patches.

The main interest in the quarry lies in the widespread, pervasive greisening, which is very rare; silicification; the development of disseminated cassiterite and sulphide mineralization. Reid and Scrivenor (1906) described some of the replacement phenomena, and Hosking (1964) and Hosking and Camm (1985), who give a full description of the quarry (Figure 5.21), proposed a complex paragenesis in which the relations between greisening and mineralization are examined in detail. They ascribe these features to permeation of the granite by fluids moving through it along a network of 'knife-edge' and microscopic fissures. Bromley and Holl (1986) consider the fracture system to be unrelated to the greisening, the impermeable cover of the granite cusp being more important in giving rise to a 'ponding' effect.

Description

The country rocks round Cameron Quarry are the semipelitic and psammitic Porthtowan Formation (formerly the Ladock Beds, a subdivision of the more extensive Upper Devonian Gramscatho Group), and although the psammites at the contact show little obvious signs of alteration, loose fragments on the ground surface show that thermal metamorphism has caused spotting of the hornfelses, with andalusite developed in pelitic bands. These features are additional to tourma-

Figure 5.21 (Opposite) Detailed map of Cameron Quarry (after Hosking and Camm, 1985).

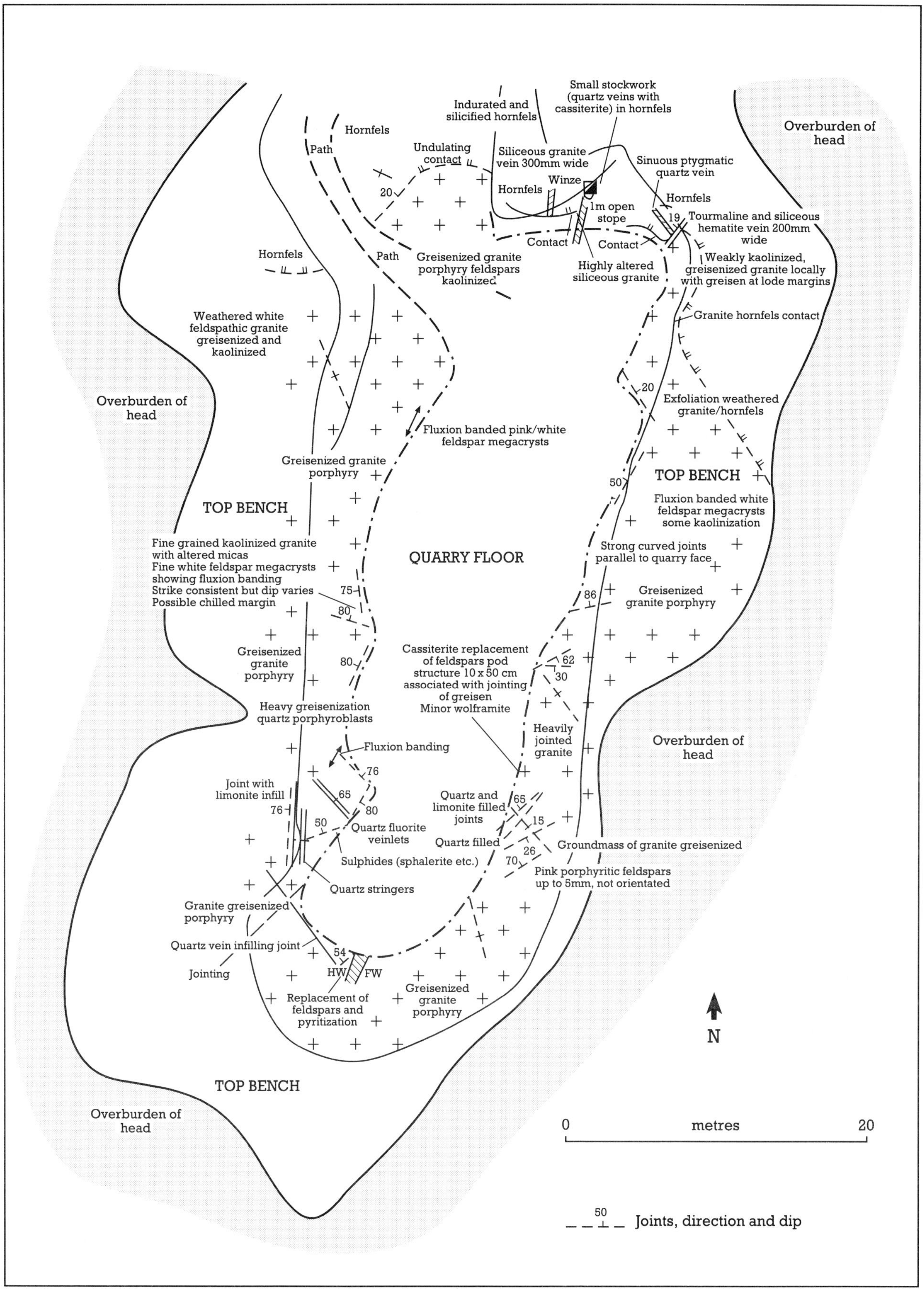

Small stockwork (quartz veins with cassiterite) in hornfels
Indurated and silicified hornfels
Hornfels
Undulating contact
Siliceous granite vein 300mm wide
Winze
Sinuous ptygmatic quartz vein
Overburden of head
Path
20
Hornfels
1m open stope
Hornfels
19
Tourmaline and siliceous hematite vein 200mm wide
Contact
Contact
Hornfels
Path
Greisenized granite porphyry feldspars kaolinized
Highly altered siliceous granite
Weakly kaolinized, greisenized granite locally with greisen at lode margins
Weathered white feldspathic granite greisenized and kaolinized
Granite hornfels contact
20
Exfoliation weathered granite/hornfels
Overburden of head
Fluxion banded pink/white feldspar megacrysts
Greisenized granite porphyry
TOP BENCH
50
Fluxion banded white feldspar megacrysts some kaolinization
TOP BENCH
Fine grained kaolinized granite with altered micas
Fine white feldspar megacrysts showing fluxion banding
Strike consistent but dip varies
Possible chilled margin
QUARRY FLOOR
Strong curved joints parallel to quarry face
75
80
86
Greisenized granite porphyry
Greisenized granite porphyry
80
Cassiterite replacement of feldspars pod structure 10 x 50 cm associated with jointing of greisen
Minor wolframite
62
30
Heavy greisenization quartz porphyroblasts
Heavily jointed granite
Overburden of head
Fluxion banding
76
Joint with limonite infill
65
76
80
50
Quartz fluorite veinlets
Quartz and limonite filled joints
65
15
Quartz filled
26
70
Groundmass of granite greisenized
Sulphides (sphalerite etc.)
Pink porphyritic feldspars up to 5mm, not orientated
Quartz stringers
Granite greisenized porphyry
Quartz vein infilling joint
54
Jointing
HW
FW
Replacement of feldspars and pyritization
Greisenized granite porphyry
N
TOP BENCH
Overburden of head
0
metres
20
50
Joints, direction and dip

linization caused by the underlying granite and which preceded greisening. The contact with the granite is seen at the northern end of the quarry (Figure 5.21), and within a metre or two of the igneous rock, which is normally a medium- to coarse-grained, poorly megacrystic biotite granite with megacrysts up to 20 mm long (Dangerfield and Hawkes, 1981; Type B, Table 5.1; Exley and Stone, 1982) has a chilled, fine-grained texture. Fine-grained contacts are not common in Cornwall, as is demonstrated at the contacts at Rinsey Cove and Porth Ledden for example. There is also a pegmatitic facies visible at the contact in places.

Massive and pervasive greisening and silicification give much of the granite an abnormally dark colour, fine grain size and glassy appearance. There has been extensive alteration of feldspar megacrysts to greisen locally, and although some have been eroded to leave hollow moulds, others, near post-greisen fractures, have been replaced by aggregates of minerals which conspicuously include cassiterite. Mineralization, in the form of disseminated copper sulphides, is best seen in the north-east and south-west corners of the quarry.

Interpretation

Almost always in south-west England (and generally elsewhere in Britain), greisening is very obviously related to a joint or fracture system. This is not the case in Cameron Quarry where, although varying in intensity, nearly all the granite has been altered, first by greisening and then by silicification. Hosking and Camm (1985) believe this permeation to have been achieved as a result of the development of a complex network of fine fractures due partly 'to contraction and partly ... to the pressure exerted by residual fractions in the magma'. Bromley and Holl (1986), on the other hand, state specifically that the quarry contains 'massive greisen not related to penecontemporaneous fractures' and presume that 'the greisening solutions were ponded beneath the impermeable carapace of tourmalinized hornfelses'. There seems to be no reason why both should not be right if the 'network of fine fractures' is on a scale approaching the microscopic and regarded as distinct from the usual type of megascopic joint system. It is certain, however, that mineralization took place in a series of steps following the influx of pulses of fluid as suggested by Halls (1987), and that it varied in degree over very short distances.

Following the early stages of alteration, during which most of the feldspar was replaced by secondary mica and quartz or dissolved to leave cavities, there was extensive cassiterite and Cu, As, Fe and Zn sulphide mineralization via open, although still very narrow, 'knife-edge' fractures, and this resulted in both disseminated deposition and infilling of the feldspar moulds. Replacement was in two stages: K-feldspar was removed before deposition of cassiterite, and wolframite and plagioclase before deposition of sulphide. Virtually the full range of Cornish mineral parageneses is represented in a small vertical span; there is, therefore, a telescoped version of the mineral zonation so well known from the famous Camborne–Redruth mining district not far to the south. The mines of the St Agnes area and Cligga Head also show this effect, but less distinctively. Detailed study here shows that mineral deposition took place in stages, as it did elsewhere, with reactivation of channelways from time to time (Hosking and Camm, 1985). Unlike Cligga Head, Cameron Quarry rocks exhibit very little kaolinization, largely because most of the feldspar had already been altered to quartz and mica, but also because less water was available at the low temperature stage following greisening.

The date of the intrusion of the St Agnes granite has not been established but the main mineralization in western Cornwall was at about 270 Ma BP (for example, Jackson *et al.*, 1982; Darbyshire and Shepherd, 1985), substantially after granite emplacement; and the pattern of events seen at both Cameron Quarry and Cligga Head has close similarities with the general chronology of the mineralized areas round Camborne and St Just.

Conclusions

This is a unique site in which can be seen the widespread effects of greisening, silicification and intense mineralization in a 'telescoped' succession in which the effects are superimposed rather than in distinct zones, all less closely related to jointing than is normal.

C19 CLIGGA HEAD AREA (SW 738536)

Highlights

This classic site contains metamorphosed and mineralized metasediments adjacent to a small granite stock deformed to show an antiform-and-synform structure. The concentration of spectacular greisening, condensed W–Sn–Cu–Fe mineralization and kaolinization into zones is determined by this structure.

Introduction

Cligga Head is a classic site for the study of greisening, mineralization and kaolinization. Described by several authors in the last century, among whom were Conybeare (1817), Sedgwick (1820), Henwood (1838, 1843), De la Beche (1839) and le Neve Foster (1877), and then by Scrivenor (1903) and Reid and Scrivenor (1906), it is situated on the north Cornish coast between Perranporth and St Agnes, and consists of a small stock of altered granite and adjoining metasediments, the former being superbly exposed in section in the westerly facing cliffs (Figure 5.22).

The granite, together with the neighbouring St Agnes stock, rises from a northerly projection of the Cornubian batholith (Bott *et al.*, 1958; Toombs, 1977) and is composed of a coarse, poorly megacrystic granite (Dangerfield and Hawkes, 1981), but the presence of lithium mica makes it a Type-D granite (Table 5.1) of Exley and Stone (1982). This mica has been variously named by Hall (1971) as protolithionite, and by Stone *et al.* (1988) as lithian siderophylite; its presence suggests that the present granite evolved by metasomatic alteration of biotite granite (Hawkes and Dangerfield, 1978; Dangerfield *et al.*, 1980; Exley and Stone, 1982; Exley *et al.*, 1983; Manning and Exley, 1984; Stone and Exley, 1986, Hawkes *et al.*, 1987). Charoy (1981) considers the albite was derived from original oligoclase by hydrothermal alteration. There is disagreement as to the nature of the granite's southern contact (Moore and Jackson, 1977; Badham, 1980).

The granite has been deformed in a way not seen elsewhere in south-west England, and Moore and Jackson (1977) have related the distribution of alteration and mineralization zones to this deformation. Hall (1971) has described the nature of the greisening and Charoy (1979, 1981, 1982) has discussed the tourmalinization, greisening and kaolinization. From fluid-inclusion studies, Jackson *et al.* (1977) have reported three phases of mineralization at temperatures below 400°C.

Description

Cligga Head itself and the westerly facing cliffs immediately to the south are formed of altered granite, which is part of an elliptical stock measuring roughly 600 m from north to south and 350 m from east to west. The westerly part has been eroded down to sea-level. As is usual in Cornubian granites, the concentration of megacrysts is somewhat variable, and those near the adjoining upper Gramscatho Group hornfelses are aligned approximately parallel with the margin. All visible contacts are steep, but while those in the north and east (seen in old mine workings) appear intrusive, that in the south is described as a fault by Moore and Jackson (1977), and as a stoped contact by Badham (1980). The hornfelses within about 30 m of the junctions are spotted.

The flat-lying joints of the granite constitute an antiform with a WSW-plunging axis about 350 m south of the Head and a synform to the south of this. The steep joints are often curved, some of them having developed into small faults. Zones of greisening are found along the walls of the primary joints and, where these are closely spaced, they merge to give pervasive greisening of the intervening rock, a feature well developed in the core of the antiform and also seen in the Cameron Quarry, St Agnes (discussed above). Hall (1971) lists Fe, Ca, F, B, Li, Mn, Rb, Sn, W and Zn as being enriched by this process, which extends as far as the southern boundary, and Na, Al, Ba, Cu and Sr as being depleted.

The mineral assemblage wolframite–cassiterite–arsenopyrite–molybdenite–mispickel, associated with a quartz gangue, also follows the joints, although some of the veins are transgressive. There are two intensely mineralized zones: one situated about 200 m south of Cligga Head (this is some 130 m wide) and the second at the southern boundary (up to 30 m wide). Iron staining is conspicuous in the vicinity of the latter. Zones of extreme kaolinization are also present in these areas, although the whole stock has been kaolinized to some extent. The adjacent

The Cornubian granite batholith (Group C sites)

The main structural elements of the Cligga Stock

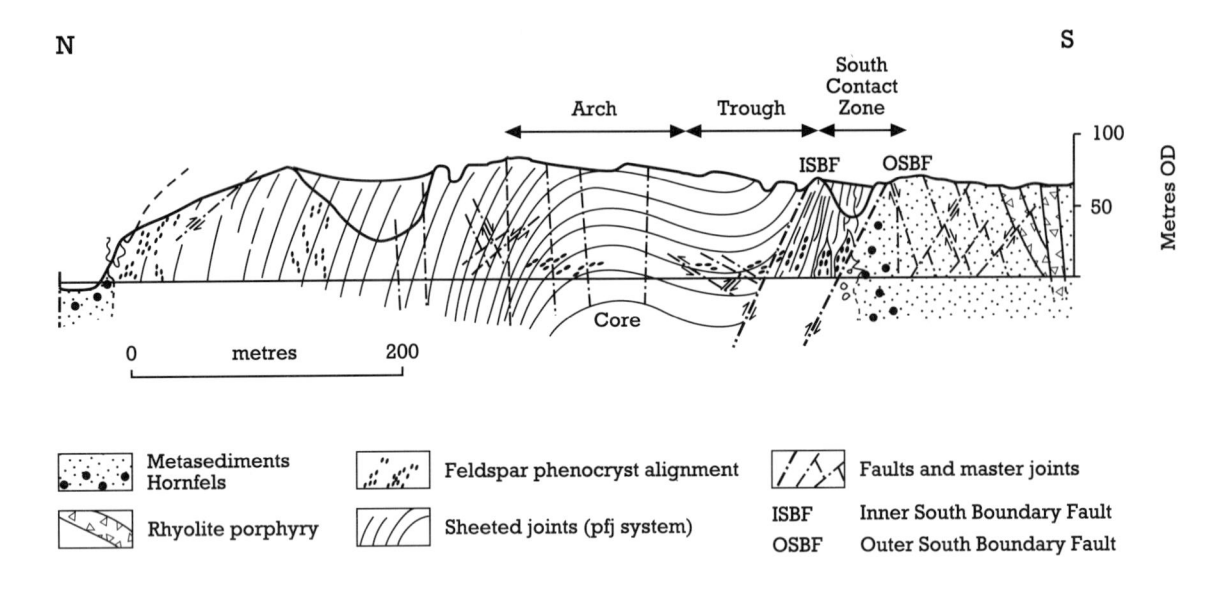

Lithology and mineralization

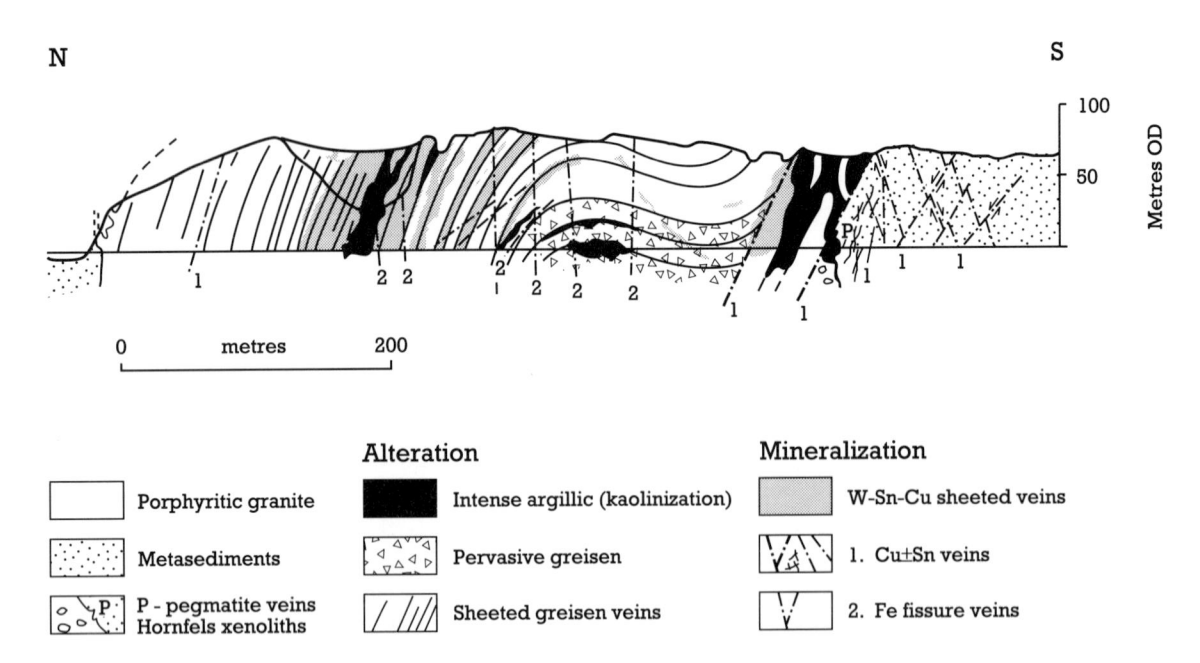

Figure 5.22 Coastal section of the Cligga Head Granite, site C19 (after Moore and Jackson, 1977).

'killas' has also been affected locally by these processes.

About 100 m south of the spotted hornfelses of the southern contact, two narrow, steeply dipping outcrops of granite-porphyry containing chalcopyrite, and believed to converge at about sea-level, strike roughly north-east inland from Hanover Cove.

Interpretation

The granite at Cligga appears, like that in the central parts of the St Austell mass, to have been a biotite granite which has been metasomatized. Apart from the lithium-rich nature of the mica, Charoy (1981) points out that the present albite has been derived from earlier oligoclase. This alteration is thought to have been brought about by a complex exchange process induced by a Li- and F-rich differentiated magma which substituted Na, Al and Li for Ca, Fe and Mg, as described in the 'Petrogenesis' section and under Tregargus Quarries. The Li-bearing mica now present has been identified as protolithionite (Hall, 1971; Charoy, 1981) or lithian siderophyllite close to zinnwaldite in composition (Stone *et al.*, 1988). The implication is that Type-E granite (Table 5.1), which brought about metasomatism, could be present in the batholith nearby.

As explained earlier in this chapter, the crystallization of the granite magma gave rise to hydrothermal fluids capable of causing metasomatism and producing pegmatites, aplites, metalliferous mineralization and eventually kaolinization. The origin of the water in these fluids was increasingly meteoric and decreasingly magmatic with time and they were associated with two periods of granite intrusion, at 290 Ma and 270 Ma BP. From the Li-rich nature of the Cligga Head Granite, it is evident that the mineralization here belongs to the second period and it therefore corresponds with the 'Main mineralization' described from the St Just area of Land's End by Jackson *et al.* (1982).

According to Moore and Jackson (1977) the development of the undulating 'floor' joints and subvertical joints and faults resulted from stress in a NNW–SSE direction, first synkinematically due to magmatic pressure and then cooling, and secondly, post-kinematically from the relaxation of stress, while Halls (1987) argues for a successsion of hydrothermal pulses. These controlled the distribution of both hydrothermal alteration and mineralization (Figure 5.22), the wide-ranging mineral assemblage indicating a condensed or telescoped sequence in which mineralized zones overlap instead of being separate, like that at Cameron Quarry but not to the same degree. The work of Jackson *et al.* (1977) led them to conclude that mineralization took place in three stages. The first was itself a two-stage process; the earlier, at a T_h range of 400–280°C, deposited cassiterite and wolframite and the later, with a T_h range of 320–240°C, deposited sulphides of Sn, Cu, As, Zn, Fe, Bi and Mo. The hydrothermal fluids involved were of low salinity (2–12 equiv. wt. % NaCl) and might also have caused the greisening. The second stage, post-faulting, resulted in quartz–hematite and quartz–pyrite–chalcopyrite assemblages from fluids with a T_h range of 260–210°C and a narrower range of low salinities (3–9 equiv. wt. % NaCl); and the final stage, leaving iron hydroxides, was effected at T_h of 150 to <70°C from fluids with salinities as low as 1–3 equiv. wt. % NaCl. Jackson *et al.* suggest that the minimum depth of formation of these deposits was 400 m, and that the last fluids might have played a part in the kaolinization process which, at similar temperatures, subsequently broke down the feldspar.

It is interesting that the salinities recorded by Jackson *et al.* (1977) are lower than those measured in the Main Mineralization of the Land's End area by Jackson *et al.* (1982) at 10–20 (and locally 40) equiv. wt. % NaCl, and that the depth of cover at 400 m minimum compares with 2800 for Land's End. Cligga Head Granite is a small granite, peripheral to the main batholith, and, as Jackson *et al.* (1982, Figure 2) suggest, salinities in the flanks of the intrusion tend to be lower than those in the main body. Jackson *et al.* (1977) do not date the later episodes of mineralization at Cligga but, by analogy with Land's End (Jackson *et al.*, 1982), they probably include Mesozoic and Cenozoic events.

Charoy (1979, 1981, 1982) has used the Cligga Head Granite in his studies of all three of the 'traditional' alteration processes in Cornwall. As regards greisening, he agrees in essence with Jackson *et al.* (1977) about the temperature and salinity of the altering fluids but, in addition, lays emphasis on the difficulties in drawing firm conclusions about processes and changes in areas where there has been repeated and successive alteration. While Hall (1971) was firm about the changes at Cligga Head, he also emphasized the difficulties in making comparisons. Charoy (1981) pointed out the importance of a shallow depth and relatively open physicochemical system in interpreting a complex site like Cligga Head. Charoy regards the tourmaline at Cligga as 'a true magmatic phase', unlike those at Porth Ledden and Roche Rock, which were a consequence of an unmixed, B-rich, late- or post-magmatic liquid giving rise to a secondary mineral. Kaolinization he regarded as a low-temperature continuation of the earlier hydrothermal processes, again agreeing with Jackson *et al.* (1977).

Conclusions

Like the St Agnes Granite, the Cligga Head Granite mass is a comparatively small projection from the Cornwall–Devon granite (the Cornubian batholith). The flat-lying joints in the granite show that it is folded and all the main joints have been affected by greisenization (see Cameron Quarry above). Cligga Head is an ideal location for the study of this process as well as metalliferous mineralization and kaolinization. The last process, the decomposition of the granite and the transformation of its feldspar crystals to the clay mineral kaolinite, was brought about by the action of the youngest and coolest of the fluids persisting from the magmatic phase which generated the granite. So here may be seen mineral veins with greisen borders and intervening zones of kaolinized granite. The full range of these phenomena, typical of south-west England, is available on this site.

Chapter 6

Post-orogenic volcanics (Group D sites)

Lithological and chemical variation

INTRODUCTION

The five sites descibed in this chapter include examples of the small-volume extrusives and intrusives that developed after the emplacement of the main granite plutons of the Cornubian batholith. Their locations are shown in Figure 6.1. Volcanism occurred during the prevalence of the 'red-bed' environment of the late Carboniferous and the early Permian Periods; it includes the mixed volcanics of the Exeter Volcanic 'Series', as well as suprabatholithic rhyolite lavas which were fed by late granite porphyry dykes. The volcanics of this period are often referred to as being Permian in age, although recalculated isotopic age dates suggest that they could also represent late Carboniferous activity. These volcanics represent the last remnant of magmatic activity that developed largely after continent–continent collision in a post-orogenic, tectonic setting.

LIST OF SITES

Rhyolitic suite:

D1 Kingsand Beach (SX 435506)

Basaltic suite:

D2 Webberton Cross Quarry (SX 875871)
D3 Posbury Clump Quarry (SX 815978)

Potassic suite – lamprophyres:

D4 Hannaborough Quarry (SS 529029)
D5 Killerton Park (SS 971005)

LITHOLOGICAL AND CHEMICAL VARIATION

Although three series of post-orogenic volcanics are generally recognized (see Chapter 2), in

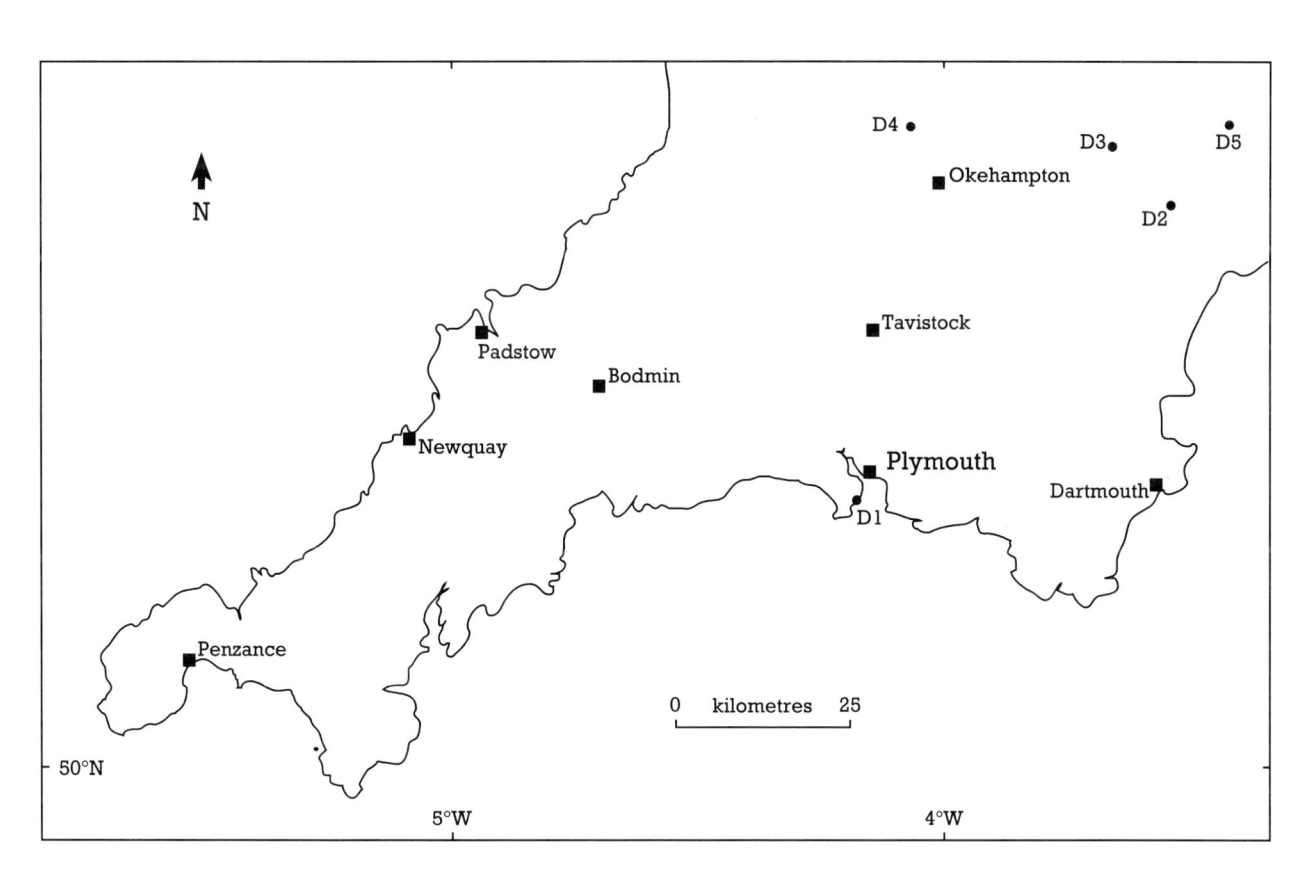

Figure 6.1 Outline map of south-west England, showing the location of Group-D sites.

Post-orogenic volcanics (Group D sites)

compositional terms they can be grouped as follows:

1. basaltic suite (in Exeter Volcanic 'Series');
2. potassic suite (dominated by minette-type lamprophyres in the Exeter Volcanic 'Series' and regionally throughout south-west England);
3. rhyolitic suite (including pebbles of acidic volcanics in 'red bed' sequences).

The petrogenesis of these small-volume volcanics and high-level intrusives has often been linked to the Cornubian granites and cross-cutting granite-porphyry dykes. Rhyolites have been interpreted as the volcanic expression of the plutonic granites (Goode, 1973; Cosgrove and Elliott, 1976), whereas the highly potassic nature of some basic lavas was considered to reflect contamination by granitic material or fluids (Tidmarsh, 1932; Knill, 1969). Although this latter feature is no longer considered significant for the production of potassic magmas, Leat *et al.* (1987) have suggested that the granites could have been derived by fractionation of a mantle-derived potassic magma that contaminated melts mainly produced by crustal anatexis.

Basaltic suite

This comprises a comagmatic, mildly alkaline series of olivine–plagioclase-phyric basalts and ophitic olivine dolerites invariably altered by post-eruptive weathering (Knill, 1969). They are characterized chemically by high incompatible-element contents (especially the LIL group), moderate light REE enrichment (La$_n$/Yb$_n$ = 6–10) and relatively evolved mafic compositions, with Ni varying between 100–200 ppm (Cosgrove, 1972; Thorpe *et al.*, 1986; Thorpe, 1987; Leat *et al.*, 1987). As seen in Figure 6.2 chondrite-normalized multi-element diagrams show progressive enrichment patterns with increasing element incompatibility (Thorpe *et al.*, 1986; Grimmer and Floyd, 1986; Leat *et al.*, 1987), together with large negative Sr anomalies (that reflect plagioclase fractionation) and minor, but variable, Nb–Ta anomalies, that Grimmer and Floyd (1986) interpreted as possible sediment or crustal contamination. Overall the basalts have a chemical composition indicative of a within-plate, continental, eruptive setting.

Potassic suite

This suite comprises all the lamprophyres (minettes) and minor trachybasalts, mafic syenites and leucitites (Knill, 1969, 1982), all of which feature abundant K-feldspar. The minettes are characterized by aligned phenocrysts of dark-rimmed phlogopitic biotite, rarer diopsidic augite, olivine and small idiomorphic apatites set in an often highly altered and reddened biotite–alkali–feldspar–Fe ore matrix (Knill, 1969; Exley *et al.*, 1982). Plagioclase may occasionally be common, but is very variable in its distribution; brown amphiboles are also recorded (Hall, 1982). Some of the lamprophyres may be vesicular, with infillings of alkali feldspar, quartz, calcite, chlorite and clays (Exley *et al.*, 1982). Sedimentary and granitic inclusions, together with a variety of xenocrysts, are relatively common (Smith, 1929).

By far the most interesting chemical feature of the potassic lavas and intrusives (Figure 6.2) are the strong enrichments in light REE and LIL elements, marked depletions of Nb, Ta and Ti (Cosgrove, 1972; Thorpe, 1987; Leat *et al.*, 1987) and high ^{87}Sr/^{86}Sr (Thorpe *et al.*, 1986). This chemical fingerprint is characteristic of subduction-related magmas and suggests the potassic lavas were derived from lithosphere subducted during the Variscan (Leat *et al.*, 1987). Similarly, Cosgrove (1972) originally suggested that the Exeter Volcanic 'Series' were generated at or near a plate margin on the basis of their similarity to high-alkaline shoshonitic rocks and the eruptive setting of such rock types elsewhere. However, although these volcanics have features similar to some magmas generated in a subduction environment, it does not necessarily imply the presence of an active subduction zone in this area during the Permian – for which there is no direct geological evidence. Instead, the particular characters could have been obtained or inherited from a mantle that had previously undergone melting in such an environment at some time in the late Palaeozoic.

Figure 6.2 Chondrite-normalized multi-element patterns for the A) basaltic and B) lamprophyric suites of the Exeter Volcanic 'Series' (data from Leat *et al.*, 1987).

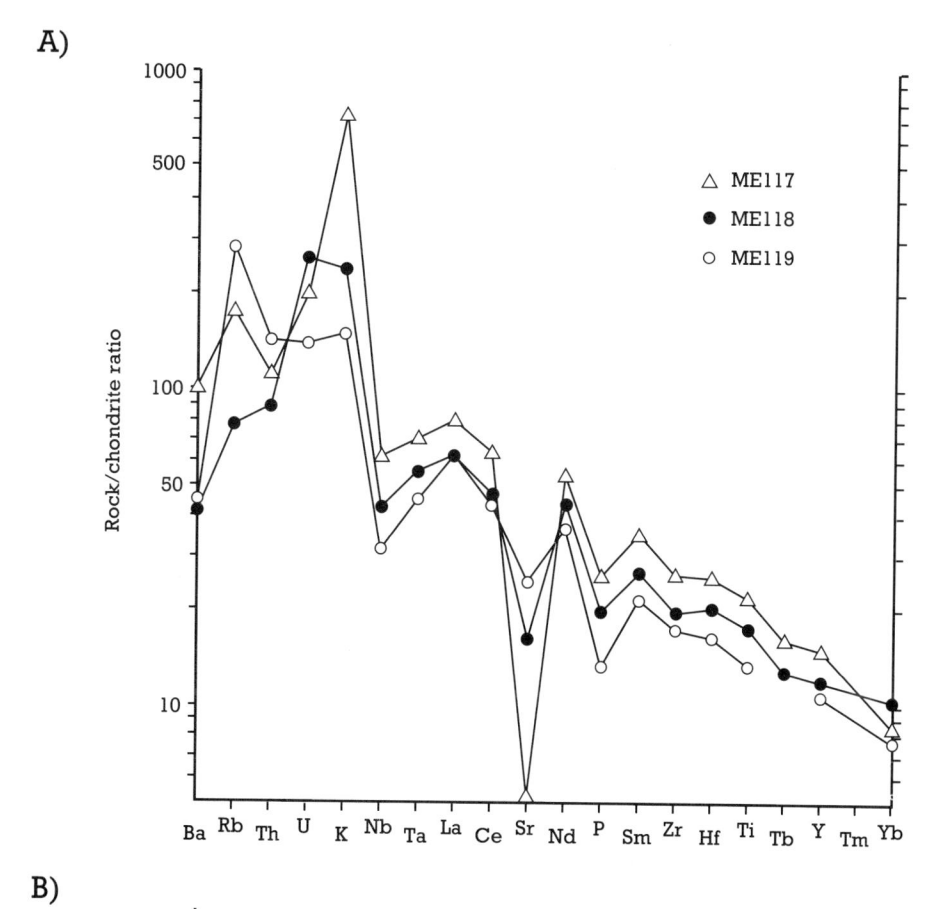

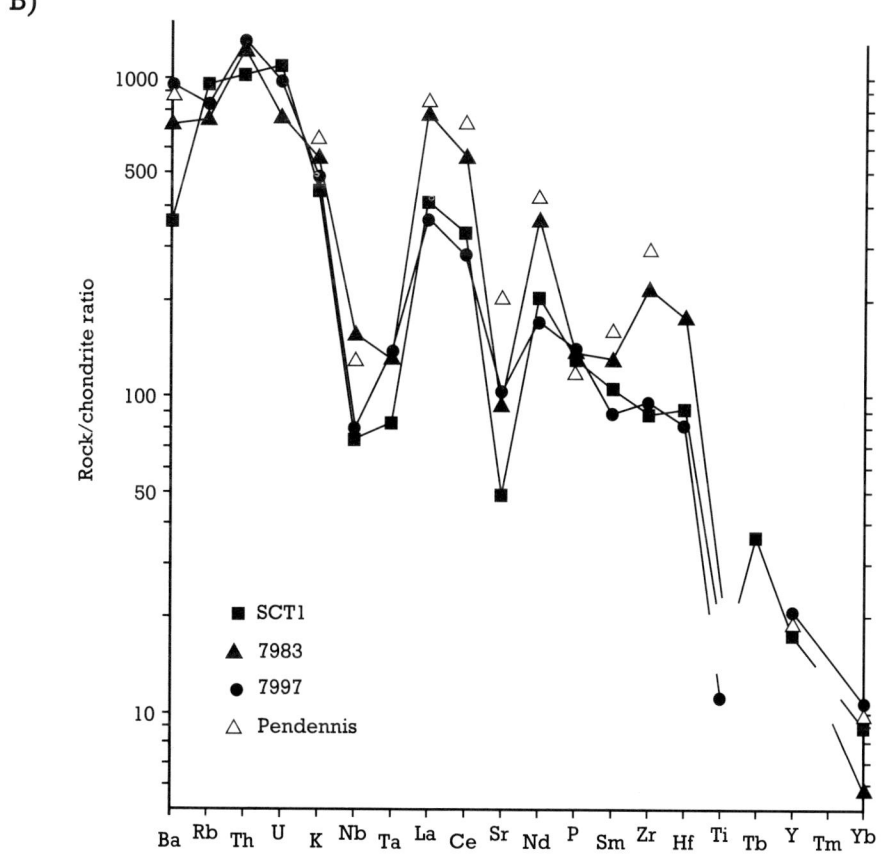

Post-orogenic volcanics (Group D sites)

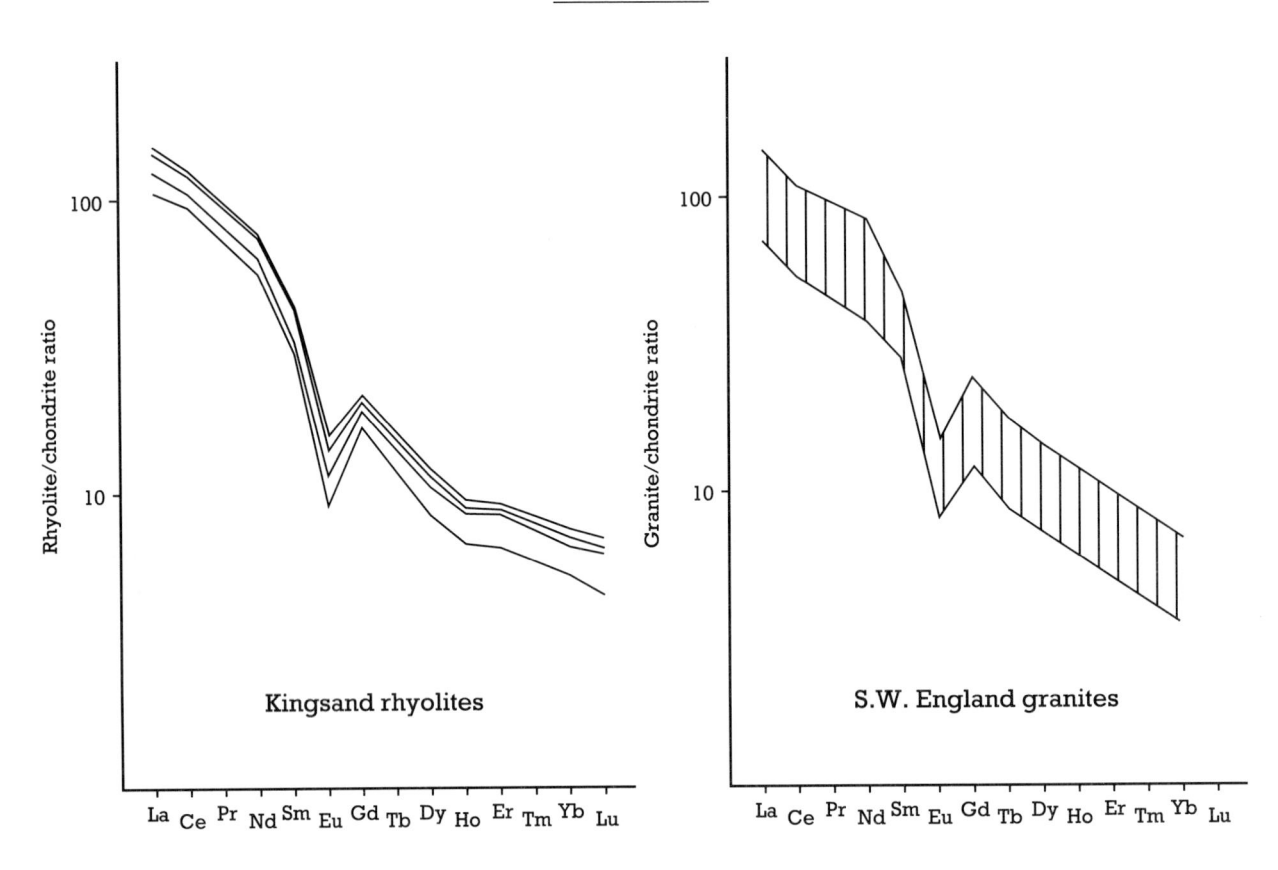

Figure 6.3 Chondrite-normalized REE patterns for Permian rhyolites (from Floyd, unpublished) and south-west England granites (data from Alderton et al., 1980).

Rhyolitic suite

Rhyolitic pebbles found in the New Red Sandstone (Laming, 1966) have similar textural, petrographic and chemical features to exposed flow-banded rhyolite lavas near Plymouth. They form the remnants of possibly extensive suprabatholithic calc-alkaline acid volcanism that was fed by the granite-porphyry dykes which cut the main granite plutons. Most commonly, they are reddened quartz–K-feldspar–biotite-phyric rhyolites with relict spherulitic textures, that have often recrystallized to a granular cryptocrystalline quartzofeldspathic matrix (Cosgrove and Elliott, 1976). Chondrite-normalized geochemical data (from Cosgrove and Elliott, 1976; Floyd, unpublished) shows strong LIL enrichment patterns, but with marked Nb, Ta, Sr and Ti negative anomalies, light REE enrichment (Figure 6.3) with negative Eu anomalies (Eu/Eu* $c.$ 0.4). All are typical features of acidic calc-alkaline rocks formed in a syncollisional tectonic setting. These rhyolites are chemically distinct from the Lower Devonian rhyolites with higher LIL/HFS ratios (Rb/Nb $c.$ 50 relative to $c.$ 5) and although REE patterns are broadly similar to, and overlap, those of the granites (Figure 6.3), they have more in common with the granite porphyries (Exley et al., 1983).

D1 KINGSAND BEACH (SX 435506)

Highlights

This locality provides the only exposure of *in situ* rhyolite of the post-Variscan volcanics.

Introduction

This site occupies the rocky platforms of a 800 m stretch of foreshore from Kingsand Beach to Sandway Point in Cawsand Bay.

Post-orogenic volcanism included both mafic dykes and flows, as well as extensive rhyolite lava flows. Apart from this site, very few actual exposures of the latter remain, although relicts of what was probably a widespread suprabatholithic rhyolite lava field, are now found as numerous altered pebbles within the local New Red Sandstone succession (Laming, 1966). An additional example of this volcanic event is the small, circular plug of flow-banded rhyolite at nearby Withnoe.

Cosgrove and Elliott (1976) showed that the Kingsand rhyolite still retained many features indicative of high-temperature rapid quenching, and that they were chemically comparable with the pebbles of rhyolite in the New Red Sandstone. Geochemical comparisons using immobile elements indicated that the rhyolites were not the volcanic equivalents of the Cornubian main granites, but were probably comagmatic with the later granite-porphyry dykes which could have acted as feeders to the lava field (Floyd, 1983). Unpublished chemical data (Floyd) show the rhyolites to be typically calc-alkaline, with chondrite-normalized negative Nb and Ta anomalies, and highly enriched in large-ion-lithophile elements. It is generally assumed that the rhyolite magma was derived by the partial melting of lower continental crust rather than of local sediments (Floyd *et al.*, 1983).

Description

An unconformable relationship between the rhyolite and the nearby Devonian is inferred at the southern end of Kingsand Beach. Separated from the Devonian sediments by a small pebble beach, the tabular rhyolite outcrop apparently overlies the deformed and near-vertical green and purple phyllites. The sediments become bleached and heavily veined with hematite near the inferred junction, possibly due to subaerial oxidative reddening of the eroded surface and the subsequent contact effects of the hot lava flow. The only other contact with sediments is near Sandway Cellar, at the north-eastern end of the outcrop, where a basal Permian conglomerate is developed. Again an actual contact is not exposed, although flow banding at the margin of the rhyolite body and in the seaweed-covered reefs dips away from the conglomerate. Back-projection of the rhyolite banding suggests that it could lie on top of the conglomerate, but equally the conglomerate could be banked against the rhyolite margin. The conglomerate was deposited rapidly as a debris flow containing many large rounded/subrounded

Figure 6.4 Flow-banded rhyolite lava of Permian age that may have formed part of the volcanic field developed above the Cornubian granite batholith. Kingsand, Devon. (Photo: P.A. Floyd.)

Post-orogenic volcanics (Group D sites)

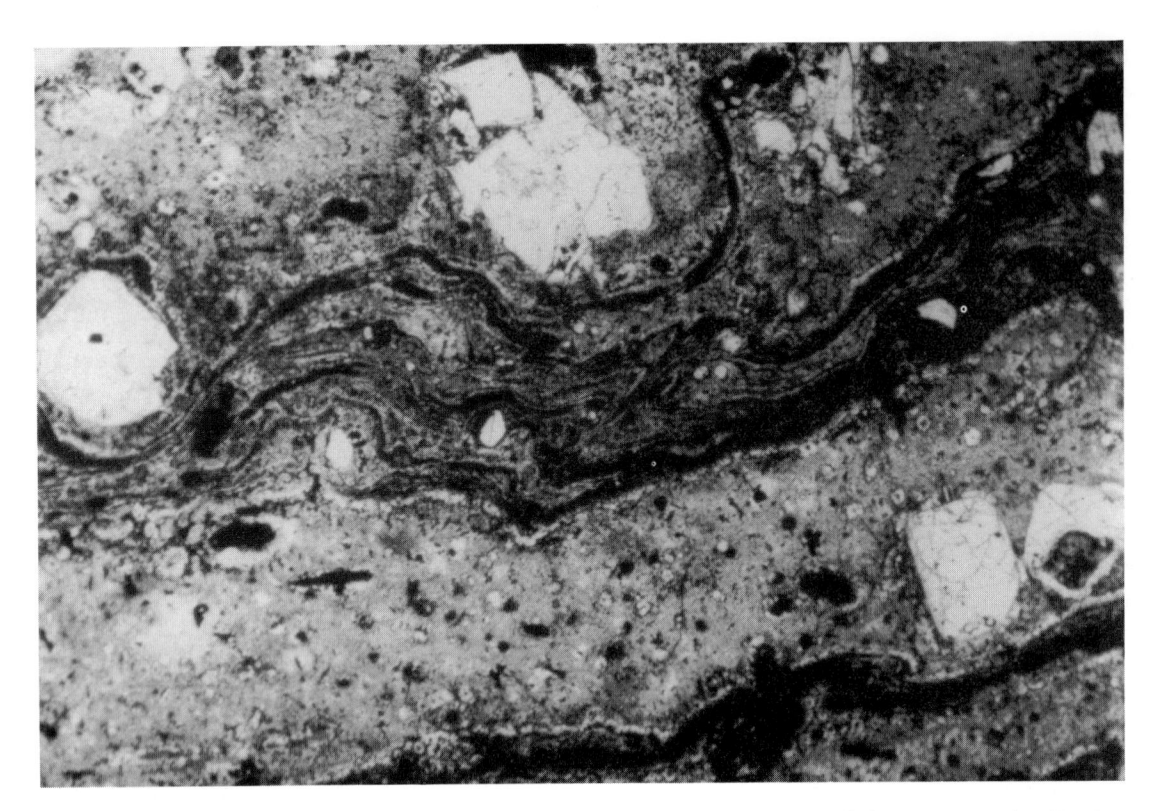

Figure 6.5 Silica phenocrysts in the flow-banded, partly devitrified matrix of the Permian rhyolite lava. Kingsand, Devon. (Photo: P.A. Floyd.)

pebbles and boulders (up to c. 1 m in diameter) of quartzite and flow-banded rhyolite. About 4 m is exposed before being succeeded by a uniform, red sandstone.

The general uniformity and tabular shape of the rhyolite along the beach, together with the lack of an extensive autobrecciated carapace, suggests that it is a lava flow, rather than a dome. Also, the lack of internal brecciated horizons indicates that the lateral extent of the outcrop probably represents a single flow. Most of the flow appears massive, with flow banding (Figure 6.4), when present, restricted to narrow, relatively uniform zones and, rarely, highly contorted on the small scale. All of the lava is pervasively reddened due to subaerial oxidation after eruption, although there are small green reduction spots (centred on spherulites) and linear zones adjacent to east–west-trending veins and fractures.

The lava contains many microscopic features indicative of rapid chilling, with a devitrified quartz–feldspar groundmass containing vestiges of glass, spherulites and high-temperature phases (Cosgrove and Elliott, 1976). The rhyolite is composed of interbedded phyric and aphyric flow units, with the former exhibiting variable proportions of quartz, feldspar (sanidine and orthoclase) and dark-red biotite phenocrysts (Figure 6.5). Chemical data on the Kingsand body (Cosgrove and Elliott, 1976; Floyd, unpublished) show it to be a typical calc-alkaline rhyolite, with high contents of large-ion-lithophile elements and chondrite-normalized negative Nb–Ta anomalies.

Interpretation

The importance of the site lies in the fact that it is the only sizeable *in situ* exposure of a supposed Permian rhyolite flow in south-west England. It also has wider regional implications, in that the rhyolites are considered to represent extensive acid volcanism genetically related to the Cornubian granite batholith. The evidence for late Carboniferous–early Permian acid volcanicity, that perhaps formed a volcanic superstructure to the high-level granite plutons, is now seen in rhyolite pebbles and feldspar crystals of extrusive origin in the New Red Sandstone. The Kingsand rhyolite is thus a solitary remnant of a once-extensive, suprabatholithic, volcanic lava field. However, comparisons based on limited chemical data suggest the rhyolites are more likely to be genetically related to the later granite-porphyry dykes than to the main granites of the batholith. The dykes could have acted as feeders to the rhyolite lavas above. Also, isotopic age dating of the granite porphyries overlaps the basal Permian at 280–270 Ma. However, the only age data on the rhyolites, quoted (Hawkes, 1981) for a pebble in a mass-flow conglomerate, adjacent to the Kingsand rhyolite, gives a Stephanian age of 295 Ma.

Other points of interest about the flow include its microscopic features and chemical characteristics. Although highly oxidized and reddened, the rhyolite flow still preserves primary features that indicate rapid quenching, together with the presence of high-temperature feldspar, and quartz and biotite phenocrysts. The rhyolite has a typical calc-alkaline composition, chondrite-normalized negative Nb–Ta anomalies and Zr/Nb ratios >10. It is chemically distinct from pre-orogenic Lower Devonian rhyolites in having higher large-ion-lithophile/high-field-strength element ratios which probably reflect derivation from a different crustal source by partial melting. When chemical plots that discriminate the nature of the eruptive environment of acidic rocks are employed (for example, Pearce *et al.*, 1984), the rhyolites apparently have a composition typical of the syncollisional tectonic environment. This chemical discrimination appears to be anomalous, as the rhyolites are clearly post-orogenic and probably later than the main granite emplacement events. The misinterpretation is an artefact of the chemical database used to construct the original tectonic discrimination plot; this assumed that the south-west England granites were typical of the syncollisional environment rather than post-collisional (Pearce *et al.*, 1984).

Conclusions

This site is unique in showing an isolated remnant of a rhyolite lava flow that was probably extruded around 295 million years ago. Apart from chemical differences these rhyolites are distinct from the Lower Devonian submarine lavas of similar composition, in being extruded after the main phase of Variscan mountain building had terminated in the late Carboniferous. It is generally considered that these rhyolites are representative of a large volcanic field situated on top of the granite batholith of south-west England.

They were probably fed by late-stage granitic dykes that cut through the main granite and surrounding country rocks. They were once much more extensive, having been eroded away, such that evidence for their existence is largely in the form of pebbles in later sedimentary deposits, particularly the red sandstones and pebble beds of the Permian Period (also seen at Kingsand). Thus Kingsand yields important evidence for the latest phase of volcanic activity at the end of the Carboniferous to earliest Permian times that followed the deformation of the Variscide mountain-building phase.

D2 WEBBERTON CROSS QUARRY (SX 875871)

Highlights

This quarry provides one of the best exposures of a basalt of the post-Variscan volcanics; the basal lava/soft-sediment contact is also well exposed.

Introduction

The site covers the disused, elongate northern quarry of the pair located in School Wood, about 300 m south-west of Webberton Cross, near Dunchideock.

The late Carboniferous–early Permian post-orogenic volcanics of the Exeter Volcanic 'Series' comprise both basalts and a mixed bag of potassic lavas that includes lamprophyres (Tidmarsh, 1932; Knill, 1969, 1982). The site is representative of the basaltic group which was considered by Ussher (1902) to occur at the base of the Permian and to be the earliest expression of post-orogenic volcanism in the area. A K/Ar whole-rock date for the basalt from Webberton Quarry gave an age of 291 Ma (recalculated from 281 Ma quoted by Miller and Mohr, 1964), that is, late Stephanian relative to the Carboniferous–Permian boundary at 286 Ma.

The petrography of the basalt and others in the Exeter area has been described by Knill (1969). Chemical data on this basalt and other members of the suite have illustrated the highly incompatible-element-enriched nature of these rocks (Cosgrove, 1972; Thorpe *et al.*, 1986; Leat *et al.*, 1987; Thorpe, 1987). As a group, the basalts form a single, cogenetic, mildly alkaline series related

Post-orogenic volcanics (Group D sites)

by minor mafic and plagioclase fractionation (Floyd, 1983; Leat *et al.*, 1987). They (including the Webberton Cross basalt) are also characterized by small, variable Nb and Ta negative anomalies on chondrite-normalized multi-element diagrams, that are often a feature of subduction-related magmas. This chemical signature was interpreted as implying that the basalts might have come from a mantle source modified by subduction-related processes (Leat *et al.*, 1987), although similar features can be produced by crustal contamination by local sediments, as suggested by Grimmer and Floyd (1986).

Description

Although the quarry walls are somewhat overgrown and mainly composed of basalt, towards the existing quarry floor can be seen the highly vesicular base of a flow resting on an intimate admixture of lava and baked sediment (Figure 6.6). The sandy sediment originally represented the surface over which the lava flowed and which became incorporated into the base of the flow. Lithological change towards the base of the lava flow is gradational, with an increase in the size, proportion and irregularity of the vesicles in the slightly flow-banded lava, together with the incoming of hematitic sandstone and siltstone fragments in greater and greater abundance. Neither the lava nor the sediments are sharp-angled blocks in this zone; they exhibit very irregular, penetrating junctions and intimately mixed relationships. This feature suggests that the lava flowed over and penetrated wet, partly consolidated sediments that were baked, and which, in turn, rapidly chilled the hot lava. Further evidence for the reaction between lava and sediment is probably provided by the neptunean dykes mentioned by Knill (1969). Although sediments overlie the basalts (eastern half of the quarry), it seems more likely that the sediment fill of these structures was extruded up from the base, rather than infilling erosional gullies in the lava flow from the top.

Most of the quarry face (15–20 m high) is apparently composed of a single lava flow, as neither interflow breccias nor reddened horizons are visible. The lava is an olivine–plagioclase-phyric alkali basalt with a matrix of granular plagioclase, calcic augite, abundant Ti–Fe oxides and minor alkali feldspar. It is often highly altered with abundant secondary serpentine, hematite, carbonate, zeolites, analcite and clays. Olivine is always replaced by serpentine and Fe-oxides. The plagioclase phenocrysts are zoned labradorites with more sodic rims; they exhibit an unusual lacy texture of vermicular Fe-oxide inclusions (Knill, 1969). The lava is variably vesicular (Figure 6.7), although away from the basal contact, the vesicles are generally small and round. They are often infilled with zeolites or white clay. The secondary assemblages are typical of very low-grade, hydrothermal oxidative alteration, which is often characteristic of ancient subaerial flows.

Interpretation

The mildly alkaline lava of this site is representative of the basaltic suite of post-orogenic, graben-related volcanics that form part of the Exeter Volcanic 'Series' of Stephanian–Permian age. According to Whittaker (1975), they are confined to the Crediton Trough and environs, which is interpreted as a small graben. In the global context of Permian rift-related magmas, they are insignificant and very small-volume extrusives (Grimmer and Floyd, 1986). However, in southwest England they are temporally associated with a potassic suite of rocks (dominated by lamprophyres) and are seen to share some of their chemical characteristics. In particular, all the Exeter Volcanic 'Series' rocks are incompatible-element-rich and have chemical features generally considered to be indicative of magmas generated in subduction zones (Grimmer and Floyd, 1986; Thorpe *et al.*, 1986; Leat *et al.*, 1987), although the chemical signature is less marked in the basalts. Previously, Cosgrove (1972) had considered that the whole suite belonged to the highly alkaline shoshonite series and that they were generated at a continental-plate margin. On the basis of their chemistry, Leat *et al.* (1987) concluded that the basalts were derived from an asthenospheric mantle source possibly modified by processes related to subduction, e.g. LIL-element enrichment, although Grimmer and Floyd (1986) stressed that a similar chemical feature could be generated by sediment contamination. Trace-element geochemistry cannot give a definitive answer in this respect, although geological inference concerning the tectonic environment suggests that there may well have been a subduction zone somewhere to

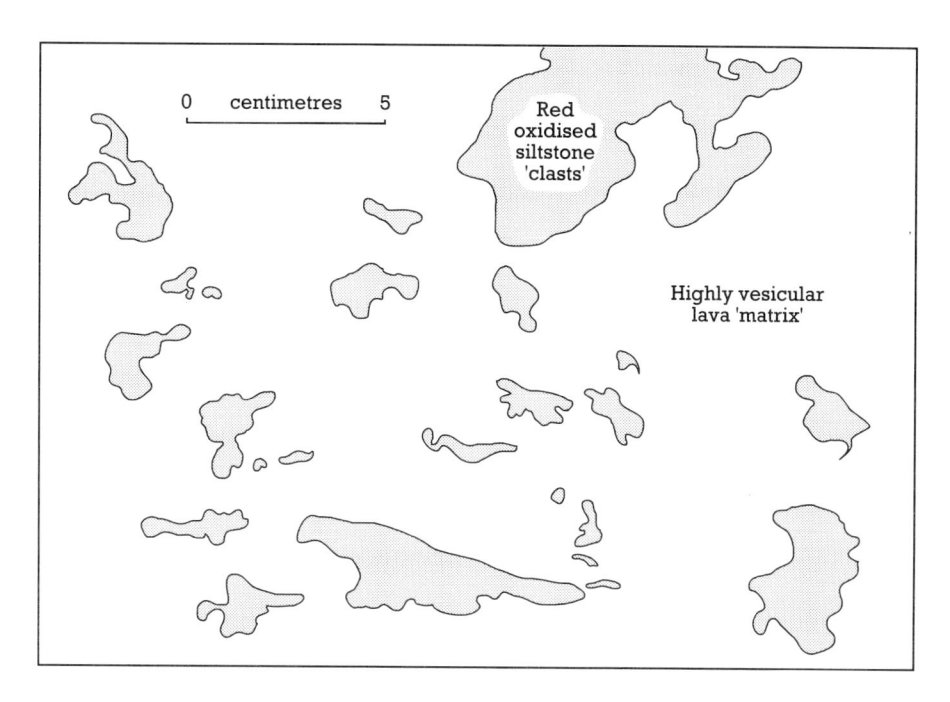

Figure 6.6 Sketch of the lava–sediment relationship at the base of a late Stephanian basalt lava flow of the Exeter Volcanic 'Series', Webberton Cross Quarry, near Exeter.

Figure 6.7 Highly amygdaloidal (vesicles infilled with white zeolites and/or clays) and oxidized subaerial basalt lava flow. Webberton Cross Quarry, Devon. (Photo: P.A. Floyd.)

Post-orogenic volcanics (Group D sites)

the south during the early Devonian (Leeder, 1982), which could have provided a modified mantle source for the later Permian volcanics.

The other major feature of this site concerns the nature of the land surface over which the basaltic lavas flowed. Sedimentological studies of the New Red Sandstone indicate that post-orogenic denudation and a tropical climate produced an aeolian desert landscape partly buried in fluvial debris (Laming, 1966). The lower contact of the basalt with the sediments indicates that the lava not only flowed over, but into, wet, partly consolidated sandy sediments, incorporating them into its base. The intimate relationships between the lava and sediment seen here are also apparent at other basalt localities, such as Stone Cross Quarry (SX 681013), although in the latter case they have now been obscured by tipping of domestic rubbish.

Conclusions

Basaltic lavas in this area erupted near the junction of the Carboniferous and Permian periods. In contrast to most of the Devonian and Carboniferous lavas described in this volume, this was not in a marine setting, but eruption occurred over an arid, rocky desert landscape under a tropical climate. At this site, however, the lavas traversed still-wet, riverine sands, fragments of which were incorporated into the base of the lava as it flowed across the irregularities of the land surface. Faint banding within the lavas indicates the direction of liquid flow. Such lavas are a feature of the Exeter–Crediton area which was probably a downfaulted trough at the time, with magma seeping up along marginal fractures. They are associated with another group of basic lavas with characteristic high potassium levels, which are known as lamprophyres. The particular chemistry of these two lava groups has been used as evidence of their initial formation from a mantle source associated at some time with a subduction zone such as found under modern oceanic island arcs. This site presents key evidence for the last phase of volcanicity that affected onshore southern Britain.

D3 POSBURY CLUMP QUARRY (SX 815978)

Highlights

This is the best section through trachybasalt lavas of the post-Variscan volcanics; autobrecciated lava tops and basal lava/soft-sediment contacts are both well seen.

Introduction

The site comprises the disused, elongate quarry on the hill about 0.4 km to the north-east of Posbury village. It is now heavily overgrown, although at the far end of the quarry and along the northern side a 15 m high face is partly accessible.

The Exeter Volcanic 'Series' was divided by Knill (1969) into a basaltic suite and a potassic suite. The latter mainly consisted of lamprophyres together with minor syenites and trachybasalts. This site is representative of the trachybasalts, although these rocks are not particularly K-rich compared with the other volcanics in the potassic suite. It was also demonstrated by Floyd (1983) that the trachybasalts around Crediton, including this site, were chemically related to the olivine basalts of Dunchideock and Silverton, analysed by Cosgrove (1972). The general field and chemical features of the basaltic suite described at Webberton Cross Quarry (discussed above) also apply here.

Description

Most of the quarry was developed in a massive, purplish-grey, variably vesicular trachybasalt lava flow. At the northern end of the quarry (by the stepway), can be seen a crude flow foliation, marked by steeply dipping trains of elongate vesicles. Their trend is not uniform across the quarry face, but varies in direction and degree of dip. Towards the quarry top, the massive lava is replaced by a reddened rubbly top to the flow. The junction is very irregular, dipping down into the massive lava, but crudely parallels the flow foliation. In detail, the junction between the rubbly flow top and massive lava is gradational and probably indicative of local autobrecciation, with cracks outlining incipient blocks in the massive portion of the flow. Spaces between the blocks in the rubble are now filled with coarse sparry calcite. At a lower level along the quarry wall, a further autobrecciated lava horizon is exposed, with the same gradational relationship. Many of the angular, purple-hearted lava blocks have highly vesicular, oxidized chilled rims and vesicle-poor centres.

Much of the rest of the quarry shows the lava flow closely associated and physically admixed with clastic sediments. This *mélange* is composed of highly vesicular, often flow-laminated, lava and baked red sandstone and siltstone. There appear to be two relationships exhibited by the lava and sediments. In the first, the lava forms angular blocks (varying from a few centimetres in width up to 1 m) mixed with smaller, red, baked sediment fragments. This suggests that the rubbly lava base flowed over a partly consolidated sediment surface, picking up fragments as it travelled forward. In other instances, the two main constituents do not form discrete blocks as such, and the lava appears rather to net vein the sediment, so that lava fingers and tubes intimately penetrate a sediment matrix. Contacts are often very irregular, cuspate, or in some cases diffuse and wispy rather than sharp. This suggests that the lava flowed over wet, unconsolidated sediments with basal lava lobes penetrating deep into an unstable sandy surface. A further feature of interest in this connection is the neptunean dyke at the northern end of the quarry. This is a vertical sediment dyke with very irregular margins, a width of *c.* 1 m at the base, and composed of structureless dolomitic sandstone. The actual mode of emplacement is not clear, although there are no sedimentary features which suggest that it was formed by the infilling of sand from the top.

Petrographically, most of the trachybasalt is highly altered, although it can be inferred from the fresher rock samples that the lava was olivine- and plagioclase-phyric, with a matrix of olivine, plagioclase, augite, K-feldspar and magnetite. Hematite, carbonate and clays are the common alteration products, with calcite, quartz and zeolites infilling the vesicles (Knill, 1969). Quartz xenocrysts, presumably derived from the sandstones over which the lava flowed, are also present. A trachytic flow texture is sometimes discernible if the rock matrix is not too altered.

Interpretation

The trachybasalt of this site probably belongs to the mildly alkaline basaltic suite of the Exeter Volcanic 'Series' with which it has petrographic and chemical affinities. The regional significance of these post-orogenic basaltic volcanics and their particular chemical signature has been discussed previously in the description of the Webberton Cross Quarry.

Apart from representing an additional example of the Exeter Volcanic 'Series' basaltic suite, the geological interest in this site concerns the lava flow itself and its relationship with the associated red-bed sediments. Since the times of Ussher (1902), when the exposure was better and less degraded, it has been one of the few sites where there is reasonable evidence remaining to demonstrate that the Permian basalts were in fact lava flows and they had autobrecciated flow tops. The trachybasalt–sediment association also provides evidence for the nature of the land surface over which the lava was flowing. This was composed of a wet, partly consolidated sand into which the base of the flow ploughed to produce an intimate admixture. The highly vesicular nature of the basal portion of the lava suggests it was charged with volatiles that, together with steam generated from the sediments, fluidized and aided penetration of the sediments.

Conclusions

Posbury Clump Quarry shows graphic evidence for volcanic eruption on to the arid Permian landscape. Here basaltic lava, part of a suite of lavas called collectively the Exeter Volcanic 'Series', was erupted over newly deposited Permian sediments. Several features characteristic of lava flows may be seen to advantage at the site: evidence of the flow direction in the form of vesicle trains (former gas bubbles), aligned crystals in what is called a flow texture, rubbly broken tops to flows and an intricate relationship with the sediments below. This last feature takes the form of an intimate mixing between the lava and sediment, so that sharp contacts do not exist. This was produced by the effusion of the lava onto or partly into wet sediments. Apart from a mechanical mixing, water flashed to steam helped to fluidize the sediment and allow lava penetration through this medium. On the other hand, more rubbly mixture of lava blocks and sediment clasts resulted where consolidated sediments were ripped up from the surface as the lava flow moved across it. This is a key site for the study of the lavas which were formed after the Variscan Orogeny and for their relationship with Permian sedimentary rocks.

D4 HANNABOROUGH QUARRY (SS 529029)

Highlights

This is one of the best exposures of a minette-type lamprophyre of the post-Variscan volcanics; an associated breccia has been variously interpreted as a vent agglomerate and as a debris-flow infill.

Introduction

The site occupies the old, disused, overgrown and partly water-filled quarry 2 km south-west of Hatherleigh. The most accessible part of the quarry is a small 5-m-high face that (photographed in better condition) is illustrated in the *British Regional Guide to South-west England* (Edmonds *et al.*, 1969; plate 8A).

The post-orogenic Stephanian–Permian volcanics of the Exeter Volcanic 'Series' can be divided petrographically and chemically into a basaltic group and a potassic group (Knill, 1969; Cosgrove, 1972). The latter includes K-feldspar-rich lamprophyres of the minette type, which are exhibited at the Hannaborough Quarry site. The site and the rock type has been described previously as representative of the lamprophyre group (Ussher, 1902; Tidmarsh, 1932; Knill, 1969), although it differs from other minette localities in being olivine-, as well as biotite-phyric. The other feature noted by previous workers is that fissures within the minette are filled with 'vent agglomerate', which quarrying in the past has left as a small knoll or peak within the northern face (Edmonds *et al.*, 1968).

The actual form and mode of emplacement of the Hannaborough minette is not clear. Tidmarsh (1932) referred to it as a lava flow, and it is generally depicted as such on geological maps, although the dominant and gently dipping 'floor joints' suggest a sheet-like intrusive body relative to the adjacent sediments. Magnetic and resistivity surveying (Edmonds *et al.*, 1968) suggest that the body is largely restricted to the quarry area, and that it is either a vertical neck (which is at variance with the subhorizontal tabular jointing) or truncated by faults.

Chemical data by Cosgrove (1972) showed that it has a composition similar to other lamprophyres in the Exeter area, being highly enriched in incompatible elements, especially those of the large-ion-lithophile group. On the basis of the highly potassic nature of these rocks generally, it was considered by earlier workers (Tidmarsh, 1932; Knill, 1969) that they were generated from basic/ultrabasic magma contaminated by K-rich fluids or materials, possibly related to the nearby Dartmoor Granite. More extensive chemical data (Cosgrove, 1972; Thorpe *et al.*, 1986; Thorpe, 1987; Leat *et al.*, 1987), however, have demonstrated that the lamprophyres exhibit a strong subduction-related chemical signature (for example, negative Nb–Ta anomalies) that is typical of many continental lamprophyric rocks.

Description

The minette is generally massive throughout, and in the present exposure does not exhibit features clearly indicative of a lava flow. However, in the vicinity of the 'vent agglomerate' peak, the minette shows concentrations of small, yellow sediment blocks that may have been derived from a thin (20–60 mm), subhorizontal, yellow, baked siltstone lens just below. The attitude of the sediment lens suggests that it might be the remnant of an interflow sediment horizon, rather than an inclusion, although this is open to question. The minette outcrop is stucturally dominated by strong, shallow-dipping (25–35°) joints, although there is no evidence to suggest they are related to the emplacement mode of the body.

One of the major features of this site is the presence of the coarse breccia that rests on the minette and has generally been considered to be a 'vent agglomerate'. A number of east–west-trending fissures in the minette are filled with the breccia, the most prominent of which can be traced for about 85 m west of the quarry (Edmonds *et al.*, 1968). Whether this breccia is really a volcanic agglomerate filling a small vent cut through the minette, is questionable. The breccia appears to cut down into the minette along a sharp, but irregular junction, which is probably an erosion surface. The breccia fragments are mainly angular, poorly sorted (<0.1–0.15 m in size), baked mudstones, laminated siltstones and fine sandstones in a reddish clay matrix. It is predominantly a chaotic, matrix-supported deposit, with only crudely developed layering near at the base, and could represent a sediment-rich debris flow. Lamprophyre clasts are virtually absent, although Tidmarsh (1932) suggested the clay matrix might have been originally devitrified glass.

The pale, purplish-grey, non-vesicular, fine-grained lamprophyre is an olivine–biotite-phyric minette that has been variably oxidized and altered. The matrix is often obscured by secondary hematite, carbonate and clays, although primary minerals, either observed or inferred, are olivine, K-feldspar, biotite, apatite and possibly plagioclase. Phenocrystic olivine is always replaced by hematite–carbonate (seen as red spots in hand specimen), whereas biotite may be corroded and partially replaced by chlorite. Rounded and apparently corroded quartz xenocrysts are also present, and these probably represent disrupted sandstone clasts incorporated by the magma upon emplacement.

Interpretation

The site is representative of the minette-type of lamprophyre of the Exeter Volcanic 'Series' potassic group. However, it is atypical in that it also contains phenocrystic olivine, unlike the more normal biotite-phyric minettes of the Exeter area. Whether the minette body here is intrusive or extrusive is not clear, although it is generally assumed to be a lava flow, as inferred from its very fine-grained nature and subhorizontal baked sediment interlayers. It has also suffered extensive oxidation, with the production of hematite that gives the body its predominantly reddish-purple colour. Comparison with similarly reddened, but proven lavas in the general area, suggest this could have been due to subaerial weathering during the Permian.

Of questionable status is the so-called 'vent agglomerate' that rests on the minette. The chaotic nature of the deposit and its composition suggest that it is a sediment-dominated debris-flow deposited on a pre-existing erosional surface of the lava, rather than a volcanic vent blasted through the body and subsequently filled with debris. Under New Red Sandstone environmental conditions, such debris flows deposited by flash floods would have been common, scouring and transporting local sediments and subaerial flows alike.

In common with the Exeter Volcanic 'Series' potassic group, the minettes of the local area (and elsewhere in South-west England), exhibit a LIL-enriched chemistry similar to lavas found in a subduction-dominated tectonic setting. In this sense they are similar to other, earlier, continental lamprophyres as seen in the Caledonides of northern Britain (Macdonald *et al.*, 1985), that were generated in a similar, post-tectonic environment. Previous ideas that the high large-ion-lithophile enrichment of these rocks was due to the specific involvement of a facies of the potassic Dartmoor Granite are no longer considered to be viable (Floyd *et al.*, 1983). More recent petrogenetic models for the potassic group involve the melting of subcontinental lithosphere previously enriched via a subduction process (Thorpe *et al.*, 1986). On the other hand, Leat *et al.* (1987) suggest that another source for Cornubian post-collision minettes could have been highly variable lithosphere (different aged crust and mantle) downthrust during ocean closure and crustal shortening. Hydrous melting of this chemically heterogeneous material would have generated incompatible-element-enriched potassic melts distributed over a wide area of the resulting fold belt. On the basis of chemical studies of potassic lavas and lamprophyres elsewhere (for example, Backinski and Scott, 1979), it seems more likely, that the specific enrichments exhibited by the Cornubian minettes were a consequence of source enrichment processes rather than subsequent involvement by the Dartmoor Granite or its fluids. It seems likely, therefore, that they were generated by the melting of mantle that had been metasomatically enriched in LIL elements during a previous subduction episode, rather than via some form of crustal contamination process.

Conclusions

This old quarry shows evidence for volcanic eruptions during the Permian Period, that is around 280 million years before the present. Here a body of rock, assumed to be a thick lava flow, occurs and is part of a suite of lavas collectively called the Exeter Volcanic 'Series'. The Hannaborough lava is a lamprophyre com-

Post-orogenic volcanics (Group D sites)

posed of abundant feldspar and mica, as well as being characteristically rich in incompatible elements. Cutting down through the lava is a body of angular blocks (a breccia), composed mainly of sedimentary rocks and a few lamprophyre clasts. In the past, this breccia has been interpreted as a vent agglomerate refilling a feeder (vent) for surface eruptions. However, this has been disputed, and current ideas indicated that the breccia represents an infill of the eroded and irregular top to the lava flow. Chemical analyses of the lamprophyres in the Exeter area have been the subject of debate, and various origins and settings have been proposed to explain their specific features. In particular, it is proposed that they were derived from a specific mantle composition and/or genesis involving the melting and contamination of crustal rocks. This site is important for chemical studies of late-stage subaerial volcanism associated with the stabilization of continental crust after the Variscan Orogeny.

D5 KILLERTON PARK AND QUARRIES (SS 971005)

Highlights

This is the best locality for the study of compositional variation within the lamprophyres

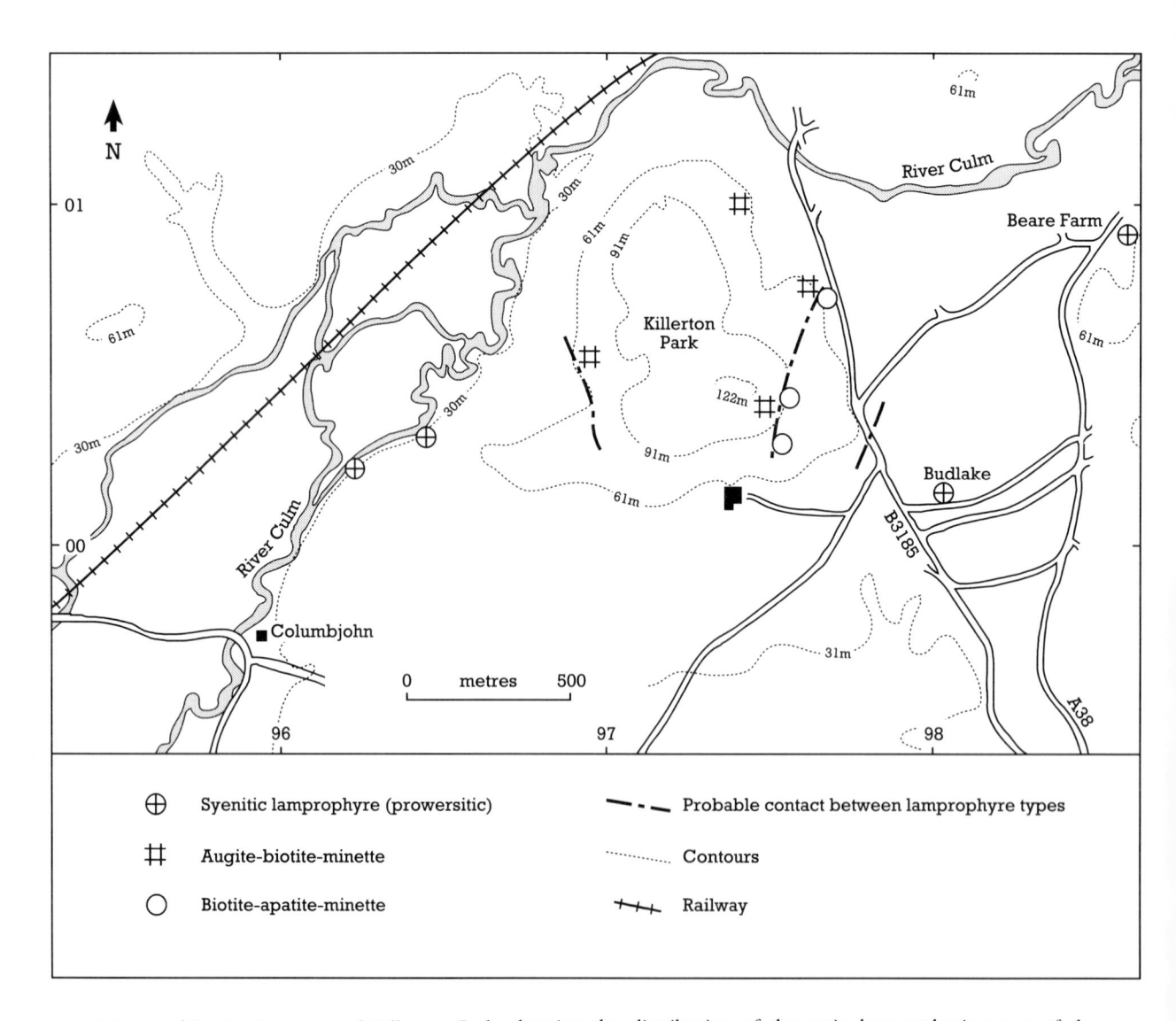

Figure 6.8 Outline map of Killerton Park, showing the distribution of the main lamprophyric types of the Exeter Volcanic 'Series' (after Knill, 1969).

of the post-Variscan volcanics; three distinct types of lamprophyres are clearly associated here.

Introduction

The lamprophyres of Killerton Park have been known for some time (Ussher, 1902), with petrographic details recorded by Tidmarsh (1932) and Knill (1969); the last placed them in the broad potassic group of the Exeter Volcanic 'Series'. A map showing the distribution of the various lamprophyric types in the area is shown in Figure 6.8.

The site covers much of Killerton Park (National Trust) and Columbjohn Wood. The disused quarry (SS 980001) at Budlake just outside the Park is also included, exposing a basaltic lava that belongs to the same volcanic province.

The few chemical analyses available (Cosgrove, 1972; Thorpe *et al.*, 1986; Leat *et al.*, 1987) exhibit the same high level of incompatible-element enrichment that is typical of the potassic group as a whole, as well as the characteristic subduction-related chemical signature (see Hannaborough Quarry above).

Description

Distribution of the lamprophyric lavas in Killerton Park (Figure 6.8) shows the close association of three main types: a biotite–apatite minette, an augite–biotite minette and a highly potassic minette or syenitic lamprophyre (Knill, 1969). The field relationships of the minettes are generally obscure, although Knill (1969) reports that the biotite–apatite-phyric variety occurs as xenoliths in the augite–biotite-phyric type. They are all assumed to be lavas fed by small fissures, but good evidence as to their mode of emplacement is lacking. In Columbjohn Wood, however, Permian red-bed sandstones were at one time seen to overlie the eroded surface of a lava flow (Knill, 1969).

One of the Killerton lavas has yielded a whole-rock K–Ar age of 291 Ma BP (Thorpe *et al.*, 1986; recalculated from Miller and Mohr, 1964) just below the Stephanian–Permian boundary and contemporaneous with the basaltic group of the Exeter Volcanic 'Series'.

One of the best exposures in the Killerton Park site is the quarry 100 m east of The Clump hillock and the surrounding area. This exposes approximately 20 m of a massive, fine-grained, blue-grey minette. Although generally considered to be a lava flow, it is poorly vesicular and does not show any internal flow features or flow boundaries. It is heavily jointed with a dominant subvertical set, the surfaces of which are reddened by hematite. There are small, apparently random differences in texture – in particular, grain-size changes from very fine-grained (almost glassy looking) to pitted and weathered coarser-grained areas. Small, brown biotite and rarer green, pyroxene phenocrysts can be seen in hand specimen, again irregularly distributed throughout the exposed body. The bulk of the lamprophyre is a biotite–pyroxene–phyric minette with a matrix of K-feldspar, biotite, titanomagnetite and apatite. Biotite may occur as both mega- and micro-phenocrysts which show different pleochroic schemes. A chemical analysis of a separated megacrystic biotite (Tidmarsh, 1932) shows it to be rich in Ti and Mg (with a chacteristically low FeO*/MgO ratio of 0.67) distinct from the Dartmoor Granite high-Fe biotites. Much of the matrix may be composed of secondary minerals such as hematite, carbonate and clay.

Interpretation

This site shows representatives of the biotite-phyric minette, which is the typical lamprophyre type of south-west England, as well as the Exeter Volcanic 'Series'. A number of other minettes are present here in close proximity to each other, although their actual contact relationships are obscure. These volcanics are always assumed to be lavas (rather than high-level intrusives), although evidence for good flow features in the present outcrops is lacking.

The specific chemical features of the Killerton Park minettes are similar to other lamprophyres from south-west England, and the same general comments concerning their chemical petrogenesis apply (see Hannaborough Quarry).

Conclusions

The sections here show various lamprophyre lavas of the Exeter Volcanic 'Series' erupted around 290 million years ago. One of the sections here shows lava buried beneath red Permian sandstone, the latter having been deposited in the equatorial desert conditions of that time. Like the

Post-orogenic volcanics (Group D sites)

site at Hannaborough (as discussed above) the rocks here are characterized by very high incompatible-element contents, and are subject to debate over their origin and the nature of the source. Their unusual composition is considered generally to reflect a specific mantle composition, although crustal contamination is also a possibility. The site shows key sections in representatives of the biotite-phyric minette, which is a typical lamprophyre of south-west England.

References

Adams, C.J.D. (1976) Geochronology of the Channel Islands and adjacent French mainland. *Journal of the Geological Society of London*, **132**, 233–50.

Agrell, S.O. (1939) Adinoles of Dinas Head, Cornwall. *Mineralogical Magazine*, **25**, 305–37.

Agrell, S.O. (1941) Dravite-bearing rocks from Dinas Head, Cornwall. *Mineralogical Magazine*, **26**, 81–93.

Alderton, D.H.M. (1978) Fluid inclusion data for lead–zinc ores from S.W. England. *Transactions of the Institution of Mining and Metallurgy, Section B*, **87**, B132–5.

Alderton, D.H.M. (1979) Luxullianite *in situ* within the St Austell granite, Cornwall – a discussion. *Mineralogical Magazine*, **43**, 441–2.

Alderton, D.H.M. and Jackson, N.J. (1978) Discordant calc-silicate bodies from the St Just aureole, Cornwall. *Mineralogical Magazine*, **42**, 427–34.

Alderton, D.H.M., Pearce, J.A. and Potts, P.J. (1980) Rare earth element mobility during granite alteration: evidence from south-west England. *Earth and Planetary Science Letters*, **49**, 149–65.

Allman-Ward, P., Halls, C., Rankin, A.H. *et al.* (1982) An intrusive hydrothermal breccia body at Wheal Remfry in the western part of the St Austell granite pluton, Cornwall, England. In *Metallization Associated with Acid Magmatism*, (ed. A.M. Evans), Wiley, Chichester, pp. 1–28.

Al-Turki, K.I. and Stone, M. (1978) Petrographic and chemical distinction between the megacrystic members of the Carnmenellis granite, Cornwall. *Proceedings of the Ussher Society*, **4**, pp. 182–9.

Amstutz, G.C. (ed.) (1974) *Spilites and Spilitic Rocks*, Springer-Verlag, Berlin, 482 pp.

Andrews, D.S. and Power, G.M. (1984) Garnetiferous phosphatic nodules within the Upper Devonian–Carboniferous Transition Group, near Boscastle, north Cornwall. *Proceedings of the Ussher Society*, **6**, 121–8.

Backinski, S.W. and Scott, R.B. (1979) Rare-earth and other trace element contents and the origin of minettes (mica-lamprophyres). *Geochimica et Cosmochimica Acta*, **43**, 93–100.

Badham, J.P.N. (1980) Late magmatic phenomena in the Cornish batholith – useful field guides for tin mineralization. *Proceedings of the Ussher Society*, **5**, 44–53.

Badham, J.P.N. (1982) Strike-slip orogens – an explanation for the Hercynides. *Journal of the Geological Society of London*, **139**, 495–506.

Badham, J.P.N. and Kirby, G.A. (1976) Ophiolites and the generation of ocean crust: data from the Lizard Complex, Cornwall. *Bullétin de la Société Géologique de France*, **7**, 885–8.

Ball, T.K. and Basham, I.R. (1979) Radioactive accessory minerals in granites from southwest England. *Proceedings of the Ussher Society*, **4**, 437–48.

Barnes, R.P. (1983) The stratigraphy of a sedimentary mélange and associated deposits in south Cornwall. *Proceedings of the Geologists' Association*, **94**, 217–29.

Barnes, R.P. (1984) Possible Lizard-derived material in the underlying Meneage Formation. *Journal of the Geological Society of London*, **141**, 79–85.

Barnes, R.P. and Andrews, J.R. (1981) Pumpellyite–

References

actinolite grade regional metamorphism in S. Cornwall. *Proceedings of the Ussher Society*, **5**, 139–46.

Barnes, R.P. and Andrews, J.R. (1984) Hot or cold emplacement of the Lizard Complex? *Journal of the Geological Society of London*, **141**, 37–9.

Barnes, R.P. and Andrews, J.R. (1986) Upper Palaeozoic ophiolite generation and obduction in south Cornwall. *Journal of the Geological Society of London*, **143**, 117–24.

Barton, D.B. (1965) *A Guide to the Mines of West Cornwall*, 2nd edn, Bradford Barton Ltd, Truro.

Behr, H.-J., Engel, W., Franke, W. *et al.* (1984) The Variscan belt in central Europe: main structures, geodynamic implications, open questions. *Tectonophysics*, **109**, 15–40.

Bonney, T.G. (1877a) On the microscopic structure of luxullianite. *Mineralogical Magazine*, **1**, 215–21.

Bonney, T.G. (1877b) On the serpentine and associated rocks of the Lizard district with notes on the chemical composition of some rocks in the Lizard district by W.H. Hudleston. *Quarterly Journal of the Geological Society of London*, **33**, 884–928.

Bonney, T.G. (1883) The hornblendic and other schists of the Lizard district, with some additions on the serpentine. *Quarterly Journal of the Geological Society of London*, **34**, 1–24.

Bonney, T.G. (1914) *The Crystalline Rocks of the Lizard. Some Notes on their History and Origin.* Cambridge.

Bonney, T.G. and McMahon, C.A. (1891) Results of an examination of the crystalline rocks of the Lizard district. *Quarterly Journal of the Geological Society of London*, **47**, 464–99.

Booth, B. (1966) Petrogenesis of the Land's End granites. Unpublished Ph.D. thesis, University of Keele.

Booth, B. (1967) Land's End granites and their relations to the granite system. *Nature, London*, **213**, 896–7.

Booth, B. (1968) Petrogenetic significance of alkali feldspar megacrysts and their inclusions in Cornubian granites. *Nature, London*, **217**, 1036–8.

Booth, B. and Exley, C.S. (1987) Petrological features of the Land's End granites. *Proceedings of the Ussher Society*, **6**, 439–46.

Bott, M.H.P. and Scott, P. (1964) Recent geophysical studies in south-west England. In *Present Views of some Aspects of the Geology of Cornwall and Devon*, (eds K.F.G. Hosking and G.J. Shrimpton), 150th Anniversary Volume, Royal Geological Society of Cornwall, pp. 25–44.

Bott, M.H.P., Day, A.A. and Masson-Smith, D. (1958) The geological interpretation of gravity and magnetic surveys in Devon and Cornwall. *Philosophical Transactions of the Royal Society of London*, **A251**, 161–91.

Bott, M.H.P., Holder, A.P., Long, R.E. *et al.* (1970) Crustal structure beneath the granites of south-west England. In *Mechanism of Igneous Intrusion*, (eds G. Newall and N. Rast), Geological Journal Special Issue No. 2, pp. 93–102.

Bowen, N.L. (1928) *The Evolution of the Igneous Rocks.* Princeton University Press, Princeton N.J., 332 pp.

Brammall, A. (1926) The Dartmoor granite. *Proceedings of the Geologists' Association*, **37**, 251–77.

Brammall, A. and Harwood, H.F. (1923) The Dartmoor granite: its mineralogy, structure and petrology. *Mineralogical Magazine*, **20**, 39–53.

Brammall, A. and Harwood, H.F. (1925) Tourmalinization in the Dartmoor granite. *Mineralogical Magazine*, **20**, 319–30.

Brammall, A. and Harwood, H.F. (1927) The temperature-range of formation for tourmaline, rutile, brookite and anatase in the Dartmoor granite. *Mineralogical Magazine*, **21**, 205–20.

Brammall, A. and Harwood, H.F. (1932) The Dartmoor granites: their genetic relationships. *Quarterly Journal of the Geological Society of London*, **88**, 171–237.

Bray, C.J. (1980) Mineralisation, greisenisation and kaolinisation at Goonbarrow china clay pit, Cornwall, UK. Unpublished D.Phil. thesis, University of Oxford.

Brazier, S., Robinson, D. and Matthews, S.C. (1979) Studies of illite crystallinity in south-west England; some preliminary results and their geological setting. *Neues Jahrbuch für Geologie und Paläontologie Monatshefte*, **11**, 641–62.

British Institutions Reflection Profiling Syndicate (BIRPS) and Etude de la Croute Continentale et Oceanique par reflection et refraction sismique (ECORS). (1986) Deep seismic reflection profiling between England, France and Ireland. *Journal of the Geological Society of London*, **143**, 45–52.

References

Bristow, C.M. (1977) A review of the evidence for the origin of the kaolin deposits in S.W. England. *Proceedings of the 8th International Kaolin Symposium and Meeting on Alunite*, Madrid–Rome, pp. 7–19.

Bristow, C.M. (1987) World kaolin: genesis, exploitation and application. *Industrial Minerals*, July, 1987.

Bristow, C.M. (1988) Ball clays, weathering and climate. *Proceedings of the 24th Forum on the Geology of Industrial Minerals*. South Carolina Geological Survey, Columbia, South Carolina, pp. 25–38.

Bristow, C.M. In press. The genesis of the china clays of south-west England – a multi-stage story. *Special Publication, Clay Minerals Society of America*.

Bristow, C.M. and Scrivener, R.C. (1984) The stratigraphy and structure of the Lower New Red Sandstone of the Exeter district. *Proceedings of the Ussher Society*, 6, 68–74.

Bristow, C.M., Exley, C.S., Highley, D.E. et al. In press. Kaolin and kaolinitic clays of Britain. *Research Report, British Geological Survey*.

Bromley, A.V. (1973) The sequence of emplacement of basic dykes in the Lizard Complex, south Cornwall. *Proceedings of the Ussher Society*, 2, p. 508.

Bromley, A.V. (1976) A new interpretation of the Lizard Complex, south Cornwall in the light of the ocean crust model. *Proceedings of the Geological Society of London*, 132, p. 114.

Bromley, A.V. (1979) Ophiolitic origin of the Lizard Complex. *Journal of the Camborne School of Mines*, 79, 25–38.

Bromley, A.V. (1989) Field guide – the Cornubian orefield. *6th International Symposium on Water–Rock Interaction, Malvern (UK)*. International Association of Geochemistry and Cosmochemistry, Camborne School of Mines, 111 pp.

Bromley, A.V. and Holl, J. (1986) Tin mineralization in south-west England. In *Mineral Processing at a Crossroads*, (eds B.A. Wills and R.W. Barley), NATO ASI Series, Series E: Applied Sciences No. 117. Martinus Nijhoff Publishers, pp. 195–262.

Brooks, M., Doody, J.J. and Al-Rawi, F.R.J. (1984) Major crustal reflectors beneath SW England. *Journal of the Geological Society of London*, 141, 97–103.

Brooks, M., Mechie, J. and Llewellyn, D.J. (1983) Geophysical investigations in the Variscides of south-west Britain. In *The Variscan fold belt in the British Isles*, (ed. P.L. Hancock), Adam Hilger, Bristol, pp. 186–97.

Burton, C.J. and Tanner, P.W.G. (1986) The stratigraphy and structure of the Devonian rocks around Liskeard, east Cornwall, with regional implications. *Journal of the Geological Society of London*, 143, 95–105.

Butcher, N.E. (1958) The Culm igneous suite near Tavistock, west Devonshire. *Abstracts of the Proceedings of the Conference of Geologists and Geomorphologists in the Southwest of England*, Exeter (1958), pp. 21–3.

Butcher, N.E. (1982) Megacrysts in volcanic rocks from Pitts Cleave Quarry, near Tavistock. *Proceedings of the Ussher Society*, 5, p. 392.

Butler, F.H. (1908) Kaolinization and other changes in west of England rocks. *Mineralogical Magazine*, 15, 128–46.

Champernowne, A. (1889) On the Ashprington Volcanic Series. *Quarterly Journal of the Geological Society of London*, 45, 369–79.

Chandler, P. and Isaac, K.P. (1982) The geological setting, geochemistry and significance of Lower Carboniferous basic volcanic rocks in central south-west England. *Proceedings of the Ussher Society*, 5, 279–88.

Chandler, P., Davey, R.F., Durrance, E.M. et al. (1984) A gravity survey of the Polyphant ultrabasic complex. *Proceedings of the Ussher Society*, 6, 116–20.

Chappell, B.W. and White, A.J.R. (1974) Two contrasting granite types. *Pacific Geology*, 8, 173–4.

Charoy, B. (1979) Définition et importance des phénomènes deutériques et des fluides associés dans les granites: conséquences métallogéniques. *Sciences de la Terre, Mémoire*, 37, 364 pp.

Charoy, B. (1981) Post-magmatic processes in south-west England and Brittany. *Proceedings of the Ussher Society*, 5, 101–15.

Charoy, B. (1982) Tourmalinization in Cornwall, England. In *Metallization Associated with Acid Magmatism*, (ed. A.M. Evans), Wiley, Chichester, pp. 63–70.

Charoy, B. (1986) The genesis of the Cornubian batholith (south-west England): the example of the Carnmenellis granite. *Journal of Petrology*, 27, 571–604.

Chaudhry, M.N. and Howie, R.A. (1973) Lithium–aluminium micas from the Meldon aplite. *Mineralogical Magazine*, 39, 289–96.

Chaudhry, M.N. and Howie, R.A. (1976) Lithium tourmaline from the Meldon aplite, Devon-

References

shire, England. *Mineralogical Magazine*, **40**, 747–51.

Chauris, L., Souet, J. and Zimmermann, J.L. (1969) Études géochronologiques et géotectoniques dans le Nord-Finistére. *Science de la Terre*, **14**(4), 329–58.

Chayes, F. (1955) Modal composition of two facies of the Carnmenellis granite. *Geological Magazine*, **92**, 364–6.

Chesher, J.A. (1968) The geology of the middle Teign valley. *Proceedings of the Ussher Society*, **2**, 15–17.

Chesher, J.A. (1969) The geology of the middle Teign valley. Unpublished Ph.D. Thesis, University of Exeter.

Chinner, G.A. and Fox, J.S. (1974) The origin of cordierite–anthophyllite rocks in the Land's End aureole. *Geological Magazine*, **111**, 397–408.

Collins, J.H. (1878) *The Hensbarrow Granite District: a Geological Description and Trade History*. Lake and Lake, Truro.

Collins, J.H. (1887) On the nature and origin of clays: the composition of kaolinite. *Mineralogical Magazine*, **7**, 205–14.

Collins, J.H. (1909) Geological features visible at the Carpalla china clay pit. *Quarterly Journal of the Geological Society of London*, **65**, 155–61.

Collins, J.H. and Coon, J.M. (1914) On the topaz rock of St Mewan Beacon, Cornwall. *Transactions of the Royal Geological Society of Cornwall*, **15**, 43–54.

Conybeare, J.J. (1817) Memoranda relative to the porphyritic veins etc. of St Agnes in Cornwall. *Transactions of the Geological Society of London*, **4**, p. 401.

Coon, J.M. (1911) China clay. *Transactions of the British Ceramic Society*, **10**, 81–92.

Coon, J.M. (1913a) China stone. *Transactions of the British Ceramic Society*, **12**, 227–38.

Coon, J.M. (1913b) On china stone, Cornish stone or petuntzyte. *Transactions of the Royal Geological Society of Cornwall*, **13**, 561–72.

Cosgrove, M.E. (1972) The geochemistry of potassium-rich Permian volcanic rocks of Devonshire, England. *Contributions to Mineralogy and Petrology*, **36**, 155–70.

Cosgrove, M.E. and Elliott, M.H. (1976) Suprabatholithic volcanism of the southwest England granites. *Proceedings of the Ussher Society*, **3**, 391–401.

Cosgrove, M.E. and Hamilton, N. (1973) Geochemical and preliminary palaeomagnetic results of the Lemail lamprophyre, Wadebridge, Cornwall. *Proceedings of the Ussher Society*, **2**, p. 482.

Coward, M.P. and McClay, K.R. (1983) Thrust tectonics of S. Devon. *Journal of the Geological Society of London*, **140**, 215–28.

Dangerfield, J. and Hawkes, J.R. (1969) Unroofing of the Dartmoor granite and possible consequences with regard to mineralization. *Proceedings of the Ussher Society*, **2**, 122–31.

Dangerfield, J. and Hawkes, J.R. (1981) The Variscan granites of S.W. England: additional information. *Proceedings of the Ussher Society*, **5**, 116–20.

Dangerfield, J., Hawkes, J.R. and Hunt, E.C. (1980) The distribution of lithium in the St Austell granite. *Proceedings of the Ussher Society*, **5**, 76–80.

Darbyshire, D.P.F. and Shepherd, T.J. (1985) Chronology of granite magmatism and associated mineralization, S.W. England. *Journal of the Geological Society of London*, **142**, 1159–77.

Darbyshire, D.P.F. and Shepherd, T.J. (1987) Chronology of magmatism in south-west England: the minor intrusions. *Proceedings of the Ussher Society*, **6**, 431–8.

Davies, G.R. (1984) Isotopic evolution of the Lizard Complex. *Journal of the Geological Society of London*, **141**, 3–14.

Day, G.A. and Williams, C.A. (1970) Gravity compilation in the N.E. Atlantic and interpretation of gravity in the Celtic Sea. *Earth and Planetary Science Letters*, **8**, 205–13.

de Albuquerque, C.A.R. (1971) Petrochemistry of a series of granitic rocks from northern Portugal. *Bulletin of the Geological Society of America*, **82**, 2783–98.

Dearman, W.R. (1959) The structure of the Culm Measures at Meldon, near Okehampton, North Devon. *Quarterly Journal of the Geological Society of London*, **115**, 65–106.

Dearman, W.R. (1962) *Dartmoor – the North-West Margin and Other Selected Areas*. Geologists' Association Guide, 33, Benham and Co. Ltd, Colchester.

Dearman, W.R. (1963) Wrench-faulting in Cornwall and south Devon. *Proceedings of the Geologists' Association*, **74**, 265–87.

Dearman, W.R. (1969) An outline of the structural geology of Cornwall. *Proceedings of the Geological Society of London*, **1654**, 33–9.

Dearman, W.R. (1971) A general view of the

References

structure of Cornubia. *Proceedings of the Ussher Society*, **2**, 220–36.

Dearman, W.R. and Butcher, N.E. (1959) Geology of the Devonian and Carboniferous rocks of the N.W. border of the Dartmoor granite. *Proceedings of the Geologists' Association*, **70**, 51–92.

Dearman, W.R., Leveridge, B.E. and Turner, R.G. (1969) Structural sequences and the ages of slates and phyllites from south-west England. *Proceedings of the Geological Society of London*, **1654**, 41–5.

Dearman, W.R., Freshney, E.C., King, A.F. et al. (1970) *The North Coast of Cornwall*. Geologists' Association Guide, 10, Benham and Co. Ltd, Colchester.

De la Beche, H.T. (1839) *Report on the Geology of Cornwall, Devon and West Somerset*. Memoir of the Geological Survey of Great Britain. Longman, Orme, Brown, Green and Longmans, London, 648 pp.

Dewey, H. (1914) The geology of north Cornwall. *Proceedings of the Geologists' Association*, **25**, 154–79.

Dewey, H. (1915) On the spilosites and adinoles from north Cornwall. *Transactions of the Royal Geological Society of Cornwall*, **15**, 71–84.

Dewey, H. and Flett, J.S. (1911) On some British pillow lavas and the rocks associated with them. *Geological Magazine*, **48**, 202–9 and 241–8.

Dineley, D. (1966) The Dartmouth Beds of Bigbury Bay, south Devon. *Quarterly Journal of the Geological Society of London*, **122**, 187–217.

Dineley, D. (1986) Cornubian quarter-century: advances in the geology of south-west England, 1960–1985. *Proceedings of the Ussher Society*, **6**, 275–90.

Dodson, M.H. and Rex, D.C. (1971) Potassium–argon ages of slates and phyllites from south-west England. *Quarterly Journal of the Geological Society of London*, **126** (for 1970), 465–99.

Duddridge, G.A. (1988) Metasomatic tourmaline at Cape Cornwall, Land's End. *Proceedings of the Ussher Society*, **7**, 93–4.

Durrance, E.M. (1985a) Lower Devonian igneous rocks of south Devon: implications for Variscan plate tectonics. *Proceedings of the Ussher Society*, **6**, 205–10.

Durrance, E.M. (1985b) A possible major Variscan thrust along the southern margin of the Bude Formation, south-west England. *Proceedings of the Ussher Society*, **6**, 173–9.

Durrance, E.M. and Bristow, C.M. (1986) Kaolinization and isostatic readjustment in south-west England. *Proceedings of the Ussher Society*, **6**, 318–22.

Durrance, E.M. and Laming, D.J.C. (eds) (1982) *The Geology of Devon*. University Press, Exeter, 346 pp.

Durrance, E.M., Bromley, A.V., Bristow, C.M. et al. (1982) Hydrothermal circulation and post-magmatic changes in granites of south-west England. *Proceedings of the Ussher Society*, **5**, 304–20.

Edmonds, E.A., McKeown, M.C. and Williams, M. (1969) *British Regional Geology: South-West England*, 3rd edn. Institute of Geological Sciences, HMSO, London.

Edmonds, E.A., Wright J.E., Beer, K.E. et al. (1968) *Geology of the Country around Okehampton* (Sheet 324). Memoir of the Geological Survey of Great Britain, HMSO, London.

Edmondson, K.M. (1970) A study of the alkali feldspars from some S.W. England granites. Unpublished M.Sc. thesis, University of Keele.

Edmunds, W.M., Andrews, J.N., Burgess, W.G. et al. (1984) The evolution of saline and thermal groundwaters in the Carnmenellis granite. *Mineralogical Magazine*, **48**, 407–24.

Edwards, J.W.F. (1984) Interpretation of seismic and gravity surveys over the eastern part of the Cornubian platform. In *Variscan Tectonics of the North Atlantic Region*, (eds D.H.W. Hutton and D.J. Sanderson), Special Publication of the Geological Society of London, No. 14, pp. 119–24.

Embrey, P.G. and Symes, R.F. (1987) *Minerals of Cornwall and Devon*. British Museum (Natural History), London.

Emmerman, R. (1977) A petrogenetic model for the origin and evolution of the Hercynian Granite Series of the Schwarzwald. *Neues Jahrbuch für Geologie und Paläontologie Abhandlungen*, **128**, 219–53.

Eskola, P. (1914) On the petrology of the Orijarvi region in south-western Finland. *Bullétin de la Commission Géologique de Finlande*, **40**, 1–274.

Exley, C.S. (1959) Magmatic differentiation and alteration in the St Austell granite. *Quarterly Journal of the Geological Society of London*, **114**, 197–230.

Exley, C.S. (1964) Some factors bearing on the

References

natural synthesis of clays in the granites of south-west England. *Clay Minerals Bulletin*, **5**, 411–26.

Exley, C.S. (1965) Some structural features of the Bodmin Moor granite mass. *Proceedings of the Ussher Society*, **1**, 157–9.

Exley, C.S. (1966) The granitic rocks of Haig Fras. *Nature, London*, **210**, 365–7.

Exley, C.S. (1976) Observations on the formation of kaolinite in the St Austell granite. *Clay Minerals*, **11**, 51–63.

Exley, C.S. and Stone, M. (1964) The granitic rocks of south-west England. In *Present Views of some Aspects of the Geology of Cornwall and Devon*, (eds K.F.G. Hosking and G.J. Shrimpton), 150th Anniversary Volume, Royal Geological Society of Cornwall, pp. 131–84.

Exley, C.S. and Stone, M. (1982) Hercynian intrusive rocks: petrogenesis. In *Igneous Rocks of the British Isles*, (ed. D.S. Sutherland), Wiley, Chichester, pp. 311–20.

Exley, C.S., Stone, M. and Floyd, P.A. (1983) Composition and petrogenesis of the Cornubian granite batholith and post-orogenic volcanic rocks in south-west England. In *The Variscan Fold Belt in the British Isles* (ed. P.L. Hancock), Adam Hilger, Bristol, pp. 153–77.

Exley, C.S., Stone, M. and Lees, G.J. (1982) Petrology of the granites and minor intrusions. In *Igneous Rocks of the British Isles*, (ed. D.S. Sutherland), Wiley, Chichester, pp. 293–302.

Fehn, U., Cathles, L.M. and Holland, H.D. (1978) Hydrothermal convection and uranium deposits in abnormally radioactive plutons. *Economic Geology*, **73**, 1556–66.

Flett, J.S. (1903) Petrography of W. Cornwall. In *Geological Survey of Great Britain, Summary of Progress*, pp. 150–62.

Flett, J.S. (1946) *Geology of the Lizard and Meneage* (Sheet 359). Memoir of the Geological Survey of Great Britain, 2nd edn, HMSO, London.

Flett, J.S. and Hill, J.B. (1912) *Geology of the Lizard and Meneage* (Sheet 359). Memoir of the Geological Survey of Great Britain, 1st edn, HMSO, London.

Floyd, P.A. (1965) Metasomatic hornfelses of the Land's End aureole at Tater-du, Cornwall. *Journal of Petrology*, **6**, 223–45.

Floyd, P.A. (1966a) Greenstone sills and metamorphic zoning in the Land's End aureole at Newlyn, Cornwall. *Proceedings of the Ussher Society*, **1**, 252–6.

Floyd, P.A. (1966b) Distribution and mobility of fluorine in contact metamorphosed and metasomatized hornfelses. *Nature, London*, **212**, 676–8.

Floyd, P.A. (1967) The role of hydration in the spatial distribution of metasomatic cations in the Land's End aureole, Cornwall. *Chemical Geology*, **2**, 147–56.

Floyd, P.A. (1975) Exotic hornfelses from the Land's End aureole. *Geological Magazine*, **112**, 315–9.

Floyd, P.A. (1976) Geochemical variation in the greenstones of S.W. England. *Journal of Petrology*, **17**, 522–45.

Floyd, P.A. (1982a) Chemical variation in Hercynian basalts relative to plate tectonics. *Journal of the Geological Society of London*, **139**, 505–20.

Floyd, P.A. (1982b) The Hercynian trough: Devonian and Carboniferous volcanism in south-western Britain. In *Igneous Rocks of the British Isles*, (ed. D.S. Sutherland), Wiley, Chichester, pp. 227–42.

Floyd, P.A. (1982c) Introduction: geological setting of Upper Palaeozoic magmatism. In *Igneous Rocks of the British Isles*, (ed. D.S. Sutherland), Wiley, Chichester, pp. 217–25.

Floyd, P.A. (1983) Composition and petrogenesis of the Lizard Complex and pre-orogenic basaltic rocks in southwest England. In *The Variscan Fold Belt in the British Isles*, (ed. P.L. Hancock), Adam Hilger, Bristol, pp. 130–52.

Floyd, P.A. (1984) Geochemical characteristics and comparison of the basic rocks of the Lizard Complex and the basaltic lavas within the Hercynian troughs of SW England. *Journal of the Geological Society of London*, **141**, 61–70.

Floyd, P.A. and Al-Samman, A.H. (1980) Primary and secondary chemical variation exhibited by some west Cornish volcanic rocks. *Proceedings of the Ussher Society*, **5**, 68–75.

Floyd, P.A. and Fuge, R. (1973) Distribution of F and Cl in some contact and regionally metamorphosed Cornish greenstones. *Proceedings of the Ussher Society*, **2**, 483–8.

Floyd, P.A. and Lees, G.J. (1972) Preliminary petrological and geochemical data on the Cudden Point greenstone. *Proceedings of the Ussher Society*, **2**, 421–3.

Floyd, P.A. and Leveridge, B.E. (1987) Tectonic environment of the Devonian Gramscatho basin, south Cornwall: framework mode and

References

geochemical evidence from turbiditic sandstones. *Journal of the Geological Society of London*, **144**, 531–42.

Floyd, P.A. and Rowbotham, G. (1979) Chemical composition of relict clinopyroxenes from the Mullion Island lavas, Cornwall. *Proceedings of the Ussher Society*, **4**, 419–29.

Floyd, P.A. and Rowbotham, G. (1982) Chemistry of primary and secondary minerals in titaniferous brown amphibole-bearing greenstones from north Cornwall. *Proceedings of the Ussher Society*, **5**, 296–303.

Floyd, P.A., Exley, C.S. and Stone, M. (1983) Variscan magmatism in south-west England – discussion and synthesis. In *The Variscan Fold Belt in the British Isles*, (ed. P.L. Hancock), Adam Hilger, Bristol, pp. 178–85.

Floyd, P.A., Lees, G.J. and Parker, A. (1976) A preliminary geochemical twist to the Lizard's new tale. *Proceedings of the Ussher Society*, **3**, 414–25.

Foster, C. Le Neve (1877) Remarks on some tin lodes in the St Agnes district. *Transactions of the Royal Geological Society of Cornwall*, **9**, 205–19.

Fox, H. (1888) On the gneissic rocks of the Lizard; with notes on the specimens, by J.J.H. Teall. *Quarterly Journal of the Geological Society of London*, **44**, 309–17.

Fox, H. (1893) Mullion Island. *Journal of the Royal Institute of Cornwall*, **12**, 34–41.

Fox, H. (1895) On a soda feldspar rock at Dinas Head, north coast of Cornwall. *Geological Magazine*, **2**, 13–20.

Fox, H. and Teall, J.J.H. (1893) On the radiolarian chert from Mullion Island. *Quarterly Journal of the Geological Society of London*, **49**, 211–20.

Fox, H. and Somervail, A. (1888) On the occurrence of porphyritic structure in some rocks of the Lizard district. *Geological Magazine*, **5**, 74–7.

Francis, H. (1970) Review of Carboniferous volcanism in England and Wales. *Journal of Earth Science, Leeds*, **8**, 41–56.

Franke, W. (1989) Tectonostratigraphic units in the Variscan belt of central Europe. In *Terranes in the Circum-Atlantic Palaeozoic Orogens*, (ed. R.D. Dallmeyer), Geological Society of America, Special Paper, No. 230, pp. 67–90.

Franke, W. and Engel, W. (1982) Variscan sedimentary basins on the continent and relations with SW England. *Proceedings of the Ussher Society*, **5**, 259–69.

Freshney, E.C. and Taylor, R.T. (1972) The Upper Carboniferous stratigraphy of north Cornwall and west Devon. *Proceedings of the Ussher Society*, **2**, 464–71.

Freshney, E.C., McKeown, M.C. and Williams, M. (1972) *Geology of the Coast between Tintagel and Bude* (Sheet 233). Memoir of the Geological Survey of Great Britain, HMSO, London.

Freshney, E.C., Edmonds, E.A., Taylor, R.T. *et al.* (1979) *Geology of the Country around Bude and Bradworthy* (Sheet 322). Memoir of the Geological Survey of Great Britain, HMSO, London.

Frey, F.A. (1969) Rare earth abundances in a high-temperature peridotite intrusion. *Geochimica et Cosmochimica Acta*, **33**, 1429–77.

Frey, F.A. (1984) Rare earth element abundances in upper mantle rocks. In *Rare Earth Element Geochemistry*, (ed. P. Henderson), Elsevier, Amsterdam, 153–203.

Gauss, G.A. and House, M.R. (1972) The Devonian successions in the Padstow area, north Cornwall. *Journal of the Geological Society of London*, **128**, 151–72.

George, M.C., Stone, M., Fejer, E.E. *et al.* (1981) Triplite from the Megiliggar Rocks, Cornwall. *Mineralogical Magazine*, **44**, 236–8.

Georget, Y., Matineau, F. and Capdevila, R. (1986) Age tardi-Hercynien et origine crustale du granite de Brignogau (Finistère, France). Conséquences sur l'interprétation des granites Nord-armoricains. *Comptes Rendus des Séances de l'Académie des Sciences, Paris*, **302**, Série II, no. 5, 237–42.

Gerrard, A.J.W. (1974) The geomorphological importance of jointing in the Dartmoor granite. In *Progress on Geomorphology*, (eds E.H. Brown and R.S. Waters), Institution of British Geographers Special Publication, No. 7, pp. 39–51.

Ghosh, P.K. (1927) Petrology of the Bodmin Moor granite (eastern part), Cornwall. *Mineralogical Magazine*, **21**, 285–309.

Ghosh, P.K. (1934) The Carnmenellis granite: its petrology, metamorphism and tectonics. *Quarterly Journal of the Geological Society of London*, **90**, 240–76.

Goldring, R. (1962) The bathyal lull: Upper Devonian and Lower Carboniferous sedimentation in the Variscan geosyncline. In *Some Aspects of the Variscan Fold Belt*, (ed. K. Coe), Manchester University Press, Manchester, pp. 75–91.

Goode, A.J.J. (1973) The mode of intrusion of

References

Cornish elvans. *Report of the Institute of Geological Sciences*, **73/7**, 8 pp.

Goode, A.J.J. and Merriman, R.J. (1987) Evidence of crystalline basement west of the Land's End granite, Cornwall. *Proceedings of the Geologists' Association*, **98**, 39–43.

Goode, A.J.J. and Taylor, R.T. (1980) Intrusive and pneumatolytic breccias in south-west England. *Report of the Institute of Geological Sciences*, **80/2**, 23 pp.

Goode, A.J.J., Hawkes, J.R., Dangerfield, J. et al. (1987) The geology, petrology and geophysics of the Land's End, Tregonning–Godolphin and St Michael's Mount Variscan granites. *British Geological Survey Open File Report*, HMSO, London.

Goode, A.J.J. and Taylor, R.T. (1988) *Geology of the Country around Penzance* (Sheet 358). Memoir of the British Geological Survey.

Grainger, P. and Witte, G. (1981) Clay mineral assemblages of Namurian shales in Devon and Cornwall. *Proceedings of the Ussher Society*, **5**, 168–78.

Green, D.H. (1964a) Petrogenesis of the high-temperature peridotite intrusion in the Lizard area, Cornwall. *Journal of Petrology*, **5**, 134–88.

Green, D.H. (1964b) Metamorphic aureole of the peridotite at the Lizard, Cornwall. *Journal of Geology*, **72**, 543–63.

Green, D.H. (1964c) A re-study and re-interpretation of the geology of the Lizard peninsula, Cornwall. In *Present Views on Some Aspects of the Geology of Cornwall and Devon*, (eds K.F.G. Hosking and G.J. Shrimpton), Transactions of the Royal Geological Society of Cornwall, 150th Anniversary Volume, pp. 87–144.

Grimmer, S. and Floyd, P.A. (1986) Geochemical features of Permian rift volcanism – a comparison of Cornubian and Oslo basic volcanics. *Proceedings of the Ussher Society*, **6**, 352–9.

Grosser, J. and Dorr, W (1986) MOR-typ-basalte im ostlichen Rheinischen Schiefergebirge. *Neues Jahrbuch für Geologie und Paläontologie Monatshefte*, **12**, 705–22.

Groves, A.W. (1931) The unroofing of the Dartmoor granite and the distribution of its detritus in the sediments of southern England. *Quarterly Journal of the Geological Society of London*, **87**, 62–9.

Hall, A. (1971) Greisenization in the granite of Cligga Head, Cornwall. *Proceedings of the Geologists' Association*, **82**, 209–30.

Hall, A. (1974) Granite porphyries in Cornwall. *Proceedings of the Ussher Society*, **3**, 115–48.

Hall, A. (1982) The Pendennis peralkaline minette. *Mineralogical Magazine*, **45**, 257–66.

Hall, A. (1988) The distribution of ammonium in granites from south-west England. *Journal of the Geological Society of London*, **145**, 37–41.

Hall, A. and Jackson, N.J. (1975) Report of the Summer Field Meeting in west Cornwall. *Proceedings of the Geologists' Association*, **86**, 95–102.

Hall, S. (1930) The geology of the Godolphin granite. A study of the coastal geology between Perranuthnoe and Looe Pool. *Proceedings of the Geologists' Association*, **41**, 117–47.

Halls, C. (1987) A mechanistic approach to the paragenetic interpretation of mineral lodes in Cornwall. *Proceedings of the Ussher Society*, **6**, 548–54.

Hampton, C.M. and Taylor, P.N. (1983) The age and nature of the basement of southern Britain: evidence from Sr and Pb isotopes in granites. *Journal of the Geological Society of London*, **140**, 499–509.

Hancock, P.L. (ed.) (1983) *The Variscan fold belt in the British Isles*. Adam Hilger, Bristol, 217 pp.

Hawkes, J.R. (1968) Igneous rocks and metamorphic aureole of the Dartmoor granite. In *Geology of the Country around Okehampton*, (eds E.A. Edmonds et al.), Memoir of the Geological Survey of the Great Britain, HMSO, London, pp. 134–43.

Hawkes, J.R. (1974) Volcanism and metallogenesis: the tin province of south-west England. *Bulétin Volcanologique*, **38**, 1125–46.

Hawkes, J.R. (1981) A tectonic 'watershed' of fundamental consequence in the post-Westphalian evolution of Cornubia. *Proceedings of the Ussher Society*, **5**, 128–31.

Hawkes, J.R. (1982) The Dartmoor granite and later volcanic rocks. In *The Geology of Devon* (eds E.M. Durrance and D.J.C. Laming), University Press, Exeter, pp. 85–116.

Hawkes, J.R. and Dangerfield, J. (1978) The Variscan granites of south-west England: a progress report. *Proceedings of the Ussher Society*, **4**, 158–71.

Hawkes, J.R., Harding, R.R. and Darbyshire, D.P.F. (1975) Petrology and Rb:Sr age of the Brannel, South Crofty and Wherry elvan dykes, Cornwall. *Bulletin of the Geological Survey of Great Britain*, **52**, 27–42.

References

Hawkes, J.R., Harris, P.M., Dangerfield, J. et al. (1987) The lithium potential of the St Austell Granite. *Report of the British Geological Survey*, **19**(4), 54 pp.

Hendricks, E.M.L. (1939) The Start–Dodman–Lizard boundary zone in relation to the Alpine structure of Cornwall. *Geological Magazine*, **76**, 385–401.

Hendricks, E.M.L., House, M.R. and Rhodes, F.H.T. (1971) Evidence bearing on the stratigraphical successions in south Cornwall. *Proceedings of the Ussher Society*, **2**, 270–5.

Henley, S. (1972) Petrogenesis of quartz-porphyry dykes in south-west England. *Nature, Physical Science*, **235**, 95–7.

Henley, S. (1974) Geochemistry and petrogenesis of elvan dykes in the Perranporth area, Cornwall. *Proceedings of the Ussher Society*, **3**, 136–45.

Henwood, W.J. (1838) *Granite of Cligga Head and the Elvan Course of the Same Neighbourhood.* 20th Annual Report of the Royal Institute of Cornwall, p. 29.

Henwood, W.J. (1843) On the metalliferous deposits of Cornwall and Devon. *Transactions of the Royal Geological Society of Cornwall*, **5**, 1–512.

Hickling, G.H. (1908) China clay: its nature and origin. *Transactions of the Institution of Mining Engineers*, **36**, 10–33.

Higgs, R. (1984) Possible wave-influenced sedimentary structures in the Bude Formation (Westphalian), south-west England and their environmental implications. *Proceedings of the Ussher Society*, **6**, 88–94.

Hill, P.I. and Manning, D.A.C. (1987) Multiple intrusions and pervasive hydrothermal alteration in the St Austell granite, Cornwall. *Proceedings of the Ussher Society*, **6**, 447–53.

Hobson, D.M. (1977) Polyphase folds from the Start Complex. *Proceedings of the Ussher Society*, **4**, 102–10.

Holder, A.P. and Bott, M.H.P. (1971) Crustal structure in the vicinity of S.W. England. *Geophysical Journal of the Royal Astronomical Society*, **23**, 465–89.

Holder, M.T. and Leveridge, B.E. (1986) A model for the tectonic evolution of south Cornwall. *Journal of the Geological Society of London*, **143**, 125–34.

Holdsworth, R.E. (1989) The Start–Perranporth line: a Devonian terrane boundary in the Variscan Orogen of SW England. *Journal of the Geological Society of London*, **146**, 419–21.

Holmes, A. (1916) The origin of igneous rocks. *Science Progress*, 67–73.

Holmes, A. (1932) The origin of igneous rocks. *Geological Magazine*, **69**, 543–58.

Holwill, F.J.W. (1966) Conglomerates, tuffs and concretionary beds in the Upper Devonian of Waterside Cove, near Goodrington Sands, Torbay. *Proceedings of the Ussher Society*, **1**, 238–41.

Hosking, K.F.G. (1964) Permo-Carboniferous and later primary mineralization of Cornwall and south-west Devon. In *Present Views of Some Aspects of the Geology of Cornwall and Devon*, (eds K.F.G. Hosking and G.J. Shrimpton), 150th Anniversary Volume, Royal Geological Society of Cornwall, pp. 201–45.

Hosking, K.F.G. and Camm, G.S. (1985) The occurrence of cassiterite and other species of economic interest in the greisenized granite porphyry of Cameron Quarry, St Agnes, Cornwall. In *High Heat Production (HHP) Granites, Hydrothermal Circulation and Ore Genesis*, Institution of Mining and Metallurgy, 517–33.

Hosking, K.F.G. and Shrimpton, G.J. (eds) (1964) *Present Views of Some Aspects of the Geology of Cornwall and Devon.* 150th Anniversary Volume, Royal Geological Society of Cornwall.

House, M.R. (1963) Devonian ammonoid successions and facies in Devon and Cornwall. *Quarterly Journal of the Geological Society of London*, **119**, 1–27.

House, M.R. (1975) Facies and time in Devonian tropical areas. *Proceedings of the Yorkshire Geological Society*, **40**, 233–88.

Howe, J.A. (1914) *A Handbook to the Collection of Kaolin, China Clay and China Stone in the Museum of Practical Geology.* HMSO, London, 271 pp.

Hutchins, P.F. (1963) The lower New Red Sandstone of the Crediton valley. *Geological Magazine*, **100**, 107–28.

Hutton, D.H.W. and Sanderson, D.J. (eds) (1984) *Variscan Tectonics of the North Atlantic Region.* Geological Society Special Publication, No. 14, Blackwell Scientific Publications, 270 pp.

Institute of Geological Sciences (1982) I.G.S. boreholes 1980. *Report of the Institute of Geological Sciences*, **81/11**, 12 pp.

Isaac, K.P. (1981) The Hercynian geology of Lydford Gorge, NW Dartmoor and its regional significance. *Proceedings of the Ussher Society*, **5**, 147–52.

Isaac, K.P. (1982) Hercynian metamorphism in

References

the Launceston–Tavistock area of SW England. *Journal of the Geological Society of London*, **139**, 366–7.

Isaac, K.P. (1985) Discussion of papers on the Hercynian back-arc marginal basin of SW England. *Journal of the Geological Society of London*, **142**, 927–9.

Isaac, K.P., Turner, P.J. and Stewart, I.J. (1982) The evolution of the Hercynides of central SW England. *Journal of the Geological Society of London*, **139**, 521–31.

Jackson, N.J. (1974) Grylls Bunny, a 'tin floor' at Botallack. *Proceedings of the Ussher Society*, **3**, 186–8.

Jackson, N.J. and Alderton, D.H.M. (1974) Discordant calc-silicate bodies in the Botallack area. *Proceedings of the Ussher Society*, **3**, 123–8.

Jackson, N.J., Moore, J.McM. and Rankin, A.H. (1977) Fluid inclusions and mineralisation at Cligga Head, Cornwall, England. *Journal of the Geological Society of London*, **134**, 343–50.

Jackson, N.J., Halliday, A.N., Sheppard, S.M.F. *et al.* (1982) Hydrothermal activity in the St Just mining district, Cornwall, England. In *Metallization Associated with Acid Magmatism*, (ed. A.M. Evans), Wiley, Chichester, pp. 139–79.

Jackson, N.J., Willis-Richards, J., Manning, D.A.C. *et al.* (1989) Evolution of the Cornubian ore field, south-west England; Part II: Mineral deposits and ore-forming processes. *Economic Geology*, **84**, 1101–33.

Jahns, R.H. and Burnham, C.W. (1969) Experimental studies of pegmatite genesis. I: A model for the derivation and crystallization of granitic pegmatites. *Economic Geology*, **64**, 843–64.

Jahns, R.H. and Tuttle, O.F. (1963) Layered pegmatite–aplite intrusions. *Special Paper of the Mineralogical Society of America*, **1**, 78–92.

Jefferies, N.L. (1984) The radioactive accessory mineral assemblage of the Carnmenellis granite, Cornwall. *Proceedings of the Ussher Society*, **6**, 35–41.

Jefferies, N.L. (1985a) The origin of sillimanite-bearing pelitic xenoliths within the Carnmenellis pluton, Cornwall. *Proceedings of the Ussher Society*, **6**, 229–36.

Jefferies, N.L. (1985b) The distribution of the rare-earth elements in the Carnmenellis pluton, Cornwall. *Mineralogical Magazine*, **49**, 495–504.

Jefferies, N.L. (1988) The distribution of uranium in the Carnmenellis pluton, Cornwall. Unpublished Ph.D. Thesis, University of Exeter.

Juteau, T. and Rocci, G. (1974) Vers une meilleure connaissance du problème des spilites à partir de données nouvelles sur le cortège spilito-kératophyrique Hercynotype. In *Spilites and Spilitic Rocks*, (ed. G.C. Amstutz), Springer-Verlag, Berlin, pp. 253–329.

Keeling, P.S. (1961) Cornish Stone. *Transactions of the British Ceramics Society*, **60**, 390–426.

Kingsbury, A.W.G. (1961) Beryllium minerals in Cornwall and Devon: helvine, genthelvite and danalite. *Mineralogical Magazine*, **32**, 921–40.

Kingsbury, A.W.G. (1964) Some minerals of special interest in south-west England. In *Present Views of Some Aspects of the Geology of Cornwall and Devon*, (eds K.F.G. Hosking and G.J. Shrimpton), 150th Anniversary Volume, Royal Geological Society of Cornwall, pp. 247–66.

Kingsbury, A.W.G. (1970) Meldon Aplite Quarry and Meldon (Railways) Quarry. In *Geological Highlights of the West Country*, (ed. W.A. Macfadyen), Butterworths, London, pp. 61–3.

Kirby, G.A. (1978) Layered gabbros in the eastern Lizard, Cornwall and their significance. *Geological Magazine*, **115**, 199–204.

Kirby, G.A. (1979a) The petrochemistry of rocks of the Lizard Complex, Cornwall. Unpublished Ph.D. Thesis, University of Southampton.

Kirby, G.A. (1979b) The Lizard Complex as an ophiolite. *Nature, London*, **282**, 58–61.

Kirby, G.A. (1984) The petrology and geochemistry of dykes of the Lizard Ophiolite Complex, Cornwall. *Journal of the Geological Society of London*, **141**, 53–9.

Knill, D.C. (1969) The Permian igneous rocks of Devon. *Bulletin of the Geological Survey of Great Britain*, **29**, 115–38.

Knill, D.C. (1982) Permian volcanism in southwestern England. In *Igneous rocks of the British Isles*, (ed. D.S. Sutherland), Wiley, Chichester, pp. 329–32.

Knox, D.A. and Jackson, N.J. (1990) Composite granite intrusions of SW Dartmoor, Devon. *Proceedings of the Ussher Society*, **7**, 246–50.

Kossmat, F. (1927) Gliederung des varistischen Gebirgsbaues. *Abhandlungen des Sächsischen Geologischen Landesamts*, **1**, 39 pp.

Krebs, W. (1977) The tectonic evolution of Variscan Meso-Europe. In *Europe from Crust*

References

to Core, (eds D.V. Ager and M. Brooks), Wiley, New York, pp. 119–42.

Lacy, E.D. (1958) Some features of the contact aureole of the Land's End granite. *Abstracts of the Proceedings of the Conference of Geologists and Geomorphologists in the south-west of England*, Exeter (1958), 14–19.

Laming, D.J.C. (1966) Imbrication, palaeocurrents and other sedimentary features in the Lower New Red Sandstone, Devonshire, England. *Journal of Sedimentary Petrology*, **36**, 940–59.

Laming, D.J.C. (1968) New Red Sandstone stratigraphy in Devon and west Somerset. *Proceedings of the Ussher Society*, **2**, 23–5.

Leake, R.C. and Styles, M.T. (1984) Borehole sections through the Traboe hornblende schists, a cumulate complex overlying the Lizard peridotite. *Journal of the Geological Society of London*, **141**, 41–52.

Leake, R.C., Styles, M.T. and Rollin, K.E. (1990) Exploration for vanadiferous magnetite and other minerals in the Lizard Complex. *Mineral Reconnaissance Programme Report, British Geological Survey*, 113.

Leat, P.T., Thompson, R.N., Morrison, M.A. *et al.* (1987) Geodynamic significance of post-Variscan intrusive and extrusive potassic magmatism in SW England. *Transactions of the Royal Society of Edinburgh: Earth Sciences*, **77**, 349–60.

Leech, J.G.C. (1929) St Austell detritals. *Proceedings of the Geologists' Association*, **40**, 139–46.

Leeder, M.R. (1982) Upper Palaeozoic basins of the British Isles – Caledonide inheritance versus Hercynian plate margin processes. *Journal of the Geological Society of London*, **139**, 479–91.

Lees, G.J., Alderton, D.H.M., Pearce, J.A. and Exley, C.S. (1978) Rare earth elements in acid rocks of SW England. *Proceedings of the Ussher Society*, **4**, p. 217 (abstract).

Le Gall, B., Le Herisse, A. and Deunff, J. (1985) New palynological data from the Gramscatho Group at the Lizard front (Cornwall): palaeogeographical and geodynamical implications. *Proceedings of the Geologists' Association*, **96**, 237–53.

Leveridge, B.E., Holder, M.T. and Day, G.A. (1984) Thrust nappe tectonics in the Devonian of south Cornwall and the western English Channel. In *Variscan Tectonics of the North Atlantic Region*, (eds D.H.W. Hutton and D.J. Sanderson), Geological Society Special Publication No. 14, Blackwell Scientific Publications, pp. 103–12.

Leveridge, B.E. and Holder, M.T. (1985) Olistostromic breccias at the Mylor/Gramscatho boundary, south Cornwall. *Proceedings of the Ussher Society*, **6**, 147–54.

Linton, D.L. (1955) The problem of tors. *Geographical Journal*, **121**, 470–87.

Lister, C.J. (1978) Luxullianite *in situ* within the St Austell granite, Cornwall. *Mineralogical Magazine*, **42**, 295–7.

Lister, C.J. (1979a) Luxullianite *in situ* within the St Austell granite, Cornwall: a reply [to discussion by D.H.M. Alderton]. *Mineralogical Magazine*, **43**, 442–3.

Lister, C.J. (1979b) Quartz-cored tourmalines from Cape Cornwall and other areas. *Proceedings of the Ussher Society*, **4**, 401–18.

Lister, C.J. (1984) Xenolith assimilation in the granites of south-west England. *Proceedings of the Ussher Society*, **6**, 46–53.

Lorenz, V. and Nicholls, I.A. (1984) Plate and intraplate processes of Hercynian Europe during the Late Palaeozoic. *Tectonophysics*, **107**, 25–56.

Lowe, H.J. (1901) The sequence of the Lizard rocks. *Transactions of the Royal Geological Society of Cornwall*, **7**, 438–66.

Lowe, H.J. (1902) The sequence of the Lizard rocks. *Transactions of the Royal Geological Society of Cornwall*, **7**, 507–34.

MacDonald, R., Thorpe, R.S., Gaskarth, J.W. *et al.* (1985) Multi-component origin of Caledonian lamprophyres of northern England. *Mineralogical Magazine*, **49**, 485–94.

Mackenzie, D.M. (1972) The mineralogy and chemistry of axinite reaction veins cutting the Meldon Aplite. *Proceedings of the Ussher Society*, **2**, 410–6.

Malpas, J. and Langdon, G.S. (1987) The Kennack Gneisses of the Lizard Complex, Cornwall: partial melts produced during ophiolite emplacement. *Canadian Journal of Earth Sciences*, **24**, 1966–73.

Manning, D.A.C. (1979) An experimental study of the effect of fluorine, in addition to water, on crystallization in the system Qz–Ab–Or and its application to Cornish granitic rocks rich in fluorine. *Proceedings of the Ussher Society*, **4**, 380–9.

Manning, D.A.C. (1981) The application of experimental studies in determining the origin of topaz–quartz–tourmaline rock and tourma-

References

line–quartz rock. *Proceedings of the Ussher Society*, **5**, 121–7.

Manning, D.A.C. (1982) An experimental study of the effects of fluorine on the crystallization of granitic melts. In *Metallization Associated with Acid Magmatism*, (ed. A.M. Evans), Wiley, Chichester, pp. 191–203.

Manning, D.A.C. and Exley, C.S. (1984) The origins of late-stage rocks in the St Austell granite – a re-interpretation. *Journal of the Geological Society of London*, **141**, 581–91.

Marshall, B. (1962) The small structures of Start Point, south Devon. *Proceedings of the Ussher Society*, **1**, 19–21.

Matte, P. (1986) Tectonics and plate tectonics model for the Variscan belt of Europe. *Tectonophysics*, **126**, 329–74.

Matthews, S.C. (1977) The Variscan fold belt in south-west England. *Neues Jahrbuch für Geologie und Paläontologie Abhandlungen*, **154**, 94–127.

Matthews, S.C. (1978) Caledonian connections of Variscan tectonism. *Zeitschrift der Deutschen Geologischen Gesellschaft*, **129**, 419–28.

Matthews, S.C. (1981) A cross-section through south-west England. In *The Variscan Orogeny in Europe*, (eds H.J. Zwart and U.F. Dorsiepen), Geologie en Mijnbouw, 60, 145–8.

McMahon, C.A. and Hutchings, W.M. (1895) Note on pseudo-spherulites. *Geological Magazine*, **2**, 257–9.

Middleton, G.V. (1960) Spilitic rocks in south-east Devonshire. *Geological Magazine*, **97**, 192–207.

Miller, J.A. and Mohr, P.A. (1964) Potassium–argon measurements on the granites and some associated rocks from south-west England. *Geological Journal*, **4**, 105–26.

Mitropoulos, P. (1982) REE patterns of the metasedimentary rocks of the Land's End granite aureole (SW England). *Chemical Geology*, **35**, 265–80.

Mitropoulos, P. (1984) REE distribution in the metabasic rocks of the Land's End granite aureole (SW England). *Mineralogical Magazine*, **48**, 495–505.

Moore, J.McM. and Jackson, N. (1977) Structure and mineralization in the Cligga granite stock, Cornwall. *Journal of the Geological Society of London*, **133**, 467–80.

Morton, R.D. and Smith, D.G.W. (1971) Differentiation and metasomatism within a Carboniferous spilite–keratophyre suite in SW England. *Mineralogical Society of Japan, Special Paper*, **1**, 127–32.

Osman, C.W. (1928) The granites of the Scilly Isles and their relation to the Dartmoor granites. *Quarterly Journal of the Geological Society of London*, **84**, 258–92.

Palmer, J. and Neilson, R.A. (1962) The origin of granite tors on Dartmoor, Devonshire. *Proceedings of the Yorkshire Geological Society*, **33**, 315–40.

Parker, A. (1970) Chemical and mineralogical analyses of some basic and ultrabasic rocks and their initial weathering products. *Reading University Geological Report*, **4**, 1–44.

Pearce, J.A., Harris, N.B.W. and Tindle, A.G. (1984) Trace element discrimination diagrams for the tectonic interpretation of granitic rocks. *Journal of Petrology*, **25**, 956–83.

Peucat, J.-J., Auvray, B., Hirbec, Y. *et al.* (1984) Granites et ciscuillements Hercyniens dans le Nord du Massif Armorican: géochronologie Rb–Sr. *Bullétin de la Société Géologique de France*, **26**(6), 1365–73.

Phillips, F.C. (1928) Metamorphism in the Upper Devonian of north Cornwall. *Geological Magazine*, **65**, 541–56.

Phillips, F.C. (1964) Metamorphic rocks of the sea floor between Start Point and Dodman Point, SW England. *Journal of the Marine Biological Association of the United Kingdom*, **44**, 655–63.

Phillips, J.A. (1876) On the so-called 'Greenstones' of western Cornwall. *Quarterly Journal of the Geological Society of London*, **32**, 155–79.

Pichavant, M. (1979) Étude expérimentale à haute température et 1 kbar du role du bore dans quelques systèmes silicates. Interet pétrologique et métallogenique. *Thèse Docteur de Spécialité, 3è cycle*. L'Institut National Polytechnique de Lorraine.

Pichavant, M. and Manning, D.A.C. (1984) Petrogenesis of tourmaline granites and topaz granites: the contribution of experimental data. *Physics of the Earth and Planetary Interiors*, **35**, 31–50.

Pinto, M.S. (1983) Geochronology of Portuguese granitoids: a contribution. *Studia Geologica Salamanticensia*, **18**, 277–306.

Pound, C.J. (1983) The sedimentology of the Lower–Middle Devonian Staddon Grits and Jennycliffe Slates on the east side of Plymouth Sound, Devon. *Proceedings of the Ussher Society*, **5**, 465–72.

Power, G.M. (1968) Chemical variation in tourmaline from south-west England. *Mineralogical Magazine*, **36**, 1078–89.

Primmer, T.J. (1982) Low-grade metamorphism

References

of the Tintagel Volcanic Formation, SW England. *Journal of the Geological Society of London*, **139**, p. 366 (abstract).

Primmer, T.J. (1983a) Low-grade metamorphism of aluminous slates from Tintagel, S.W. England. *Journal of the Geological Society of London*, **140**, p. 959 (abstract).

Primmer, T.J. (1983b) Low-grade regional metamorphism across the Perranporth–Pentewan Line, Cornwall. *Proceedings of the Ussher Society*, **5**, 421–7.

Rattey, P.R. and Sanderson, D.J. (1984) The structure of S.W. Cornwall and its bearing on the emplacement of the Lizard Complex. *Journal of the Geological Society of London*, **141**, 87–95

Read, H.H. (1949) A contemplation of time in plutonism (Presidential Address). *Quarterly Journal of the Geological Society of London*, **105**, 101–56.

Reavy, R.J. (1989) Structural controls on metamorphism and syntectonic magmatism: the Portuguese Hercynian collision belt. *Journal of the Geological Society of London*, **146**, 649–57.

Reid, C. and Flett, J.S. (1907) *Geology of the Land's End District* (Sheets 351 and 358). Memoir of the Geological Survey of Great Britain, HMSO, London.

Reid, C. and Dewey, H. (1908) The origin of the pillow lavas near Port Isaac in Cornwall. *Quarterly Journal of the Geological Society of London*, **64**, 264–72.

Reid, C. and Scrivenor, J.B. (1906) *The Geology of the Country near Newquay* (Sheet 346). Memoir of the Geological Survey of Great Britain, HMSO, London.

Reid, C., Barrow, G. and Dewey, H. (1910) *Geology of the Country around Padstow and Camelford* (Sheets 335 and 336). Memoir of the Geological Survey of Great Britain, HMSO, London.

Reid, C., Barrow, G., Sherlock, R.L. et al. (1911) *Geology of the Country around Tavistock and Launceston* (Sheet 337). Memoir of the Geological Survey of Great Britain, HMSO, London.

Reid, C., Barrow, G., Sherlock, R.L. et al. (1912) *The Geology of Dartmoor* (Sheet 338). Memoir of the Geological Survey of Great Britain, HMSO, London.

Reynolds, D.L. (1946) The sequence of geochemical changes leading to granitization. *Quarterly Journal of the Geological Society of London*, **102**, 389–446.

Reynolds, D.L. (1947) Hercynian Fe–Mg metasomatism in Cornwall. *Geological Magazine*, **84**, 33–50.

Reynolds, D.L. (1954) Fluidization as a geological process and its bearing on the problem of intrusive granites. *American Journal of Science*, **252**, 577–613.

Rice-Birchall, B. and Floyd, P.A. (1988) Geochemistry and source characteristics of the Tintagel Volcanic Formation. *Proceedings of the Ussher Society*, **7**, 52–5.

Richardson, W.A. (1923) A micrometric study of the St Austell granite (Cornwall). *Quarterly Journal of the Geological Society of London*, **79**, 546–76.

Richter, D. (1965) Observations on volcanicity and tectonics of the Torquay area. *Proceedings of the Ussher Society*, **1**, 189–90.

Richter, D. (1967) Sedimentology and facies of the Meadfoot Beds (Lower Devonian) in south-east Devon (England). *Geologische Rundschau*, **56**, 543–61.

Rivalenti, G., Garuti, G., Rossi, A. et al. (1981) Existence of different peridotite types and of a layered igneous complex in the Ivrea Zone of the western Alps. *Journal of Petrology*, **22**, 127–53.

Robinson, D. (1981) Metamorphic rocks from an intermediate facies series juxtaposed at the Start boundary, southwest England. *Geological Magazine*, **118**, 297–301.

Robinson, D. and Read, D. (1981) Metamorphism and mineral chemistry of greenschists from Trebarwith Strand, Cornwall. *Proceedings of the Ussher Society*, **5**, 132–8.

Robinson, D. and Sexton, D. (1987) Geochemistry of the Tintagel Volcanic Formation. *Proceedings of the Ussher Society*, **6**, 523–8.

Rock, N.M.S. (1984) Nature and origin of calc-alkaline lamprophyres: minettes, vogesites, kersantites and spessartites. *Transactions of the Royal Society of Edinburgh: Earth Sciences*, **74**, 193–227.

Rollin, K.E. (1986) Geophysical surveys on the Lizard Complex, Cornwall. *Journal of the Geological Society of London*, **143**, 437–46.

Rothstein, A.T.V. (1977) The distribution and origin of primary textures in the Lizard peridotite, Cornwall. *Proceedings of the Geologists' Association*, **88**, 93–105.

Rutley, F. (1878) *The Eruptive Rocks of Brent Tor and its Neighbourhood*. Memoir of the Geological Survey of Great Britain, HMSO, London, 56 pp.

Sadler, P.M. (1973) An interpretation of new

References

stratigraphic evidence from south Cornwall. *Proceedings of the Ussher Society*, **2**, 535–50.

Sadler, P.M. (1974) An appraisal of the 'Lizard–Dodman–Start Thrust' concept. *Proceedings of the Ussher Society*, **3**, 71–81.

Sandeman, H.A.I. (1988) A field, petrographical and geochemical investigation of the Kennack Gneiss, Lizard peninsula, south-west England. Unpublished MSc. Thesis, Memorial University, Newfoundland.

Sanders, L.D. (1955) Structural observations on south-east Lizard. *Geological Magazine*, **92**, 231–40.

Sanderson, D.J. (1973) The development of folds oblique to the regional trend. *Tectonophysics*, **16**, 55–70.

Sanderson, D.J. and Dearman, W.R. (1973) Structural zones of the Variscan fold belt in SW England, their location and development. *Journal of the Geological Society of London*, **129**, 527–36.

Scrivenor, J.B. (1903) The granite and greisen of Cligga Head (western Cornwall). *Quarterly Journal of the Geological Society of London*, **59**, 142–59.

Sedgwick, A. (1820) On the physical structure of those formations which are immediately associated with the primitive ridge of Devonshire and Cornwall. *Transactions of the Cambridge Philosophical Society*, **1**, 89–146.

Selwood, E.B. (1961) The Upper Devonian and Lower Carboniferous stratigraphy of Boscastle and Tintagel, Cornwall. *Geological Magazine*, **98**, 161–7.

Selwood, E.B. (1971) Successions at the Devonian–Carboniferous boundary between Boscastle and Dartmoor. *Proceedings of the Ussher Society*, **2**, 275–85.

Selwood, E.B. (1974) The age of the Upper Palaeozoic volcanics between Bodmin Moor and Dartmoor. *Proceedings of the Ussher Society*, **3**, 63–70.

Selwood, E.B. and Durrance, E.M. (1982) The Devonian rocks. In *The Geology of Devon*, (eds E.M. Durrance and D.J.C. Laming), University Press, Exeter, pp. 15–41.

Selwood, E.B. and Thomas, J.M. (1984) A re-interpretation of the Meldon anticline in the Belstone area. *Proceedings of the Ussher Society*, **6**, 75–81.

Selwood, E.B. and Thomas, J.M. (1986a) Upper Palaeozoic successions and nappe structures in north Cornwall. *Journal of the Geological Society of London*, **143**, 75–82.

Selwood, E.B. and Thomas, J.M. (1986b) Variscan facies and structure in central SW England. *Journal of the Geological Society of London*, **143**, 199–207.

Selwood, E.B., Stewart, I.J. and Thomas, J.M. (1985) Upper Palaeozoic sediments and structure in north Cornwall – a re-interpretation. *Proceedings of the Geologists' Association*, **96**, 129–41.

Shackleton, R.M., Ries, A.C. and Coward, M.P. (1982) An interpretation of the Variscan structures in SW England. *Journal of the Geological Society of London*, **139**, 533–41.

Shail, R. and Floyd, P.A. (1988) An evaluation of flysch provenance – example from the Gramscatho Group of S. Cornwall. *Proceedings of the Ussher Society*, **7**, 62–6.

Shannon, W.G. and Annis, L.G. (1933) The igneous intrusions of the Stoke Fleming area, south Devon. *Proceedings of the Geologists' Association*, **42**, 53–61.

Shepherd, T.J., Miller, M.F., Scrivener, R.C. *et al.* (1985) Hydrothermal fluid evolution in relation to mineralization in southwest England with special reference to the Dartmoor–Bodmin area. In *High Heat Production (HHP) Granites, Hydrothermal Circulation and Ore Genesis*, Institution of Mining and Metallurgy, pp. 345–64.

Sheppard, S.M.F. (1977) The Cornubian batholith, SW England: D/H and $^{18}O/^{16}O$ studies of kaolinite and other alteration minerals. *Journal of the Geological Society of London*, **133**, 573–91.

Simpson, P.R., Plant, J.A. and Cope, M.J. (1976) Uranium abundance and distribution in some granites from northern Scotland and south-west England as indicators of uranium provinces. In *Geology, Mining and Extractive Processing of Uranium*, (ed. M.J. Jones), Institution of Mining and Metallurgy, pp. 126–39.

Simpson, P.R., Brown, G.C., Plant, J.A. *et al.* (1979) Uranium mineralization and granite magmatism in the British Isles. *Philosophical Transactions of the Royal Society*, **A291**, 385–412.

Smith, H.G. (1929) Some features of Cornish lamprophyres. *Proceedings of the Geologists' Association*, **40**, 260–8.

Somervail, A. (1888) On a remarkable dyke in the serpentine of the Lizard. *Geological Magazine*, **5**, 553–5.

Štemprok, M. (1986) Petrology and geochemistry

References

of the Czechoslovak part of the Krušné hory Mountains granite pluton. *Sborník Geologickych Věd: Ložiskova Geologie, Mineralogie*, **27**, 111–56.

Stewart, I.J. (1981) The Trekelland thrust. *Proceedings of the Ussher Society*, **5**, 163–7.

Stone, M. (1966) Fold structures in the Mylor Beds, near Porthleven, Cornwall. *Geological Magazine*, **103**, 440–60.

Stone, M. (1968) A study of the Praa Sands elvan and its bearing on the origin of elvans. *Proceedings of the Ussher Society*, **2**, 37–42.

Stone, M. (1969) Nature and origin of banding in the granitic sheets, Tremearne, Porthleven, Cornwall. *Geological Magazine*, **106**, 142–58.

Stone, M. (1975) Structure and petrology of the Tregonning–Godolphin granite, Cornwall. *Proceedings of the Geologists' Association*, **86**, 155–70.

Stone, M. (1979) Textures of some Cornish granites. *Proceedings of the Ussher Society*, **4**, 370–9.

Stone, M. (1984) Textural evolution of lithium mica granites in the Cornubian batholith. *Proceedings of the Geologists' Association*, **95**, 29–41.

Stone, M. (1987) Geochemistry and origin of the Carnmenellis pluton, Cornwall: further considerations. *Proceedings of the Ussher Society*, **6**, 454–60.

Stone, M. (1988) The significance of almandine garnets in the Lundy and Dartmoor granites. *Mineralogical Magazine*, **52**, 651–8.

Stone, M. and Austin, W.G.C. (1961) The metasomatic origin of the potash feldspar megacrysts in the granites of south-west England. *Journal of Geology*, **69**, 464–72.

Stone, M. and Exley, C.S. (1968) South-western granites and alkali feldspar megacrysts. *Nature, London*, **222**, 555–6.

Stone, M. and Exley, C.S. (1978) A cluster analysis of chemical data from the granites of SW England. *Proceedings of the Ussher Society*, **4**, 172–81.

Stone, M. and Exley, C.S. (1984) Emplacement of the Porthmeor granite pluton, west Cornwall. *Proceedings of the Ussher Society*, **6**, 42–5.

Stone, M. and Exley, C.S. (1986) High heat production granites of south-west England and their associated mineralization: a review. *Transactions of the Institution of Mining and Metallurgy, Section B*, **95**, B25–B36.

Stone, M. and George, M.C. (1979) Amblygonite in leucogranites of the Tregonning–Godolphin granite, Cornwall. *Mineralogical Magazine*, **42**, 151–2.

Stone, M. and Lambert, J.L.M. (1956) Shear folding in the Mylor Beds, near Porthleven, Cornwall. *Geological Magazine*, **93**, 331–5.

Stone, M., Exley, C.S. and George, M.C. (1988) Composition of trioctahedral micas in the Cornubian batholith. *Mineralogical Magazine*, **52**, 175–92.

Streckeisen, A. (1976) To each plutonic rock its proper name. *Earth Science Reviews*, **12**, 1–33.

Strong, D.F., Stevens, R.K., Malpas, J. *et al.* (1975) A new tale for the Lizard. *Proceedings of the Ussher Society*, **3**, p. 252 (abstract).

Styles, M.T. and Kirby, G.A. (1980) New investigations of the Lizard Complex, Cornwall, England and a discussion of an ophiolite model. In *Proceedings of the International Ophiolite Symposium, Cyprus, 1979*, (ed. A. Panayiotou), 517–26.

Styles, M.T. and Rundle, C.C. (1984) The Rb–Sr isochron age of the Kennack Gneiss and its bearing on the age of the Lizard Complex, Cornwall. *Journal of the Geological Society of London*, **141**, 15–19.

Sutherland, D.S. (ed.) (1982) *Igneous Rocks of the British Isles*. Wiley, Chichester, 645 pp.

Tammemagi, H.Y. and Smith, N.L. (1975) A radiogeological study of the granites of SW England. *Journal of the Geological Society of London*, **131**, 415–27.

Taylor, R.T. and Wilson, A.C. (1975) Notes on some igneous rocks of west Cornwall. *Proceedings of the Ussher Society*, **3**, 255–62.

Teall, J.J.H. (1886) The metamorphosis of the Lizard gabbros. *Geological Magazine*, **3**, 481–9.

Teall, J.J.H. (1887) On the origin of certain banded gneisses. *Geological Magazine*, **4**, 484–93.

Teall, J.J.H. (1888) *British Petrography*. Dulau, London, 469 pp.

Thorpe, R.S. (1987) Permian K-rich volcanic rocks of Devon: petrogenesis, tectonic setting and geological significance. *Transactions of the Royal Society of Edinburgh: Earth Sciences*, **77**, 361–6.

Thorpe, R.S., Cosgrove, M.E. and Van Calsteren, P.W.C. (1986) Rare-earth element, Sr- and Nd-isotope evidence for petrogenesis of Permian basaltic and K-rich volcanic rocks from south-west England. *Mineralogical Magazine*, **50**, 481–90.

References

Tidmarsh, W.G. (1932) The Permian lavas of Devon. *Quarterly Journal of the Geological Society of London*, **88**, 712–75.

Tilley, C.E. (1923) The petrology of the metamorphosed rocks of the Start area (south Devon). *Quarterly Journal of the Geological Society of London*, **79**, 172–204.

Tilley, C.E. (1925) Petrographical notes on some chloritoid rocks. II: Chloritoid phyllites of the Tintagel area, north Cornwall. *Geological Magazine*, **62**, 314–8.

Tilley, C.E. (1935) Metasomatism associated with the greenstone–hornfelses of Kenidjack and Botallack, Cornwall. *Mineralogical Magazine*, **24**, 181–202.

Tilley, C.E. (1937) Anthophyllite–cordierite granulites of the Lizard. *Geological Magazine*, **74**, 300–9.

Tilley, C.E. and Flett, J.S. (1930) Hornfelses from Kenidjack, Cornwall. *Geological Survey of Great Britain, Summary of Progress* (for 1929), **2**, 24–41.

Toombs, J.M.C. (1977) A study of the space form of the Cornubian granite batholith and its application to detailed gravity surveys in Cornwall. *Mineral Reconnaissance Programme Report, Institute of Geological Sciences*, **11**, 16 pp.

Tunbridge, I.P. (1983) The Middle Devonian shoreline in North Devon, England. *Journal of the Geological Society of London*, **140**, 147–58.

Tunbridge, I.P. (1986) Mid-Devonian tectonics and sedimentation in the Bristol Channel area. *Journal of the Geological Society of London*, **143**, 107–15.

Turner, P.J. (1982) The anatomy of a thrust: a study of the Greystone Thrust Complex, east Cornwall. *Proceedings of the Ussher Society*, **5**, 270–8.

Turner, R.E., Taylor, R.T., Goode, A.J.J. et al. (1979) Palynological evidence for the age of the Mylor Slates, Mount Wellington, Cornwall. *Proceedings of the Ussher Society*, **4**, 274–83.

Turpin, L., Velde, D. and Pinte, G. (1988) Geochemical comparison between minettes and kersantites from the western European Hercynian orogen: trace element and Pb–Sr–Nd isotope constraints on their origin. *Earth and Planetary Science Letters*, **87**, 73–86.

Tuttle, O.F. and Bowen, N.L. (1958) Origin of granite in the light of experimental studies in the system $NaAlSi_3O_8$-$KAlSi_3O_8$–SiO_2–H_2O. *Geological Society of America Memoir*, **74**, 1–153.

Ussher, W.A.E. (1892) The British Culm Measures. Part II: Mechanical effects produced by the presence of the Dartmoor granite upon the surrounding strata. *Proceedings of the Somerset Archaeological and Natural History Society*, **38**, 181–219.

Ussher, W.A.E. (1902) *Geology of the Country around Exeter* (Sheet 325). Memoir of the Geological Survey of Great Britain, HMSO, London.

Ussher, W.A.E. (1904) *Geology of the Country around Kingsbridge and Salcombe* (Sheets 355 and 356). Memoir of the Geological Survey of Great Britain, HMSO, London.

Ussher, W.A.E. (1907) *Geology of the Country around Plymouth and Liskeard* (Sheet 348). Memoir of the Geological Survey of Great Britain, HMSO, London.

Ussher, W.A.E. (1912) *Geology of the Country around Ivybridge and Modbury* (Sheet 349). Memoir of the Geological Survey of Great Britain, HMSO, London.

Ussher, W.A.E. (1913) *Geology of the Country around Newton Abbot* (Sheet 339). Memoir of the Geological Survey of Great Britain, HMSO, London.

Ussher, W.A.E., Barrow, G. and MacAlister, D.A. (1909) *Geology of the Country around Bodmin and St Austell* (Sheet 347). Memoir of the Geological Survey of Great Britain, HMSO, London.

Vallance, T.G. (1960) Concerning spilites. *Proceedings of the Linnean Society of New South Wales*, **85**, 8–52.

Vallance, T.G. (1965) On the chemistry of pillow lavas and the origin of spilites. *Mineralogical Magazine*, **34**, 471–81.

Vallance, T.G. (1967) Mafic rock alteration and isochemical development of some cordierite–anthophyllite rocks. *Journal of Petrology*, **8**, 84–96.

Van Marcke de Lummen, G. (1985) Mineralogy and geochemistry of skarn deposits in the Land's End aureole, Cornwall. *Proceedings of the Ussher Society*, **6**, 211–7.

Van Marcke de Lummen, G. (1986) Geochemical evolution of the Land's End granite (southwest England) in relation to tin potential in the light of data from western marginal areas. *Proceedings of the Ussher Society*, **6**, 398–404.

Vearncombe, J.R. (1980) The Lizard ophiolite and two phases of suboceanic deformation. In *Proceedings of the International Ophiolite Symposium, Cyprus, 1979*, (ed. A. Panayiotou), pp. 527–37.

References

Vernon, R.H. (1986) K-feldspar megacrysts in granites – phenocrysts not porphyroblasts. *Earth-Science Reviews*, **23**, 1–63.

Von Knorring, O. and Condliffe, E. (1984) On the occurrence of niobium–tantalum and other rare-element minerals in the Meldon aplite, Devonshire. *Mineralogical Magazine*, **48**, 443–8.

Waters, R.A. (1970) The Variscan structure of eastern Dartmoor. *Proceedings of the Ussher Society*, **2**, 191–7.

Watson, J.V., Fowler, M.B., Plant, J.A. *et al.* (1984) Variscan–Caledonian comparisons: late orogenic granites. *Proceedings of the Ussher Society*, **6**, 2–12.

Webb, P.C., Tindle, A.G., Barritt, S.D. *et al.* (1985) Radiothermal granites of the United Kingdom: comparison of fractionation patterns and variations of heat production for selected granites. In *High Heat Production (HHP) Granites, Hydrothermal Circulation and Ore Genesis*, Institution of Mining and Metallurgy, p. 409–24.

Webby, B.D. (1966) Middle–Upper Devonian palaeogeography of north Devon and west Somerset. *Palaeogeography, Palaeoclimatology, Palaeoecology*, **2**, 27–46.

Wedepohl, K.H., Meyer, K. and Muecke, G.K. (1983) Chemical composition and genetic relations of metavolcanic rocks from the Rhenohercynian belt of north-west Germany. In *Intracontinental Fold Belts*, (eds H. Martin and F.W. Eder), Springer-Verlag, Berlin, pp. 231–56.

Wells, M.K. (1946) A contribution to the study of luxullianite. *Mineralogical Magazine*, **27**, 186–94.

Wells, P.R.A. (1977) Pyroxene thermometry in simple and complex systems. *Contributions to Mineralogy and Petrology*, **62**, 129–39.

Whiteley, M.J. (1984) Shallow-water Dinantian sediments in south-east Cornwall. *Proceedings of the Ussher Society*, **6**, 137–41.

Whitley, N. (1849) On the remains of ancient volcanoes on the north coast of Cornwall in the parish of St Minver. *30th Annual Report of the Royal Institution of Cornwall*, HMSO, London, p. 47.

Whittaker, A. (1975) A postulated post-Hercynian rift valley system in southern Britain. *Geological Magazine*, **112**, 137–49.

Wilkinson, J.J. and Knight, R.R.W. (1989) Palynological evidence from the Porthleven area, south Cornwall: implications for Devonian stratigraphy and Hercynian structural evolution. *Journal of the Geological Society of London*, **146**, 739–42.

Williams, C.T. and Floyd, P.A. (1981) The localized distribution of U and other incompatible elements in spilitic pillow lavas. *Contributions to Mineralogy and Petrology*, **78**, 111–17.

Willis-Richards, J. and Jackson, N.J. (1989) Evolution of the Cornubian ore field, south-west England; Part I: Batholith modelling and ore distribution. *Economic Geology*, **84**, 1078–1100.

Wilson, I.R. and Floyd, P.A. (1974) Distribution of uranium in the Land's End granite and aureole, and various greenstones from Cornwall. *Proceedings of the Ussher Society*, **3**, 166–76.

Wood, D.A. (1980) The application of Th–Hf–Ta diagram to problems of tectonomagmatic classification and to establish the nature of crustal contamination of basaltic lavas of the British Tertiary Volcanic Province. *Earth and Planetary Science Letters*, **50**, 11–30.

Worth, R.H. (1920) The geology of the Meldon valleys near Okehampton, on the northern verge of Dartmoor. *Quarterly Journal of the Geological Society of London*, **75** (for 1919), 77–118.

Ziegler, P.A. (1986) Geodynamic model for the Palaeozoic crustal consolidation of western and central Europe. *Tectonophysics*, **126**, 303–28.

Zwart, H.J. and Dornsiepen, U.F. (1978) The tectonic framework of central Europe. *Geologie en Mijnbouw*, **57**, 627–54.

Glossary

This glossary contains simple explanations of the more important technical terms used in Chapters 1 and 2 and in the 'Introduction, Highlights' and 'Conclusions' sections of Chapters 3 to 6. The particular context of the igneous rocks of southwest England is reflected in these explanations. Stratigraphical terms are omitted as they are placed in context within the figures and tables. With reference to igneous rocks the following grain sizes generally apply: coarse-grained – strictly over 5 mm (in practice, often over 3 mm); medium-grained – 1–5(3) mm; fine-grained – under 1 mm.

Bold-face indicates a further glossary entry.

Acid: light-coloured **igneous rocks** relatively enriched in silica (SiO_2, nominally over 65%); also known as **felsic** and silicic.
Actinolite: non-aluminous amphibole; see **ferromagnesian** minerals.
Adinole: see **metasomatism**.
Agglomerate: a **volcaniclastic rock** composed of large, often angular, rock and mineral fragments.
Albite: sodic plagioclase **feldspar**.
Alkali feldspar: see **feldspar**.
Allochthon: rock unit not in its place of formation.
Allotriomorphic: in igneous rocks, textural term referring to the majority of minerals having poor crystal shape.
Amphiboles: see **ferromagnesian** minerals.
Anatexis: partial melting, especially in the context of progressive melting of sediments to form **granite**.
Andesine: sodic/calcic plagioclase **feldspar**.
Andesite: a fine-grained, **intermediate igneous rock** consisting mainly of plagioclase **feldspar** and **ferromagnesian** minerals; usually a lava.
Aplite: fine-grained, often light-coloured intrusive rock usually of **acid** composition; found in veins and narrow **dykes**.
Aphyric: non-**porphyritic**.
Argillite: a general term for fine-grained clay-rich **clastic** sedimentary rocks.
Augite: a **pyroxene**; see **ferromagnesian** minerals.
Aureole: the envelope of **metamorphic rocks** adjacent to an **igneous** intrusion.
Autochthon: rock unit at its place of formation.
Basalt: a fine-grained, **basic igneous rock** consisting largely of the minerals: plagioclase **feldspar** and **ferromagnesian** minerals, especially **pyroxenes** and olivine; usually a lava or a **dyke**.
Basic: dark-coloured **igneous rocks** relatively enriched in MgO, FeO, CaO, etc. – the 'bases' of early chemistry; SiO_2 relatively low (nominally 45–52%).
Basification: the **metasomatic** process by which a rock is enriched in the **basic** components without melting.
Biotite: a brown mica; see **ferromagnesian** minerals.
Breccia: a **volcaniclastic** or sedimentary or fault-related rock composed of very large, angular rock fragments.
Calc-flinta: **metamorphosed** calcareous mudstone.
Clast: a fragment.
Clinopyroxene: see **pyroxenes**.
Compatible elements: elements which prefer the solid to the **magma** during crystallization and melting.

Glossary

Diapir: a body, e.g. of **igneous rock/magma**, that has risen in consequence of its lower density and/or greater plasticity; often cylindrical and steep sided.

Diorite: a coarse-grained, relatively lime-rich, **intermediate igneous rock**, consisting largely of plagioclase **feldspar** and various **ferromagnesian** minerals; often in large intrusions (batholiths/plutons).

Dolerite: a medium-grained, **basic igneous rock**, consisting largely of plagioclase **feldspar** and the **ferromagnesian** minerals: **pyroxene** and olivine; found in **dykes** or **sills**.

Dunite: an **ultrabasic igneous rock** consisting largely of the **ferromagnesian** mineral, olivine.

Dyke: a sheet-like body of **igneous rock** which cross-cuts the structure of the rocks it intrudes; often steeply inclined.

Elvan: a Cornish term for **quartz–feldspar porphyry**; usually a **dyke**.

Facies: an assemblage of rocks and/or minerals and/or fossils which is characteristic of a particular environment and/or process.

Feldspathization: a **metasomatic** process by which the (alkali) **feldspar** content of the rock is increased.

Feldspars: a series of aluminosilicate minerals between calcium/sodium-rich plagioclase and potassium/sodium-rich alkali feldspar, e.g. orthoclase and microcline; the most abundant minerals in the Earth's crust.

Felsic: **igneous rocks** (usually) which are rich in **feldspars** and **quartz**.

Ferromagnesians: silicate minerals enriched in iron and/or magnesium (and/or lime, etc.) e.g. olivine, **pyroxenes** (augite, hypersthene), amphiboles (hornblende, actinolite, tremolite), micas (biotite, muscovite).

Flaser: an intensely laminated shearing texture in **metamorphosed igneous rocks**.

Fluidization: the mobilization that results from the passage of a fluid (usually a gas) through a granular solid.

Flysch: an assemblage (**facies**) of **clastic** sedimentary rocks characterized by **turbidites**.

Gabbro: a coarse-grained, **basic igneous rock** consisting largely of plagioclase **feldspar** and the **ferromagnesian** minerals: **pyroxene** and olivine; usually occurs in large intrusions (batholiths/plutons).

Gneiss: a coarse-grained, often banded **metamorphic rock**.

Graben: a fault-bounded rift or trough.

Granite: a coarse-grained, **acid igneous rock** consisting largely of alkali **feldspar** and **quartz**; usually in large intrusions (batholiths/plutons).

Granitization: the **metasomatic** process by which rock is converted to **granite** without melting.

Granodiorite: a coarse-grained **acid igneous rock** intermediate between **granite** and **diorite**.

Granulite: a granular **metamorphic rock**.

Greenschists: general term for foliated, green coloured, **metamorphic** rocks of low grade **basic** or **ultrabasic** composition.

Greenstone: general term for massive **basic igneous rocks** that have been (partly) **metamorphosed**.

Greisen: a rock, often a **granite**, that now consists of **quartz** and **mica** because its feldspars have been broken down by **hydrothermal** activity.

Greywacke: a poorly sorted, **clastic** sedimentary rock composed of fragments of rocks and crystals set in a clay-rich matrix.

Harzburgite: a coarse-grained, **ultrabasic igneous rock** consisting largely of the **ferromagnesian** minerals olivine and **orthopyroxene**.

Hornblende: a **ferromagnesian** mineral.

Hornfels: a hard, massive **metamorphic rock** formed by the thermal effects of hot **magma**.

Hyaloclastites: **volcaniclastic rocks** composed largely of glassy fragments.

Hydrothermal alteration: mineralogical and chemical changes in rocks brought about by the penetration of hot, aqueous fluids.

Hypidiomorphic: in igneous rocks, textural term referring to a mixture of well and poorly shaped crystals.

Idiomorphic: in igneous rocks, textural term referring to the majority of minerals having good crystal shape.

Igneous rocks: formed by the consolidation (usually crystallization) of a **magma**.

Incompatible elements: those elements which prefer magma to the solid during crystallization and melting.

Intermediate: **igneous rocks** intermediate in composition between **acid** and **basic**.

Initial ratio: the **isotopic** ratio of a rock before radioactive decay.

Isotopes: forms of a chemical element that differ only in the number of neutrons within their atomic nuclei.

I-type granite: formed by the partial melting of

Glossary

an **igneous** source rock.

Kaolinization: the process by which **feldspars** are broken down to kaolinite by waters of **magmatic** and/or **meteoric** origin.

Keratophyre: **metamorphosed trachyte**.

Lamprophyre: a group of **igneous rocks** of very varied composition, but often rich in potassic minerals; usually in **dykes**.

Lherzolite: a coarse-grained **ultrabasic igneous rock** consisting largely of the **ferromagnesian** minerals olivine and **pyroxene**.

LIL: the large-ion-lithophile group of elements – Cs, Ba, Rb, Th, U, K, La, Ce, Sr and Nd.

Magma: molten rock; referred to as lava when erupted at the Earth's surface.

Mafic: see **basic** and **ferromagnesian**.

Metamorphic rocks: rocks whose texture and mineralogy have been changed in the solid by heat and/or pressure (i.e. without melting).

Metasomatism: the fluid-assisted modification of bulk-rock chemistry that may often occur during metamorphism e.g. 1. sodic metasomatism to form **adinole**, an **albite**-rich **hornfels** at **dolerite** margins, and 2. widespread potassic metasomatism in the contact **aureoles** of **granite** intrusions.

Meteoric water: water derived directly from the atmosphere.

Micas: a group of **ferromagnesian** minerals.

Microcline: a potassium-rich **feldspar**.

Migmatite: a mixed rock, i.e. with both **igneous** and **metamorphic** components.

Minette: a **lamprophyre** rich in the **ferromagnesian** mineral biotite mica.

Moho: the **Mohorovičič** seismic discontinuity between the Earth's crust and mantle.

MORB: mid-ocean ridge **basalt**.

Muscovite: a mica; see **ferromagnesian** minerals.

Nappe: a rock unit which has been displaced laterally by Earth movements.

Obduction: the thrusting of (part of) one lithospheric plate over another, e.g. during plate convergence.

Oligoclase: a sodic plagioclase **feldspar**.

Olistostrome: a rock unit that has gravity glided from its original setting.

Olivine: a **ferromagnesian** mineral.

Ophiolite: a genetic association of mainly **basic** and **ultrabasic igneous rocks** thought to represent ancient oceanic crust.

Ophitic: an **igneous** texture where well-formed plagioclase crystals are enclosed within the **ferromagnesian** mineral pyroxene; characteristic of **dolerites**.

Orthoclase: a potassic **feldspar**.

Orthopyroxene: see **pyroxene**.

Palingenesis: the 're-birth', i.e. re-intrusion, of a **granite** after **anatexis**.

Parautochthonous: close to its place of formation.

Pegmatites: exceptionally coarse-grained **acid** (common) or **basic igneous rocks**.

Pelite: a general term for fine-grained, clay-rich, **clastic** sedimentary rocks; often applied to **metamorphosed** mudstones.

Peridotite: a coarse-grained, **ultrabasic igneous rock** consisting largely of the **ferromagnesian** minerals **pyroxene** and olivine.

Perthite: potassium **feldspar** crystals containing veins and pods of albite.

Phyllite: a **metamorphic rock** rich in the **ferromagnesian** mineral mica that is texturally intermediate between **slate** and **schist**.

...-phyric: denotes the type(s) of the larger crystals in a **porphyritic igneous rock**.

Picrite: a fine-grained **ultrabasic igneous rock** rich in the **ferromagnesian** mineral olivine; usually a lava, or cumulate associated with small intrusions.

Plagioclase: a feldspar.

Porphyroclastic: where fragments of large crystals remain within a fine-grained, sheared matrix.

Porphyritic: **igneous rocks** in which the large crystals (megacrysts, phenocrysts) occur in a matrix of finer crystals and/or glass.

Pneumatolytic: strictly, implies a gas phase; see **hydrothermal alteration**.

Pyroclastics: rocks composed largely of fragments of **igneous rocks** formed during volcanic eruptions.

Pyroxene: a **ferromagnesian** mineral; orthopyroxenes being Ca-poor, whereas clinopyroxenes are Ca-rich.

Quartz: a mineral composed entirely of silica (SiO_2).

REE: the fifteen rare-earth elements or lanthanides of Period 6 of the Periodic Table e.g. La, Ce, Pr, Nd, Sm.

Rhyolite: a fine-grained to glassy **acid igneous rock**; usually a lava.

Rudaceous: applied to very coarse-grained, **clastic** sedimentary rocks (e.g. conglomerate).

Schist: a **metamorphic rock** with well-developed platy and/or linear fabric.

Serpentinite: a **metamorphosed ultrabasic igneous rock** that now consists largely of the **ferromagnesian** mineral serpentine.

Glossary

Sial: that part of the Earth rich in silica and alumina, i.e. the continental crust.

Sill: a sheet-like body of **igneous rock** which in general does not cross-cut the structure of the rocks which it intrudes; often gently inclined.

Sima: the Earth's oceanic crust and (upper) mantle, i.e. rich in silica and magnesium.

Spilite: low-grade **metamorphosed basalt**.

Slate: clay-rich **clastic** sediments (usually) that have developed a pronounced parting (cleavage) under Earth pressures; usually of low **metamorphic** grade.

Stoping: the breaking off and envelopment of blocks of country rock (**xenoliths**) by intruding **magma**.

S-type: **granite** derived from partial melting (**anatexis**) of originally sedimentary rocks.

Subduction: the sinking of (part of) one lithospheric plate under another at a convergent plate boundary.

Syenite: a coarse-grained **intermediate igneous rock** consisting largely of alkali **feldspar** and various **ferromagnesian** minerals.

Tholeiite: **basalt** relatively deficient in alkalis.

Tourmalinization: the **metasomatic** process which gives rise to tourmaline as a replacement for **feldspars** and **micas** in **granite**, etc.

Trachybasalt: **basalt** enriched in alkalis and containing both plagioclase and alkali **feldspar**.

Trachyte: a fine-grained, **intermediate igneous rock** consisting largely of alkali **feldspar** and various **ferromagnesian** minerals; often a lava with a pronounced flow alignment.

Transition elements: some elements within Periods 4 and 5 of the Periodic Table, e.g. Sc, Ti, V, Cr, Mn, Fe, Co, Ni; characteristic of **ferromagnesian** and oxide minerals.

Tremolite: a non-aluminous amphibole; see **ferromagnesian** minerals.

Troctolite: a coarse-grained **basic igneous rock** consisting largely of plagioclase **feldspar** and the **ferromagnesian** mineral olivine.

Tuff: consolidated volcanic ash.

Turbidite: **clastic** sedimentary rock containing structures formed during its deposition from subaqueous turbidity currents; often of **greywacke** composition.

Ultrabasic: dark-coloured, coarse-grained **igneous** rocks consisting entirely of **ferromagnesian** minerals.

Ultramafic: see **ultrabasic**.

Vesicles: gas bubble cavities within consolidated lavas.

Volcaniclastic rocks: sedimentary rocks composed largely of volcanic rock fragments.

Xenoliths(crysts): exotic rock (crystal) fragments incorporated within **magma** (and the resultant **igneous rock**).

Zeolites: a group of low-temperature, hydrous, Ca, Na, K-bearing, aluminosilicate minerals.

Index

Page numbers in **bold** type refer to figures and page numbers in *italic* type refer to tables.

A-type granite 146–8
Actinolite 59, 85, 87, 90, 112
Adinole 92, 100, 104, 112, 114, 137
Agglomerate 84, 124, 127, 222, 223, 224
Albite 59, 73, 75, 83, 85, 110, 130, 150
Alkali basalts **28**, 72, 81, 106, 109, 122, 132, 218
Alkali feldspar 212, 218
Amblygonite 196
Ammonium concentration 157
Amphibole
  Lizard Complex 36, 49, 53, 73
  post-orogenic rocks 212
  pre-orogenic rocks 83, 93, 95, 104
Amphibolite 42, 45, 50, 52, 54, 60, 61–2, 64–5
Amphibolite facies 39, 41, 58
Analcite 218
Anatexis 52
Andalusite 112, 169, 170, 175, 177, 180
Anorthosite 62
Anthophyllite 98, 100
Apatite 26
  orogenic granite 170, 191
  post-orogenic rocks 212, 223, 225
  pre-orogenic rocks 86, 112, 116, 120
Aplite 156
  De Lank Quarries 168
  Megiliggar Rocks 196

Meldon Quarries 198–200
Porthmeor Cove 182
Rinsey Cove 176
Arch Zawn 86, 87
Archean rocks 61
Arsenic mineralization 5
Arsenopyrite 205
Augen 66, 70
Augite 39, 57, 83, 87, 212, 218, 225
Augite picrite 83, 117
Aureole 89, 90, 94, 95, 97, 98
Autunite 189
Axinite 87, 93, 101, 105

B-type granite 148–9
  Birch Tor 166
  Burrator Quarries 173
  De Lank Quarries 167
  Cape Cornwall 180
  Haytor Rocks 162, **163**, 164
  Porthmeor Cove 182
  Leusdon Common 172
  Luxulyan Quarry 169, 170
Back-arc basin model 12
Baggy Sandstone **19**
Barley House **28**
Barras Nose 124, **125**
Barras Nose Formation 122
Barton **28**
Basal Unit 36
Basalt 22
  Lizard Complex 41
  post-orogenic 212, 217–20
  pre-orogenic 80–1, 106, 122, 127, 132, 133

Rhenohercynian 11
Basaltic dykes 3, 39–40, 53, 57, 58, **82**
Basic microgranite *see* A-type granite
Basin models, Rhenohercynian 12
Bass Point **34**
Bastite serpentinite 36, 55, 64, 70
Batholith *see* Cornubian granite batholith
Beacon Quarry 202–4
Beare Farm **28**
Bellever Tor 165
Belowda Beacon 23
Bessy's Cove 86, **87**
Bickington Thrust **15**
Biotite
  orogenic granite 165, 166, 169
  post-orogenic rocks 212, 215, 222, 225
  pre-orogenic rocks 57, 86, 93, 104, 120
Biotite granite *see* B-type granite; C-type granite
Birch Tor 165–7
Black Cove 33, 73–5
Black Head **34**
Blackdown Nappe **15**, 22, 128
Blackdown Thrust **15**
Blue granite 164, 166
Bodmin Moor Granite 23, **147**
  age *24*, 168
  De Lank Quarries 167–9

249

Index

Bodmin Moor Granite (*cont.*)
  history of study 151, 152
  outcrop **4**
Bohemian Massif 12
Boscastle Nappe **15**, **123**
Bossiney Haven 122–7
Botallack Head 97–102
Botallack Mine 98, **99**
Botallackite 100
Boudinage 54, 124
Brannel Elvan *24*
Brent Tor 22, 127–9
Budlake **28**
Budlake Quarry 224
Burrator Quarries 173–6

C-type granite 149–50
  Carn Grey Rock **170**
  De Lank Quarries 168
  Lower Penquite 168
  Rinsey Cove 176
Cadgwith **34**
Calcite 50, 71, 93, 98, 105, 212, 221
Cameron Quarry 202–4
Cape Cornwall 178–82
Carbonate banks 13, **21**
Carboniferous
  plutonic events 22–7
  tectonic setting 13
  volcanic events 21–2
Carn Brea 23
Carn Grey Rock 187–8
Carn Marth 23
  granite age *24*
Carne Formation **16**, **17**
Carnmenellis Granite 23, *24*, **147**
  age 168
  history of study 152
  outcrop **4**
Carrick Du 91–5
Carrick Luz **34**, 66, 68, 69
Carrick Thrust 16
Cassiterite 202, **203**, 204, 205, 206
Castle-an-Dinas 23
  granite age *24*
Castle-an-Dinas Mine 160
Cawsand Bay 214
Chalcopyrite 101
Chemical analyses
  aureole 98
  granite *145*, 148, *149*, 157
  Lizard Complex
    amphibolite 45
    basaltic dykes **40**, 58, 59, 61
    mélange metabasalts 42–4
  post-orogenic rocks 212, **213**, 214, 217–18

pre-orogenic rocks
  basalt 80–1, **82**, 93
  dolerite 83, 88, 115, 116, 117
  ultramafics **84**, 118–19, 120
Chert 70, 71, 106, **131**
Chiastolite 177, 180
China clay 186
China stone 189
Chipley 20
Chipley Quarries 109–11
Chlorite
  Lizard Complex 49, 54, 64, 73
  post-orogenic rocks 212, 223
  pre-orogenic rocks 83, 85, 86, 87, 90, 91, 101, 110, **138**
Chromite 36
Chromite serpentinite 36, 70
Chrysotile 38
Chudleigh Nappe 14, **15**, 109
Church Cove **34**
Chynhalls Point **34**
Clicker Tor 20
Clicker Tor Quarry 117–19
  ultramafics 83
Cligga Head 23
Cligga Head Granite **147**, 205–7
Clinopyroxene
  Lizard Complex 49, 55, 64, 67, 71
  pre-orogenic rocks 86, 87, 115, 117, 120
Clinozoisite 73, 105
Clodgy Point 91–5
Columbjohn Wood **28**, **224**, 225
Columnar joints 132, **134**
Composite intrusion 52, 53
Conglomerate 215
Conodonts 70
Contact metamorphism 41
  Dartmoor Granite 135
  Land's End Granite 89, 90, 91, 92, 94, 95, 97–8, 102–4, 178–82
  Tregonning–Godolphin Granite 177
  see also Hornfels
Continental crust xii, 101, 102
Convective circulation 161–2
Cooling cracks 92, **93**, 108
Copper
  mineralization 5, 204
  mining history 98–100, 102
Cordierite
  Lizard Complex 45

orogenic granite 170, 173, 175, 177, 180
  pre-orogenic rocks 90, 92, 98
Cornish *mélange* 18
Cornish stone 189
Cornubian batholith
  age 23, 24
  chemical analysis 24, *145*, 148, *149*
  classification by type
    A 146–8
    B 148–9
    C 149–50
    D 150
    E 150
    F 150
    others 150–1
  dimensions 22–3
  geophysical character 23
  kaolinization 26–7
  mineralization phases 3, 26, 159–62
  outcrop **4**, 23
  petrogenetic models
    1980s model 157–9
    Dartmoor 151–3
    St Austell 153–7
  petrography *144*
  source 23–4, 171
  unroofing 27
Corundum 175, 180
Coverack Cove 33, **34**, 37, 41, 54–8
Crackington Formation **123**
Crediton Trough 28
Crockham Quarry 135
Crousa Downs Unit 36
Crousa Gabbro 38, 39, 57, 66
Crousa Gravels **34**
Cudden Point 86–9
Cummingtonite 90, 98, 100
Cumulates
  Lizard Complex 37, 41, 55, 57, **63**
  pre-orogenic rocks 83, 86, 117, 118, 119, 120
  see also Traboe Cumulate Complex

D-type granite 150
  Cligga Head 205
  St Mewan Beacon **170**, 191
  Wheal Martyn **170**, 185
Dartmoor Granite 23
  age 24, 168
  aureole 135
  Birch Tor 165–7
  Burrator Quarries 173–6
  Haytor Rocks 162–5

250

Index

Dartmoor Granite (*cont.*)
  history of study 151, 152
  Leusdon Common 172–3
  Meldon Quarries 198–200
  mineralization 160
  outcrop 4
Dartmoor petrogenetic model 151–3
Dartmouth Slate 18, **19**
De Lank Quarries 167–9
De Narrow Zawn **99**, 100
Dean Point **34**
Debris flow 215–16, 223
Deformation phases 13–14, 37
  soft sediment features 85, 86
Delabole Slates 122
Denbury Unit **15**
Devonian xii
  tectonic setting 13
  volcanic events 13
    central SW England 18–20
    Lizard and W Cornwall 16–18
Diabase 85
Diapir 35
Diaspore 100, **101**
Dinas Head 112–14
Diopside 95, 96, 100, 104, 105
Diorite 146
Dodman Phyllite **16**
Dodman Thrust **16**
Dolerite 20, 22
  Lizard Complex 41, 58
  post-orogenic 212
  pre-orogenic 83, 85, 86, 95, 97, 112, 130, 133, 135, 135–7
Dolor Point 55
Dunite 55, 62, **63**
Dunite serpentinite 36, 37, 55, 64, 69–70
Dunsmoor **28**
Dykes
  basaltic 3, 39–40, 57, 58, **82**, 214
  granitic 182, **184**, 198, 200
  see also Neptunean dykes; Sheeted dykes

E-type granite 150
  Megiliggar Rocks 194
  Rinsey Cove 176
  Tregargus Quarries 188
Elender Cove 33, 73–5
Elvan dykes 26, 151, 156–7, 167, 168, 169, 200
Energy source, hot-dry rock 3
Epidiorite 126
Epidote
  Lizard Complex 45, 73, 75

pre-orogenic rocks 85, 90, 93, 98, 101, 110
Equigranular Li-mica granite *see* E-type granite
Exeter Volcanic Series **28**, 29, 217, 220, 222, 223, 225
Exotic terrane 75
Explosion breccia 157
Extrusive activity *see* Volcanism

F-type granite 150
  Tregargus Quarries **170**, 188
F1 deformation event 13
F2 deformation event 13–14
Facies relationships **21**
Fault breccia 62
Faults 61, 62, 65
Felsite 82
Fine biotite granite *see* C-type granite
Flaser gabbro 51, 67, **68**
Flow banding 215
Fluid inclusion studies 160, 192
Fluorapatite 150
Fluorite 150, 170, 186
Fluorite granite *see* F-type granite
Fluxion banding 47
Flysch 21
Folding 65
Foliation 53, 65, 67
Folly Rocks 200–2
Forder Green Thrust **15**

Gabbro 3, 22
  Lizard Complex 38–9, 41, 48, 50, 53, 55, 56, 67, **68**
Garnet 45, 63, 98, 101, 105
Geevor, vein age *24*
George's Cove 54
Geothermometry 66
Gew Graze **34**
Gilbertite 189
Gneiss
  Kennack **34**, 36, 40–1, 46, 47, **49**
  Man of War 35, 44, 45
Godolphin Granite 195
  see also Tregonning–Godolphin Granite
Godrevy Cove **34**
Goldenpoint 169
Goonhilly Down Unit 36
Gramscatho Basin
  development 16–18, 94

Gramscatho Formation 79
Gramscatho Group 16, **17**, 18, **35**, 44
Granite, Cornubian
  age 23, 24
  chemical analysis 24, *145*, 148, *149*
  classification by type
    A 146–8
    B 148–9
    C 149–50
    D 150
    E 150
    F 150
    others 150–1
  dimension 22–3
  geophysical character 23
  kaolinization 26–7
  mineralization phases 3, 26, 159–62
  outcrop 4
  petrogenetic models
    1980s model 157–9
    Dartmoor 151–3
    St Austell 153–7
  petrography *144*
  source 23–4, 171
  unroofing 27
Granite porphyry (elvan)
  dykes 7, 26, 156–7, 167, 168, 200, 215
Granodiorite 146
Granophyre 83
Granulite 41, 65
Greenschist
  Start Complex 3, 18, 44, 73, 74, 75, **79**
  Tintagel Volcanic Formation 22
Greenschist facies
  metamorphism 5, 39, 124, 126
Greenslade **28**
Greensplat 186
Greenstone **16**, 22
  Cudden Point 86, 88
  Dinas Head 112, 114
  Gurnard's Head 95
  Lizard Complex 41
  Mullion Island 70
  Penlee Point 90
  Pitts Cleave Quarry 132–3, 135
  Porthleven 85
  Rump's Point 109
  Ryecroft Quarry 137
  Trevone Bay 114, 115
  Trusham Quarry 136

251

Index

Greisenization 5
  Cameron Quarry 202, 204
  Cligga Head 205
Greystone Nappe 81, 130
Greystone Quarry 130–2
Greystone Thrust **15**, 130
Gullastem 126
Gunheath clay pit 160
Gurnard's Head 95–7
Gurrington Slate **19**
Gurrington Slate Formation 109
Gwavas Sill 89
Gwenter 37

Haig Fras 23
Hangman Grit **19**
Hannaborough Quarry **28**, 222–4
Hantergantick Quarry 168
Harzburgite 37, 55
Haytor Rocks 162–5
  history of granite study 152
Heathfield Nappe 128
Hematite 218, 221, 223, 225
Hematitization 69
Hemerdon Ball 23
  granite age 24
Hingston Down granite 23, *24*
Holmead **28**
Hornblende 57, 69
Hornblende diorite 39, 59
Hornblende schists 5, 18, **34**, 41, 44, 50, 62
  Landewednack type 35, 36, 41, 44, 45, 49, 50–1, 53, 62, 65
  Traboe type 35, 36, 41, 62, 65
Hornfels
  Botallack 97–8, 100, 101, 102
  Burrator Quarries 175
  Cameron Quarry 202
  Rinsey Cove 177
  Tater-du 104, 105
Housel Bay **34**
Hyaloclastite 80, 127, 129
Hybridization 52
Hydrothermal alteration 57, 90–1, 95, 218
Hyner Slate **19**

I-type granite 12, 13
Idocrase 101
Ilfracombe Slate **19**
Ilmenite
  Lizard Complex 39, 57, 71
  pre-orogenic rocks 83, 90, 98, 100, 104, 112

Internappe Flysch **15**
Intracratonic strike-slip basin model 12
Intrusive activity xii, **4**
  composite 52, 53
  see also Plutonism
Iron mineralization 5
Isochron ages 55
Isoclinal folds 65
Isotopes
  dating 13, **14**, *24*, 47, 55
  fluid source studies 98

K/Ar ages **14**, 217, 225
Kaerusite 115, 116
Kaolinite 93, 161
Kaolinization 5, 26–7, 153, 161–2, 171, 186–7, 189, 190, 205
Kate Brook Slate Formation **19**, 173–5
Kate Brook Unit **15**, 173–5
Kenidjack Cliff 98, **99**
Kennack Gneiss **34**, 36, 40–1, 46, 47, **49**, 53
Kennack Granite 47
Kennack Sands 33, **34**, 46–9, 52–3
Keratophyre 12
  pre-orogenic 82
Killas 180
Killerton **28**
Killerton Park 224–6
Kingsand 29
Kingsand Beach 214–17
Kingsteignton Volcanic Group **19**, 20
Kit Hill granite 23, *24*
Knowle Quarry **28**
Krušné hory Mountains 12
Kyanite 45
Kynance Cove 33, **34**, 53–4

Ladabie Bank 23
Lamprophyre **27**, 28
  post-orogenic 7, 212
  Hannaborough Quarry 222–4
  Killerton Park 224–6
  Webberton Cross Quarry 217
Landewednack Hornblende Schist 35, 36, 41, 44, 45, 49, 50–1, 53, 62, 65
Land's End Granite 23, **147**
  age *24*, 168
  aureole 89, 90, 94, 95, 97, 98
  contact metamorphism 178–82

  outcrop **4**
  satellite plutons 182–5
Lankidden 33, 66–70
Large-ion-lithophile (LIL) elements 46, 115, 212, 214, 215, 222
Lava tubes 70, 110
Lavas
  outcrops **4**
  see also Basalt; Exeter Volcanic Series; Rhyolite; Trachybasalt
Lawarnack Cove 54
Lead mineralization 5
Legereath Zawn 195
Leggan Cove 59
Lepidolite 178
Leptites 102
Leucitite 212
Leucogabbro 65
Leucogranite 156
  Megiliggar Rocks 194, 196
  Porthmeor Cove 182, **184**
  Rinsey Cove 176
Leucoxenization 93, 112
Leusdon Common 172–3
Lherzolite 55
Limonite 120
Lit-par-lit intrusion 45, 47
Lithium mica 150, 154–5, 171, 185
  see also Lepidolite; Zinnwaldite
Lithium-mica granite *see* D-type granite; E-type granite
Lithostratigraphic relationships **19**
Little Cudden 87
Lizard Boundary Thrust **16**
Lizard Complex 3, **16**
  history of study 35–6
  lithology 33–5
    basalt dykes 39–40, 55, 59
    gabbro 38–9, 55, **68**, 69
    hornblende schist 41, 44, 50, 62
    Kennack Gneiss 40–1, 46, 47
    mélange metabasalt 41–4
    ophiolite 5, 18, 41, 54
    peridotite **4**, 36–8, 50, 53, 54, 55, 67
  structure 35–6
Lizard Point **33**, **34**, 44–6
Lizardite 38
Löllingite 189
Looe Valley Fault **118**
Lower Penquite 168

Index

Lowland Point 34
Luxton Nodular Limestone 19
Luxullianite 171, 172, 193
Luxulyan Quarry 169–72
Lynton Beds **19**

Magnetite 39, 110, 117, 120, 124, 221
Main Thrust **15**
Man of War Gneiss 35, 44, 45
Manacle Point **34**, 59
Mantle rock 54
Massif Central 13
Meadfoot Group **16**, **17**, 18, 191
Meadfoot Slate 18, **19**
Meadowend **28**
Megacrystic biotite granite *see* B-type granite
Megacrystic lithium-mica granite *see* D-type granite
Megiliggar Rocks 185, 194–8
Mélange metabasalts 41–4, **81**
Meldon Aplite 24, **147**
Meldon Aplite Quarries 198–9
Meldon Calcareous Group 199
Meldon Chert Formation 198, 199
Meldon Shales and Quartzite Formation 198, 199
Meldon volcanics 21
Meneage 41
Meneage Formation 62
Meneage *Mélange* **34**
Metabasalt 41–4, 108
Metadolerite 86, 114, 116, 130, 135, 135–6, 137
Metagabbro 41, 87
Metamorphism
  contact 41
    Dartmoor Granite 135
    Land's End Granite 89, 90, 91, 92, 94, 95, 97–8, 102-4, 178–82
    Tregonning–Godolphin Granite 177
    see also Hornfels
  regional
    amphibolite facies 39, 41, 58
    greenschist facies 5, 39, 86, 124, 126
    prehnite–pumpellyite facies 5, 42, 71, 72, 80, 114, 117, 130
Metaperidotite 120
Metapyroxenite 41
Metasomatism
  dolerite 112, 114

Land's End Granite aureole 95, 98, 101–2
Mica 22, 63, 73, 90, 130
  see also Biotite; Lithium mica; Muscovite
Microgranite 168
  see also A-type granite
Mid-German Crystalline Rise 12
Mid-ocean ridge basalt (MORB) 3, 12, 18, 42, 59
Midland Valley 3
Migmatization 52, 172, 173
Milepost Slate Formation 117, **118**
Mineralization 3, 5, 26, 153, 159–62
  Cligga Head 205
Minette
  post-orogenic 212
    Hannaborough Quarry 222–4
    Killerton Park 225
Mining history 98–100
Minverite 115
Mispickel 205
Mohorovičić discontinuity (Moho) 23, 54, 55, 58
Moldanubian zone 12
Molybdenite 205
Monazite 26
Morte Slate **19**
Mullion Island 18, 33, **34**, 41, 70–3
Muscovite 86, 104, 191
Mylonite and mylonitic fabric 51, 55, 60, 69
Mylor Slate Formation **16**, **17**, 18, **34**, **35**
  Cape Cornwall 178–82
  Carrick Du 91
  Cudden Point 86
  Megiliggar Rocks 194, 195
  Porthleven 84
  Porthmeor Cove 182
  Praa Sands 200–1
  Rinsey Cove 177

Nanpean 189
Nappes 13
Nare Point **34**, 41
Nd isotope ratios 24
Neptunean dykes 218, 221
Net vein complex 47, 85, 221
New Red Sandstone 215, 220
Newton Abbot Nappe **15**
Nordon Slate **19**
Normannia High 12

Obduction evidence 5
Ocean basin subduction model 12
Oceanic crust xii, 5, 44, 46, 53, 58, 61, 72, 75
  see also Ophiolite
Old Lizard Head 'Series' 35, 44, 62, 65
Old Red Sandstone volcanics 3
Olivine
  Lizard Complex 36, 38, 55, 57
  post-orogenic rocks 212, 222, 223
  pre-orogenic rocks 83, 115, 117
Ophiolite 3, 18, 36, 41, 46, 54, 57, 60, **82**
Ordovician rocks 61
Orthoclase 215
Orthogneiss 35
Orthopyroxene 36, 49, 51, 53, 55, 59
Outcrop map 4

Pargasite 36
Parn Voose **34**, 46, **50**, 51
Pb isotopic ratios 157
Pedn Tierre 65
Pegmatites 156
  Cape Cornwall 180, 181
  Coverack Cove 55
  Cudden Point 86
  Kennack Sands 48
  Megiliggar Rocks 194, 196
  Porthmeor 182
Pen Voose Cove **34**
Pendower Formation **16**, **17**
Penlee Point 89–91
Penlee Sill 89, 90
Penolva Quarry 90
Pentire Pillow Lava Group 106, 111
Pentire Point 20, 105–9
Pentreath 54
Penwith Peninsula 100, 101, 104, 182
Peridotite
  Coverack 55
  Kennack Sands 48
  Kynance Cove 53
  Lankidden 66, 67
  Lizard Complex 3, 36–8, 53
  The Balk 50
Permian xii
  conglomerate 215
  volcanism 27–8
Perprean Cove **34**
Perranporth-Pentewan Line 18
Petherwin Beds 19

Index

Petherwin Nappe **15**
Phacoids 104, 116, 124, 127
Phosphatic nodules 124
Pickwell Down Sandstone **19**
Picrite 83, **118**
Pigeonite 87
Pillow breccia 41, 80, 129
Pillow lava 71, **72**
  Lizard Complex 41–4, 70
  pre-orogenic 5, 80–2
    Botallock Head 97, 101
    Carrick Du 91–2
    Brent Tor 127, 129
    Chipley Quarry 109–11
    Gurnard's Head 95
    Pentire Point 106–9
Pilton Shale **19**
Piskies Cove 86, **87**
Pistil Ogo 44, 45
Pitchblende 100
Pitts Cleave Quarry 132–5
Plagioclase
  Lizard Complex 36, 38, 48, 55, 57, 67, 71
  post-orogenic rocks 218
  pre-orogenic rocks 83, 110, **138**
Plagiogranite 59
Plastic deformation 51, 64
Pleonaste 100, **101**
Plutonism 22–7
  geological setting 3
Plymouth Limestone 20
Pol Cornick **34**
Pol Gwarra 64–5
Polbarrow 46, 49, 53
Polbream Cove 45
Poldhu Cove **34**
Poldowurian 66
Polgwidden **50**
Pollurian Cove **34**, 62
Polpeor 45, 52
Poltesco 52
Polyphant 119–22
  ultramafics 83
Polyphant Complex 20, 119, 120
Polyphant stone 119
Port Isaac Nappe 14, **15**, 112, 115, **123**
Porth Ledden 97–102, 156, 193
  contact metamorphism 178, **179**, 180
Porthallow Cove 33, **34**, 45, 61, 66
Porthallow Granite Gneiss 63
Porthcew 176–8
Porthkerris Cove 33, **34**, 41, 52, 62, **64**, 65–6

Porthleven 84–6
Porthmeor Cove 182–5
Porthmissen beach 114
Porthoustock Point 33, **34**, 58–61
Porthtowan Formation **16**, 202
Portscatho Formation **16**, **17**, 44
Portwrinkle Fault **118**
Posbury **28**
Posbury Clump Quarry 220–2
Potassic rocks 212, 217
Praa Sands 200–2
Prawle Point 33, 73–5
Predannack 41
Prehnite–pumpellyite facies metamorphism 5, 42, 71, 80, 114, 117, 130
Priest's Cove 178, **179**, 180
Proterobases 83, 115, 120
Prussia Cove 86–9
Pumpellyite 83
Pumpellyite facies see Prehnite–pumpellyite facies
Pyrite 73
Pyroxene 36, 37, 49, 55, 67, **138**
Pyroxenite 49, 62, **63**, **64**

Quarry granite 164, 166
Quartz in vesicles 71
Quartz keratophyre 12
Quartz–topaz rock 150, 156, 191
Quartz–tourmaline rock 150, 156, 164, 165, 180, 191, 192–4

Raddon **28**
Radiolarian chert 22, 71
Radiometric dating 13, **14**, 24, 47, 55
Radon 3
Rare earth elements see REE analysis; REE chemistry
Rb/Sr ages 24, 25, 168, 169, 175, 180, 200–1
Reaction zones 54
Red Cross 28
Red Sea model 59, 61
Reduction spots 216
REE analysis, granite 157, **158**
REE chemistry **39**
  alkali basalt 110
  amphibolite 45
  basaltic dykes **40**, 58, 59, 61
  dolerite 83
  gabbro 40
  granite 148, 157, **158**

  greenstone 115
  hornfels 98, 101
  Mélange metabasalts 42–4
  ophiolite dykes 53
  peridotite 37, **39**
  pillow lavas 93–4
  post-orogenic rocks 212, **213**, 214
Regional metamorphism
  amphibolite facies 39, 41, 58
  greenschist facies 5, 39, 86, 124, 126
  prehnite–pumpellyite facies 5, 42, 71, 72, 80, 114, 117, 130
Relict texture 86
Restite 52, 157, 165, 166, 169
Rhenohercynian zone
  European setting 11, 12
  SW England setting 13–16
Rhyodacite 82
Rhyolite 29
  post-orogenic 214–17
  pre-orogenic 82
Rinsey Cove 173, 176–8
Rip-up clasts 84
Roche Rock 156, 192–4
Roof pendants 176
Rora Slate **19**
Roseland 41, 44
Roseland Breccia Formation **16**, **17**, 18, 63, 65, 70
Rosenun Slate Formation **118**
Rumps Point 105–9
Rutile 108
Ryecroft Quarry 137–9

S-type granite 13, 23, 157, 166
St Agnes Beacon 23
St Agnes Granite 202
St Austell Granite 23, **147**
  age 24, 168
  Carn Grey Rock 187–8
  classification by type **170**
  evolution 25
  history of study 152, 153–4
  Luxulyan Quarry 169–72
  outcrop **4**
  St Mewan Beacon 191–2
  Tregargus Quarries 188–91
  Wheal Martyn 185–7
St Austell petrogenetic model 153–7
St Mewan Beacon 156, 191–2
St Michael's Mount 23
Salite 115
Sandway Cellar 215
Sanidine 215
Sapphire **179**, 181

Index

Saussuritization 57, 69
Saxothuringian zone 11
Schists
  hornblende 5, 18, **34**, 41, 44, 50, 62
  see also Landewednack-type; Traboe-type
  Start Complex 44, 73
Schlieren 168, 169, 195
  pyroxenite **63**, 65, 69
Schorl 91, 192, **193**
Schwarzwald Massif 13
Scilly Isles Granite 23
  history of study 152
Seamounts 13
Seismic reflectors 23
Serpentinite 36, 37, 55, 64, 69–70
Serpentinization 37–8, 49
Sharp Tor 173
Shearing and shear zones
  Lizard Complex 45, 51, 55, 57, 63, 66, 69
  pre-orogenic rocks 87, 104, 109
Sheeted dyke complex 39, 58, 60
Sillimanite 45
Sills
  Botallack Head 97
  Cudden Point 86, 88
  Greystone Quarry 130, 132
  Lizard Point 45
  Penlee Point 89–90
  Pitts Cleave Quarry 133
  Porthleven 84
  Rumps point 109
  Ryecroft Quarry 137–8
  Trevone Bay 115–16
  Trusham Quarry 136
Skarn minerals 98, 100, 104–5
Slickensides 62, 117, 168
Sm/Nd ages 55
Smith's Cliff 124, **125**, 126
South Crofty, vein age *24*
Spencecombe **28**
Spernic Cove 67
Sphene
  Lizard Complex 73
  pre-orogenic rocks 86, 90, 98, 101, 104, 110, 112
Spherulitic texture 214
Spilite 80
  Rhenohercynian 12
  see also Pillow lava
Spilite-keratophyre association 3, 5
Spilosite 112, 114

Spinel 36, 37, 55, 83, 101
Sr isotope ratios 24, 157, 212
Staddon Grit 18, **19**, **118**
Standon Hill 173
Start Complex 5, 73
  greenschist 3, 18, 44, **74**, 75, **79**
Stilpnomelane 117
Stockscheider 180
Stone Cross Quarry 220
Stone Quarry **28**
Structural elements **14**, **15**
Subduction model 12
Sulphide mineralization 50, 101, 110, 202
Summit tor 164
Suns 162, 164, 165, 192
SWAT seismic survey 23
Syenite 212, 220

Talc 49, 54
Tater-du 102–6
Tectonic setting
  European region 11–13
  SW England 13–16
Teign Valley 83
Teign Valley Unit **15**
Tempellow Slate Formation **118**
The Balk 50, 52
The Crowns 98, 100
Thermal metamorphism *see* Contact metamorphism
Thermal spotting 87
Tholeiite
  Lizard Complex 40, 42, 71, 72, 73
  pre-orogenic 80, 88, 93, 95
Thorne **28**
Thorny Cliff 48
Thrust sheets 61, 65
Thrust tectonics 13–14
Tin
  mineralization 5
  modelling 3
  mining history 98–100
Tintagel Head 122–7
Tintagel Volcanic Formation 21, 22, 84, 122, 124
Titanaugite 83
Titanomagnetite 225
Topaz 150, 190, 191
Tor 164, 165–6
Tourmaline 101, 112, 150, 151, 175, 180, 181
  veins 162, 164
Tourmalinization 5, 171–2, 175, 176, 177, 178, 189, 193

Traboe Cumulate Complex **34**, 41, **42**, 57, 61, 62, 66
Traboe Hornblende Schist 35, 36, 41, 62, 65
Trace elements 58, 117, **147**
Trachybasalt 212, 220
Trambley Cove Formation 122
Trebarwith Strand 122, **123**
Tredorn Nappe 14, **15**, 122, **123**
Tregarden 169
Tregargus Quarries 162, 188–91
Tregonning–Godolphin Granite **4**, 23, *24*, 155, 156
  Megiliggar Rocks 194, 195, 197
  Rinsey Cove 176–8
Trekelland Thrust **15**, **123**
Trelan gabbro **34**, 39
Tremearne Beach **197**
Tremolite serpentinite 36
Trequean Cliff 195
Trevone Bay 114–17
Trevose Head 112–14
Trevose Slate Formation 108–9
Troctolite 41, 54, 55, **58**
Trusham Quarry 135–7
Tubbs Mill lavas 18, **43**, 44
Tuffs 44, 84, 124, 126, 129
Tungsten mineralization 5
Turwell Point **34**

Ultramafics, pre-orogenic 83, 117, 120
Upcott Slate **19**
Uralite 112
Uraninite 26
Uranium
  mineralization 72, 108
  mining history 100, 102

Variscan fold belt 11
Variscan mountain building orogen xii
  regional setting 11–13
Vellan Drang 44, 45
Vellan Head **34**
Vent agglomerate 222, 223, 224
Venton Cove 157
Vesicles 71, 75, 110, 114, 218, **219**, 220, 221
Volcaniclastics 4, 22
  Lizard Complex 45
  pre-orogenic 84, 124, 126, 128

Index

Volcanism
  Carboniferous 21–2
  Devonian
    central SW England 18–20
    Lizard and W Cornwall 16–18
  geological setting 3
    relation to structure 14
  post-orogenic 7, 27–9
  pre-orogenic 5

Webberton Cross Quarry 217–20
Westown **28**
Whale Rock 50, 52, 53
Wheal Cock Mine 100, **101**
Wheal Martyn 185–7
Wheal Owles Mine 98–100

Wheal Remfry 157
Wherry Elvan *24*
Whiteway Slate **19**
Willapark Thrust **15**, 122, **123**
Wishford Farm **28**
Withnoe 215
Wolfram mineralization 5, 204, 205, 206
Woolsgrove **28**

Xenoliths 23, *24*, **57**, 102
  Birch Tor 165, 166
  Botallack Head 97, 101
  Cape Cornwall 180
  Coverack Cove 55
  De Lank Quarries 167, 168, 169
  Haytor Rocks 162, 164
  Luxulyan Quarry 170, 171
  Parn Voose 51
  Rinsey Cove 177
Xenotime 26

Zawn a Bal **99**, 100, 101
Zawn Gamper 104
Zawns 97
Zeunerite 100
Zinc mineralization 5
Zinnwaldite 150, 171, 177, 178, 186, 188, 189, 190, 194
Zircon 26
Zirconolite 108
Zoisite 105